Studies in Big Data

Volume 105

Series Editor

Janusz Kacprzyk, Polish Academy of Sciences, Warsaw, Poland

The series "Studies in Big Data" (SBD) publishes new developments and advances in the various areas of Big Data- quickly and with a high quality. The intent is to cover the theory, research, development, and applications of Big Data, as embedded in the fields of engineering, computer science, physics, economics and life sciences. The books of the series refer to the analysis and understanding of large, complex, and/or distributed data sets generated from recent digital sources coming from sensors or other physical instruments as well as simulations, crowd sourcing, social networks or other internet transactions, such as emails or video click streams and other. The series contains monographs, lecture notes and edited volumes in Big Data spanning the areas of computational intelligence including neural networks, evolutionary computation, soft computing, fuzzy systems, as well as artificial intelligence, data mining, modern statistics and Operations research, as well as self-organizing systems. Of particular value to both the contributors and the readership are the short publication timeframe and the world-wide distribution, which enable both wide and rapid dissemination of research output.

The books of this series are reviewed in a single blind peer review process.

Indexed by SCOPUS, EI Compendex, SCIMAGO and zbMATH.

All books published in the series are submitted for consideration in Web of Science.

More information about this series at https://link.springer.com/bookseries/11970

Khaled R. Ahmed · Henry Hexmoor
Editors

Blockchain and Deep Learning

Future Trends and Enabling Technologies

 Springer

Editors
Khaled R. Ahmed
School of Computing
Southern Illinois University
Carbondale, IL, USA

Henry Hexmoor
Computer Science
Southern Illinois University
Carbondale, IL, USA

ISSN 2197-6503 ISSN 2197-6511 (electronic)
Studies in Big Data
ISBN 978-3-030-95421-5 ISBN 978-3-030-95419-2 (eBook)
https://doi.org/10.1007/978-3-030-95419-2

This Springer imprint is published by the registered company Springer Nature Switzerland AG
The registered company address is: Gewerbestrasse 11, 6330 Cham, Switzerland

Preface

This book is divided into three distinct sections containing contemporary contributed articles.

The first section of this collection contains five chapters that offer themes on emerging foundational technologies contributing to enabling technologies our focal topics. The first article embodies the concept of ShareCert on authentication of certificates for the blockchain and addresses security and privacy for usage of blockchain, whereas smart contracts enable modeling dynamic transactions via automated, logic-controlled contracts.

Our second article discuss semantic web and blockchain as dual frameworks that offer libraries and information centers capabilities to store data in a dispersed and tamper-resistant setting. These frameworks contribute to streamlined procurement of varieties of document types, which expand the range of electronic and digitized material maintained at the libraries.

The proof-of-work consensus mechanism along with scalability is the focus of our third article with an emphasis on scalability. To overcome the block size limitation of consensus mechanism known as proof of work, scalability solutions are sought. They contrasted common consensus algorithms and performed extensive simulation-based analyses and illustrated effects different block sizes and provide guidance for scalability.

Our fourth article begins with a review of traditional legal contracts among businesses. This discussion is then carried on to the blockchain and automation of legal contracts as the software-enabled smart contracts. The smart contract code is responsible for facilitating, verifying, and enforcing the negotiation or performance of a transaction or an agreement. Unlike traditional contracts, smart contracts enable automated execution and tracking of contract terms. This is ideal for codifying recurring interactions among businesses and ease the burden of cumbersome monitoring and verification of traditional contracts.

Our fifth and final paper in this section offers MyEtherWallet designed for secure public communication that relies on smart contracts. They used the solidity language from the Ethereum blockchain platform. Blockchain and cryptographic transaction management ameliorates frauds and corruption. Throughout, arguments are made about the Indian food distribution infrastructure. The introduction of MyEtherWallet would greatly empower Indian citizens.

We witness abundant evidence that blockchain and deep learning are foundational contemporary technologies and serve as tools in myriad of domains and a wide range of use cases. The second section contains a group of papers outlining applications and offer a more pragmatic perspective with nine articles.

We start with the first article addressing intelligent transportation system (ITS) and vehicular IoT application. Emerging communication and complex control systems in ITS require a decentralized and secure methodology to assure inclusion of delay-sensitive and mission-critical applications and that naturally invites the blockchain framework. Internet of vehicles is outlined as a class of cyber physical systems interfacing physically embodied sensory and control devices on modern vehicles as well as devices on roadside and other transportation infrastructure with corresponding cyber environments that remotely monitor and control parameters in the real, physical environment. Timely and efficient transportation requires foresight into existing patterns as well as expected future changes. Machine learning and deep learning are proven to be powerful prediction tools and they are suggested of benefit in this article. For a world where vehicles must work together cooperatively, pooling their capabilities and resources, several vehicular scenarios are exemplified in detail. One such scenario is ride sharing where blockchain offers increased measures of security and privacy. They conclude that blockchain and machine learning bear promising benefits for the intelligent transportation.

Ideas about blockchain enhanced IoT is offered in our second article. IoT is often associated with end user devices that are equipped with limited power and tiny computational resources making them so called lightweight devices and located at the edges of the network. Blockchain offers rapid connectivity and processing powers at designated multiple levels of network interior and away from the edge. Nuances of IoT and blockchain in a few domains of application such as e-health and agriculture are outlined and generally extrapolated for other domains.

The third article in this section offers application of machine learning for distorted speech. They focus on a speech impairment known as aphasia. They discuss machine learning approaches and sketch deep neural net layouts that aid in determination and improvements in aphasia. Authors offer empirical comparisons among a half dozen specific algorithms and suggest that much further exploration of machine learning is required for aphasia.

Our fourth article is on generative adversarial networks (GAN) as a type of deep learning. Here, networks compete as generators and discriminators. The bulk of this article is a review and a broad survey of GAN in the literature.

Our fifth article focuses on biological data integration surrounding protein patterns. They examined heterogeneous biological databases and introduced a semantic mediator-based system for proteomics data integration. This facilities access and navigation within data sets with their user tools to efficiently form and manipulate user-friendly data base queries. Their query sophistication is demonstrated on four key sources of heterogeneous proteomic data.

The sixth article explores the adoption of blockchain technology in Oman and examines barriers and enablers and suggests a vision for the adoption of blockchain technology. Along with numerous interviews, they review their 2017 symposium that featured stakeholders from both public and private sectors. In 2018, a governmental group was formed dubbed Blockchain Solutions and Services (BSS). After outlining many of adoption theories they conclude that blockchain is inevitable and offers benefits that include rapid transactions, transparency, and improved security. Although there is greater awareness and encouraging, adoption is somewhat lagging.

The seventh addition in this section outlines advancement and challenges in deep Learning, big data in bioinformatics. Each topic is discussed in some depth and deep learning algorithms are reviewed. They argue that large industries are poised to produce increasing number of tools for handling colossal amounts of data available and yet expected proliferation of data.

The eighth paper examines deep learning for medical informatics and public health. Typical data may include clinical images, outcome prediction, determining the relationships among genotypes and phenotypes and patient treatment information. After reviewing a few common neural network models, they point out to the heterogeneity of health data and that it naturally spans over disparate domains and disciplines. As harkened by authors, this requires active cooperation among stakeholders from allied disciplines for a coordinated data handling and coherent applications of deep learning methods.

Our ninth addition explores exoplanet hunting using deep learning approach of convolutional neural networks. The NASA Kepler space telescope observations are used for data to be analyzed. Convolutional neural networks are utilized for automatically categorizing Kepler transiting planet data.

The final third section of our collection contains a single article that is forward looking and sketches new horizons.

The article examines the prospectus of the blockchain technology. This article outlines the blockchain origin and details various components of blockchain as a primer. Cryptographic nature of transactions is highlighted as well as smart contracts that appeal to automating contractual obligations among businesses. The highlighted promising application area is manufacturing supply chain and early adoption by prominent organizations such Walmart lights up a bright horizon for widespread adoption of blockchain in the private and the public sector.

We hope we have provided a useful collection of articles that serve as technological primers as well as guides to key current topics and approaches.

Carbondale, USA

Khaled R. Ahmed
Henry Hexmoor

Contents

Enabling Technologies

ShareCert: Sharing and Authenticating Certificates and Credentials on Blockchain

Vasista Sai Venkata Durga Prasad Lodagala and Pallav Kumar Baruah

Abstract The process of certificate and credential verification of every candidate is an essential step in the recruitment phase of every organization. Today, most of this work is usually outsourced to external agencies. Owing to the presence of third parties and lack of transparency in the verification process, there is a possibility of fraud or undetected error creeping in at every stage. The recruiter is at the risk of hiring a fraudulent candidate and the candidate at the risk of facing a rejection due to inefficient verification. Moreover, the existing procedure for this verification takes around 30 days per candidate. Today, blockchain technology assisted with smart contracts have applications ranging from Supply Chain Management to Digital identity, Banking, Insurance and many (any) other scenarios where multiple entities form a network to trade or query any kind of assets. Using smart contracts, a decentralized data store, content-hash based verification and blockchain, we've developed a design for decentralized applications to address the existing issue of certificate and credential verification. These decentralized applications bring about transparency, privacy, provenance tracking and trust to the procedure of sharing and authenticating credentials and certificates. We demonstrate the efficacy of the proposed design based on the results from our implementation on Ethereum and Hyperledger Fabric. In terms of efficiency, our implementations support near real-time verification of certificates and credentials, thereby providing a faster, secure and transparent verification mechanism.

Keywords Certificate verification · Blockchain · Smart contracts · Ethereum · Hyperledger composer · Hyperledger fabric

V. S. V. D. P. Lodagala (✉) · P. K. Baruah
Department of Mathematics and Computer Science, Sri Sathya Sai Institute of Higher Learning, Andhra Pradesh, India

P. K. Baruah
e-mail: pkbaruah@sssihl.edu.in

1 Introduction

Blockchain technology is one of the few innovations in computer science that has surpassed the fame of its initial application which is the bitcoin cryptocurrency. Though Bitcoin cryptocurrency was the first to introduce blockchain technology as its underlying strength, the varied applications of blockchains is eventually surpassing the buzz around cryptocurrencies. The education sector has tremendous potential to adopt blockchain technology in its daily functioning. In this work we present the application of blockchain technology assisted with smart contracts and decentralized data-stores to efficiently solve the problem of certificate verification.

1.1 Motivation

A key phase of running an organization efficiently is the recruitment process. An important aspect of this process is to verify if the certificates produced by the candidate are authentic. An overview of the existing process of educational certificate verification can be found from the pipeline presented below (Fig. 1).

   This entire procedure involves participation of third-party organizations. The existing procedure is slow and involves limited participation from the candidate in the validation process.

1.2 Limitations of the Existing System

We discuss the limitations of the existing methodology from the perspective of each of the three entities.

Fig. 1 The existing verification pipeline

(A) Candidate:

A.1.   After the submission of certificates to the employer, there is no participation of the candidate in the verification procedure.

A.2.   It is not transparent. The entire verification procedure is carried out by the employer and the external agencies that work for the employer. In general, the details of the verification are kept confidential by the employer.

A.3.   Because the candidate has no say in the verification procedure, there are no means by which he/she can ascertain the legitimacy of the verification process.

A.4.   The current methodology doesn't consider the privacy concerns of the candidate. Since a candidate's personal data such as certificates and transcripts are handled by external agencies, it could lead to data breaches. Again, the candidate would have no idea who has initiated the privacy breach.

(B) University:

B.1.   With the total number of candidates who have graduated from the university increasing every year, the number of potential verifications that needs to be done increase with each passing year.

B.2.   As the current procedure isn't completely digital, the university would need to employ additional manpower to handle the verification.

B.3.   There could be internal fraud happening within the university that could facilitate validation of false records and certificates.

B.4.   The process involves a lot of human handlers which renders it error prone.

(C) Employer:

C.1.   The existing validation methodology is an overhead for the employer at the time of recruitment of candidates into the organization.

C.2.   This method isn't cost effective as it involves costs of transportation, verification fee, etc. With many candidates to be recruited, a company would need to invest a considerable amount to finance the verification process.

C.3.   The current procedure for verification takes around 30 days per candidate, which is a considerable amount of time for which the employer needs to put the recruitment on hold.

C.4.   The employer needs to trust middlemen which are the agencies that perform the actual verification.

As we can notice, the entire validation procedure contains risks and limitations for all the three entities. The core threats in the current system are lack of privacy, lack of transparency, trusting middlemen and centralization of verification procedure (only the university is involved in the final verification process). Also, on average the existing procedure takes up to 30 days for credential verification per candidate.

1.3 Intuitive Solution

Knowing that the existing methodology is time-consuming, it is a no-brainer that making the entire process digitally driven would drastically bring down the verification time. A possible solution in this direction would be to use a centralized database to Create, Read, Update and Delete records (CRUD operations) (Fig. 2).

Though the solution of using a centralized database helps in improving the speed of verification, it fails to address most of the limitations of the existing methodology. This is because the solution using a centralized database doesn't guarantee content privacy and transparency of the verification process as the verification process can still be outsourced to external agencies.

Another major threat for the traditional database model is the threat of cyber-attacks. For instance, February of 2018 witnessed one of the biggest Distributed Denial of Service (DDoS) attacks on GitHub. It was a Memcached DDoS attack, so no botnets were involved. The attackers leveraged the amplification effect of a popular database caching system known as Memcached. By flooding Memcached servers with spoofed requests, the attackers were able to amplify their attack by a magnitude by about 50,000 times. At its peak, this attack saw incoming traffic at a rate of 1.3 terabytes per second (TBps), sending packets at a rate of 126.9 million per second [3].

At this juncture, we propose to resolve this verification issue in an efficient manner using a blockchain, smart contracts and a decentralized data store, all of which are detailed about in the following section.

CENTRALIZED DATABASES VS. BLOCKCHAIN

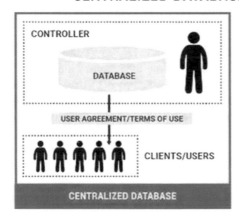 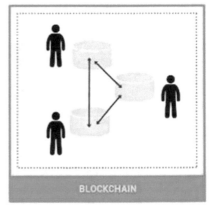

Fig. 2 Blockchain is maintained by a peer-to-peer network which decentralizes control [5]

2 Introduction to Blockchain and Smart Contracts

In this section we introduce blockchain, smart contracts and decentralized data stores which form the fundamental structure for developing our solution to share and authenticate certificates and credentials in a decentralized, transparent and privacy preserving manner.

Decentralization of power and transparency in governance are the fundamental principles that win the trust and confidence of the participants of any organization or society. Blockchain technology by design, weaves both these principles together with the additional features such as preserving user privacy and immutable record keeping.

Blockchain was first popularized by Satoshi Nakamoto through the introduction of Bitcoin [9]. The strength of blockchain lies in the fact that it is immutable and provides privacy and transparency. Now, the idea of having both privacy and transparency at the same time might seem counter intuitive. But here, the privacy preservation refers to anonymizing the users(entities) that are a part of the network. On the other hand, transparency is with respect to transactions. All the transactions made on blockchain are generally made public to its participants. So blockchain ensures that transaction data is transparent to all but at the same time doesn't reveal the sender and receiver's identity explicitly.

In layman terms, blockchain can be thought of as a secure immutable way of storing any type of data ensuring privacy and data integrity. A more technical definition would be: Blockchain is a distributed ledger technology with some or all entities in the network having the permission to write to the ledger, some or all entities in the network having to agree upon the distributed ledger and all the entities in the network having the permission to read or query the ledger. Simply put, blockchain is a decentralized, distributed and immutable database preserving user privacy.

A blockchain can also be thought of as a linked list of hash pointers [1]. A hash pointer points to the location where the data is stored and contains the hash to the value of the data. A block in a blockchain is simply a bundle of transactions that are validated by the peers in the decentralized network. Blockchain is also referred to as ledger. This ledger is updated and kept consistent among the peers through a consensus mechanism on the decentralized network. Each block in the blockchain contains the cryptographic hash of its previous block. This method of chaining blocks using cryptographic hashing is what provides blockchain its immutability. This is because, if an adversary has to alter the contents of a transaction in a block that already exists on several peers, it will also lead to the modification of the next block. Once a block is modified, the hash of the block changes, and would not match with the hash stored in the subsequent block in the ledger. In order to succeed, the attacker needs to modify all subsequent blocks that are added to the blockchain. This is computationally infeasible, especially in a blockchain network that implements Proof-of-Work (PoW) as the underlying consensus mechanism.

2.1 Computation—Record Keeping—Storage

In today's world where data is becoming increasingly valuable, privacy breaches are not quite uncommon. Assisted with an efficient record keeping mechanism that is immutable and preserves user privacy, blockchain technology offers a trustable way to handle sensitive user data. These properties bring value to user data. So, we could refer to user data as assets that are a part of the blockchain. Essentially it is the trade of these assets that brings about blockchain applications in varied fields such as banking, insurance, health care, social networking, credential authentication, voting, land ownership records, file sharing, etc.

Though blockchain can host the data digest (hash of the data) necessary for the functioning of the mentioned applications, we need efficient programs to perform computations on this data. But if these programs are not a part of the blockchain, then the entire exercise of keeping the hash of the data on blockchain becomes pointless. Programs stored on the ledger are usually referred to as "smart contracts". Smart contracts contain the business logic that is agreed upon by all the parties involved in the business network. This business logic is executed once certain transactions are triggered by certain entities in the network. The code embedded in a smart contract is executed by every node in the blockchain providing redundancy in computation leading to very high reliability and almost zero downtime.

As discussed earlier, blockchain is an effective tool for record keeping. Record keeping as known by all, just involves storing the details of an asset rather than the asset itself. Similarly, when blockchain is used as a record keeper of user data, decentralized data stores are used to store the data itself. We emphasize on decentralized data stores because centralization of data could lead to multiple issues such as: single point of failure, Denial of Service (DoS) attacks and data tampering [3].

Essentially, while smart contracts execute transactions (computation), blockchain logs these transactions (record keeping) and decentralized data stores such as the Interplanetary Filesystem (IPFS) [2] provide the necessary space for data (storage).

We'll now delve into certain details of the blockchain frameworks which we've used to develop our solution.

2.2 Frameworks

We've developed a design to solve the certificate verification problem and implemented it across two popular blockchain platforms namely, Ethereum and Hyperledger Fabric.

Before we discuss our design and implementations, we introduce the purpose and functionality of Ethereum and Hyperledger Fabric.

2.2.1 Ethereum

While the application of blockchain to bitcoin and other cryptocurrencies is treated as Blockchain1.0 [10], the Ethereum platform co-founded by Vitalik Buterin brought about a revolution in the way blockchain is applied for computation [11]. This revolution brought about by Ethereum is referred to as Blockchain 2.0 [10]. Ethereum supports distributed data storage and computation, along with providing a native cryptocurrency Ether, as a part of its implementation. The greatest contribution by the Ethereum platform is that it utilizes blockchain technology to develop decentralized applications. Such decentralized applications, which have smart contracts as the basis, are executed on every node in the Ethereum network without any possibility of downtime, censorship or third-party interference. Consensus is reached among nodes using the Proof-of-work algorithm to update the global state upon execution of smart contracts. Ethereum embodies the idea of single blockchain running any arbitrary computation. Embedding business or transaction logic as a part of smart contracts ensures its correct execution, thereby preventing any possibility of fraud. Smart contracts help in significantly reducing the business costs involved and helps run de-risked businesses.

Contracts are located at a specific address in the chain and transactions can be sent to this address to initiate state change. A contract in Ethereum includes code as well as data. In addition, contract accounts can also pass messages to each other to execute necessary contract functions or access data. These contracts are executed by the Ethereum Virtual Machine (EVM) which is run by every node that is a part of the Ethereum network. Contract code is stored in the Ethereum blockchain in a binary format called the EVM bytecode. To facilitate contract writing, many high-level languages are provided. Contracts written in such languages are then compiled into EVM bytecode which can be stored on the blockchain. Also, an intermediary JavaScript Object Notation (JSON) file can be created from the contract bytecode to be deployed locally. JSON is chosen because it can describe the functionality of the contract and remain human-readable. The most common language for writing smart contracts for the Ethereum platform is Solidity.

It is the EVM that executes the smart contracts. **Gas** is the unit which is used to measure the EVM resource usage. Gas usage in a transaction depends on the number of instructions to be executed by the EVM and the storage space used by the transaction. Each operation code in the EVM has fixed units of gas associated with it. The miners who are executing the transaction are spending some hardware resources for computation. To provide incentive for the miners, a transaction fee is paid as in the case of bitcoin. Unlike bitcoin, here the transaction fee is computed based on a gas price fixed by the user before submitting the transaction.

User defines two parameters before submitting a transaction. They are:

- Start Gas (in units): Maximum units of gas, the originator of the transaction is willing to spend.
- Gas Price (in ether): per unit gas price the originator is willing to pay.

Fee processing: Before a transaction gets validated, the maximum amount of ether the user is willing to spend is kept in an escrow. So, the escrow contains the amount of ether which equals the product of Start Gas and Gas Price. Fee that is paid to the successful miner is determined as the product of Gas Used and Gas Price. Any additional ether is refunded to the user.

Successful transaction executions update the Ethereum state, which includes the balances of the Externally Owned Account (EOA), and the balances, code and data of the contract accounts. In case the initially specified amount of ether isn't sufficient for the computation, an *Out of gas exception* is specified to the originator and the transaction is rolled back. However, the ether in the escrow is given to the miner as a reward for the computation performed.

2.2.2 Hyperledger Fabric

Hyperledger Fabric is one among the many open source blockchain platforms being developed by The Linux Foundation under the umbrella of Hyperledger projects. As stated in the Hyperledger project, only an open-source collaboration can ensure the transparency, longevity, interoperability and support required to bring blockchain technologies forward to mainstream commercial adoption [4].

Fabric aims to provide the base architecture for custom enterprise networks. It is modular, allowing for pluggable implementation of blockchain components, such as smart contracts (referred to as chain code), cryptographic algorithms, consensus protocols and data store solutions. It can be deployed using container technologies and provides a secure chain code execution environment that protects the network against malicious code. It is a robust and flexible blockchain network architecture with enterprise-ready security, scalability, confidentiality and performance.

Key design features of Fabric are:

- It provides the necessary support to develop private permissioned blockchains, as required by most of the business applications.
- In order to improve the privacy and confidentiality of assets and transactions, Fabric supports the notion of channels. Only those entities that are a part of the channel can view the transactions recorded on the ledger.
- It provides identity services in the form of Certificate Authorities and Membership Service Providers (MSPs).
- The immutable, shared ledger records the entire transaction history for each channel and supports SQL-like queries for auditing and resolution of disputes.
- The network designers can choose the specific consensus, identity management and encryption algorithms to plug-in to suit their needs.
- Transaction processing in most of the blockchain platforms is order-execute. That is transactions are ordered and blocks are formed first, following which the transactions are executed and the validation of transactions are performed by the rest of the peers. This leads to significant downtime because forming the block itself takes time in case of consensus mechanisms such as Proof-of-Work. Fabric on

the other hand adopts a execute-order-validate strategy, which aids in improving the transaction processing speed.

The Fabric architecture comprises peer nodes, ordering nodes, client applications and chain codes that can be written in any of the supported Fabric Software Development Kits (SDKs). Currently, there are SDKs for Go, Node.js, Java and Python. Fabric also offers a certificate authority service which, following its modularity principle, can again be replaced by custom implementations.

There are two places where data is stored in a Fabric network. While one is the immutable ledger (blockchain), the other more important place is the world state. The world state holds the current value of the attributes of a business object as a unique ledger state. That's useful because programs usually require the current value of an object. It would be cumbersome to traverse the entire blockchain to calculate an object's current value—you just get it directly from the world state. This state is stored in the form of key-value pairs.

Nodes in a fabric network have distinctly assigned roles. There are 3 types of nodes in a fabric network. They are peers, endorsers and orderers. All the peer nodes are committers and are responsible for maintaining the ledger by committing transactions. Some nodes are also responsible for simulating transactions. They execute the chaincodes and endorse the result. So they are called endorsers. A peer can be an endorser for certain transactions and just a committer for others. The responsibility for ordering the transactions in a block is on the orderers. The roles of committers and orderers are generally unified in common blockchain architectures.

3 Literature Survey

In the literature we find two works which aim at sharing credentials using blockchain. While one of the designs is by a set of two organizations named Serokell and Disciplina, the other implementation has been developed at the Massachusetts Institute of Technology (MIT)'s Media Lab and bears the name Blockcerts.

3.1 Disciplina

Disciplina is a platform that is designed to act as a decentralized ledger and has a special regard to privacy mechanisms and data disclosure [6]. As a platform, Disciplina aims to transform the way educational records are generated, stored and accessed.

The major requirements for a platform like Disciplina are:

i The platform should support storage of large quantities of private data such as grades, assignments, solutions, etc.
ii Only buyers/employers must be the entities to whom data can be disclosed to.

iii   There should be no third-party interference in the verification process and the platform must support fairness of data.

The design choice for using a hybrid blockchain has been made based on the following rationale. Though we can use a public ledger to store data after encryption, it suffers from incentive and scalability issues because nodes in the network must be able to store the records provided by educational institutions from all over the world. Therefore, the use of public blockchains for the verification procedure is economically unjustified.

3.1.1   Architecture

There are four entities involved in the entire validation process. They are namely, *candidates, educators, recruiters and witnesses.* A private chain is maintained between the candidates and educators. A public chain is maintained between the recruiters and witnesses.

We must note that the private layer is maintained by each educator independently. All the interactions between the educator and candidates are treated as transactions on the private chain. This private chain can be accessed by the candidate through a web or mobile application (Fig. 3).

The verification process from the platform's perspective is as follows:

i   Candidate would enroll in a course of his or her choice after choosing an educator.
ii   As the course progresses, the educator gives assignments to the candidate, which are to be completed by the candidate, in order to receive a valid score.
iii   Upon acquiring the assignment from the educator, the candidate submits a signed solution of the assignment to the educator. This candidate-educator communication happens completely off chain.
iv   After grading the solution locally, the educator transfers the score along with its hash onto the private blockchain.
v   After the course completion, the candidate acquires a final score, which is also added onto the educator's chain.

Though there exists a private chain maintained by the educators, there is no possibility of tampering with the transactions as all the private transactions are made publicly verifiable. In this way a second public layer of blockchain is introduced.

This public part consists of *witnesses*. These are the nodes that witness a private block produced by the educator. This verification of private blocks is done by the witnesses by writing the authentication information of every private block onto the public chain. This data on the public chain can be used in the future by any arbitrary verifier to substantiate the proof of transaction inclusion given to it by the candidate or the educator.

Also, witnesses process public information that is issued by the educators. This information for example could be that an educator has stopped offering a particular course. Recruiters can buy candidate data from educators.

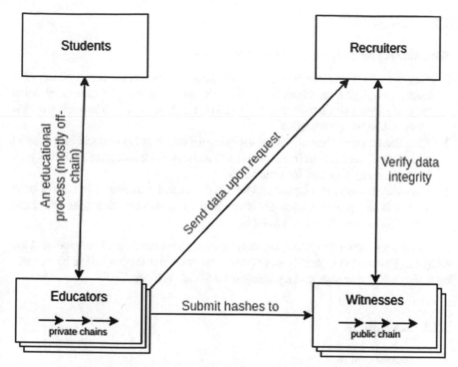

Fig. 3 The overall verification procedure on Disciplina [6]

3.2 Blockcerts

It is an open infrastructure for sharing academic credentials on the Bitcoin blockchain. Blockcerts provides a decentralized credentialing system. The bitcoin blockchain acts as a provider of trust, and credentials are tamper-resistant and verifiable. Blockcerts can be used in the context of academic, professional and workforce credentialing [7].

There are four components involved in the Blockcerts implementation:

i   Issuer: Digitized academic certificates are created by the universities and can contain an array of attributes describing the individual's skill sets and achievements. These certificates are registered on the bitcoin blockchain.

ii Certificate: open badges is the standard followed to create certificates.

iii Verifier: Any node can verify if a certificate is tamper-proof and whether it has been issued by a particular institution and to the specified user.

iv Wallet: is used by individual entities to safely store their certificates and share with other entities such as an employer.

3.2.1 Design

The following three repositories form the architecture:

1   Cert-schema: This describes the data standard to create and manage digital certificates. The digital certificate is a JSON document having the necessary fields required for cert-issuer code to be stored on blockchain. The schema is like that of open badges specification.
2   Cert-issuer: Here, the hash of the digital certificate is generated and broadcasted as a bitcoin transaction, from the issuer's address to the recipient's address with the hash being a part of the return field.
3   Cert-viewer: used to display and verify the digital certificates that have been issued. It also provides users the permission to request certificates and generate a new ID on the bitcoin blockchain.

This design uses asymmetric key cryptography to authenticate the issuer and the recipient. Revocation of certificates is done through a transaction on the blockchain by setting a flag that says that a particular certificate is invalid.

3.2.2 Advantages

- Candidates own their records.
- Vendor independent verification of blockcerts.
- Portability of certificates: because open badges standard has been used.
- Private: transactions are private and only a digital fingerprint is stored on the blockchain.

Having explained the existing solutions now we will describe our solution for the certificate and credential verification in the next section.

4   ShareCert: Design and Implementations

Keeping in view the functionality and features that blockchain technology provides, we can conclude that sharing and authenticating certificates and credentials on blockchain is a perfect application of smart contracts. With the introduction of blockchains, user privacy is maintained with the help of Public Key Identities (PKI). As any node can view the transaction history, transparency is brought into the verification procedure. Dependency on the services of middlemen is eliminated with the help of smart contracts.

With the adoption of blockchain technology to design an efficient solution, we make the entire verification process transparent. Our implementation focuses on bringing the responsibility onto the candidate to initiate the verification procedure.

This reduces the overhead on the employer and the university to take up the verification procedure. Also, no middlemen need to be entrusted with the responsibility of taking up the verification process.

The following points provide an overview of our design:

- Certificates are cryptographically secured and shared between the three entities through a decentralized data store. Each data exchange is recorded as a transaction on the blockchain for verification.
- Once recorded on the blockchain, transactions cannot be tampered with and are distributed across the nodes involved in the validation process.
- Verification of certificates and credentials doesn't involve any central authority or middlemen. Verification is done in near real-time through the smart contract used to build the decentralized application.
- A salient feature of our proposed model: The interaction between the university and the employer is now made optional. The complete verification is done through a transaction triggered by the candidate.

The smart contract which forms the backbone of our decentralized application is what interacts with the underlying blockchain. Though the decentralized data store contains the certificate, the certificate is encrypted and then uploaded to the decentralized data store. This ensures that the privacy of the candidate's data is taken into consideration. Even though the decentralized data store is public, with anybody having access to the files, this kind of encryption mechanism, prevents access to the contents of the file.

We emphasize that there is no need for any sort of interaction between the employer and the university as a part of the verification procedure, other than the initial registration of the employer with the university. The reason for the same would be clearer as we go through the implementations. All the transactions are logged on the blockchain which functions as an immutable record and can be used for audit purposes anytime using simple queries. Also, the transactions are visible to all the entities in either of the designs. The three entities would need to agree upon the smart contract that would be used as a part of the verification procedure (Fig. 4).

Also, we've not restricted the universities to follow any particular format because there is no set format that is followed by all the universities while issuing certificates. So, our design is flexible to cater to the varied formats in which universities issue certificates.

We implemented the above design on two different blockchain platforms namely, Ethereum and Hyperledger Fabric. The choice for these two platforms has been made after considering the features offered by these two platforms and the architecture of the two platforms. Ethereum is a public permission less blockchain, with any node having the permission to be a part of the blockchain and participate in the consensus procedure. Hyperledger Composer essentially is a Software Development Kit (SDK) used to interact with Hyperledger Fabric. Hyperledger Fabric is a platform that supports to maintain any kind of ledgers, public-permissioned, public-permission less, private-permissioned and private-permission less blockchains. But

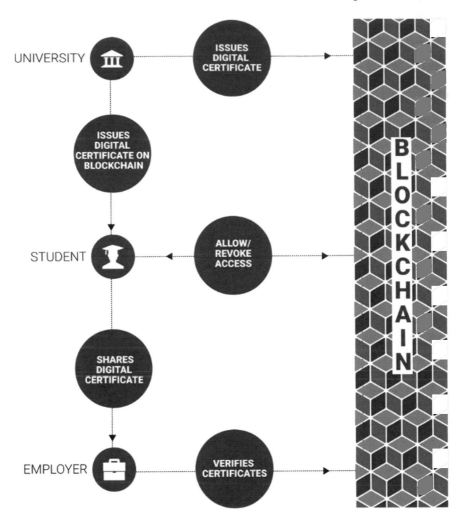

Fig. 4 Similar to our design. In our case the sharing of certificates among entities is encrypted and is over a decentralized data store [8]

among developers, Fabric is popular in maintaining private-permissioned and private-permission less blockchains.

We'll now explore the two implementations in greater detail by describing how the design has been implemented on each of these platforms.

4.1   Implementation on Ethereum

Being the most popular smart contract platform, Ethereum popularized the use of smart contracts for business application purposes. The smart contract used for the underlying decentralized application was first written and tested on Remix IDE which is available online. The smart contract was published through a transaction on the Ethereum network. As the Ethereum main net is not used for development purposes, we've deployed our smart contract onto the Ropsten test net.

The three entities (candidate, university and the employer) interact with each other using a browser-based decentralized application developed using ReactJS, which in turn provides the interface for interaction with the smart contract. The actual interaction of the decentralized application with the underlying smart contract is facilitated with the help of web3.js, which is a collection of libraries that supports interaction with local or remote Ethereum nodes, using a HTTP or IPC connection.

Openpgp.js which is a JavaScript library supporting asymmetric key cryptography is used for the encryption of the certificates before uploading to IPFS. This ensures that no participant other than the designated candidate is given the permission to view the contents of the certificate without access to the private key.

Storage on Ethereum is expensive as the EVM resource usage needs to be paid for. So, we've chosen the InterPlanetary FileSystem (IPFS) as the decentralized data store for uploading our encrypted certificates. Ipfs.js is the JavaScript library used by the decentralized application to upload the encrypted content directly onto IPFS (Fig. 5).

A simple JavaScript library named download.js has been used to facilitate the download of files from IPFS. This reduces the time taken to download the file from the daemon or the browser interface provided by IPFS.

The novelty provided by our implementation is that the employer can verify the authenticity of the certificate without any need for interaction with the university. This saves around 30 days of time spent for verification, as the verification supported by application is done on a near real-time basis.

4.1.1   IPFS Content based Addressing

In order to understand the details of the verification procedure, it is critical that we get to know how IPFS (InterPlanetary FileSystem) stores data. In our design and implementations, we chose IPFS as our decentralized data store.

InterPlaneary FileSystem or IPFS is a decentralized peer-to-peer data store that offers storage and API to interact with smart contracts that are a part of the decentralized application. In the age of client–server-based internet where data is retrieved via location addresses, IPFS brings about a revolution by introducing content addressability.

One of the greatest advantages of IPFS is its content based addressing which gives it the data de-duplication property. When a file is uploaded onto IPFS, it is passed

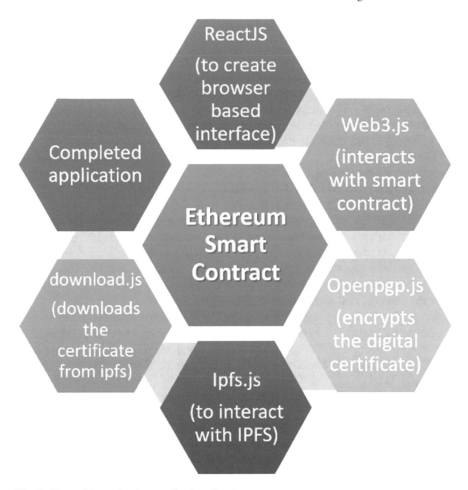

Fig. 5 Flow of the entire decentralized application

through a hash function which produces a unique digest for the file. The digest is
then converted into a Content Identifier (CID). CID is what the IPFS node searches
for in the distributed hash table (DHT) to retrieve the file. So, even if another node
adds the same file onto IPFS, the CID is not going to change because of the design
of IPFS which uses content-based hashing. Also, given a CID or IPFS hash, we can
retrieve the file from the IPFS gateway. This is because, as we mentioned earlier,
IPFS uses content-hashes for addressing as opposed to the regular location-based
addressing.

4.1.2   Stepwise Process of Certificate Verification on Ethereum

i.   (Transaction 1—Request): Candidate requests the certificate from the university using a transaction, by calling a function named *sendRequest* as defined in the smart contract. This function takes the candidate's name, registration number, a field called metadata and the public and private keys generated by Openpgp.js as input. Of course, care has been taken that no entity other than the candidate has access to the private key. The private key is given as an input here, so that the candidate has a log of it and can access it later from the smart contract, for decryption purposes. The field meta data is used to specify any other relevant data that the candidate wishes to provide. All the fields are of type string. Upon successful execution of this transaction, a request ID is generated for this request.

ii. (Transaction 2—Response): The university reviews the request. As a part of the review, the university first decides if the request made by the candidate is a genuine request or not. This can be done by the university by examining the details provided by the candidate as a part of the request and cross-examining with their records. If the request is found to be not genuine, the university marks it as a rogue request and sends the response. This check is essential because, Ethereum being a public blockchain, any person who has access to the portal with an Ethereum address can make a request (It costs them Ether though, making it infeasible to attack on a large scale). If the university deems the request to be genuine, it generates a response after encrypting the certificate using the candidate's public key and uploading the file onto IPFS. Also, it computes the IPFS hash of the certificate without encryption (not uploading the unencrypted certificate, just computing the IPFS hash) and updates the only-Hash field. This field is important as a part of the verification procedure. It then sends the response through a transaction, by calling the *sendResponse* function as defined in the smart contract. The function takes, request ID, a boolean input indicating if the request is genuine, the IPFS hash of the encrypted certificate and the IPFS hash of the unencrypted certificate as the inputs.

iii. Candidate downloads the encrypted file using download.js and decrypts it using his or her private key. The candidate can do so because anyone with the ipfs hash of a file, can download that particular file from the IPFS gateway. At the same time, we must note that this doesn't create any privacy issue because, as we know, the certificate uploaded by the university has been encrypted with the candidate's public key.

iv. (Transaction 3—Registration Request): The employer must register with the university before recruiting any of the students from the university. Without this, the employer won't be able to verify the certificates submitted by the candidate as a part of the recruitment procedure. This registration is done by the employer initiating a *sendRegRequest* transaction. This function takes the organization name, the country of origin, the public and private keys of the employer as inputs. Of course, care has been taken that no entity other than the employer has access to the private key. The private key is given as an input here, so that the employer has a log of it and can access it later from the smart contract for decryption purposes. After the execution of this transaction, an employerID is generated and the employer waits for the registration approval from the university.

v. (Transaction 4—Registration Response): The university reviews the registration request and decides to approve or reject the registration of the employer. It then generates a transaction called *employerRegistration* which takes in the employerID and a boolean input that indicates approval or rejection of the employer's registration. Only employers whose registration is successful can take part in the certificate verification.

vi. (Transaction 5—Application): Candidate encrypts the certificate using the public key of the employer, following which he or she uploads the file onto IPFS. Candidate then generates a transaction called *sendProof* which takes as inputs the candidate's name, registration number, metadata, requestID for which the candidate got his certificate from the university, employerID and the IPFS hash of the uploaded file (encrypted with the employer's public key). This transaction is the point where the candidate applies for the recruitment process. Upon successful execution of this transaction, a proofID is generated.

vii. (Transaction 6—Verification): The employer receives the candidate's application and the relevant details as submitted as a part of the *sendProof* transaction. The employer then decrypts the certificate uploaded by the candidate using his/her private key. Using the API provided by the ipfs.js library, the employer computes the hash of this decrypted file(which is the certificate), without actually uploading the certificate onto IPFS. This hash is provided as the input to the *verifyProof* transaction along with the proofID. The *verifyProof* function compares this hash with the onlyHash field which was updated by the university (computed as a part of the response transaction). If both these hashes match, then the candidate is indeed submitting the correct certificates as a part of the recruitment. This leads to successful verification. If the hashes don't match, then the verification is deemed unsuccessful, and the candidate is informed about the same. We must notice the fact that this verification is taking place through the smart contract itself without any need for interaction between the university and the employer, apart from the initial registration.

This seven-step verification procedure is what is used in the implementation on Ethereum. As we can see from the steps mentioned above, this way of verification eliminates the direct interaction between employer and university. This is a great benefit because it makes the entire verification process driven by the candidate directly and is a completely transparent process. Also, we must note that the registration for an employer is done only once and upon successful registration the employer can subsequently receive multiple applications from the candidates. So, in fact the number of steps often are less than seven.

As a part of this implementation, three separate portals (web pages) have been developed for the Candidates, University and the Employer. Essentially, these three portals interacting with the smart contract is what constitutes the implementation of our design on Ethereum.

4.2   Implementation on Hyperledger Composer

Hyperledger Composer is a framework developed under the umbrella of open source Hyperledger projects by The Linux Foundation. This framework is helpful in accelerating the development of blockchain based business applications on Hyperledger Fabric. Like Fabric, Composer supports a modular approach in building decentralized applications. Though Composer is a separate project being developed by The Linux Foundation, it can be thought of as a Software Development Kit (SDK), which has the potential to build decentralized applications in a faster and more efficient manner.

One must realize the difference between the other SDKs such as node SDK, python SDK etc., that are supported by Fabric and Composer. Compared to those SDKs, Composer brings ease to the process of application development by auto-generating the chain code based on the configuration and business logic specified by the developer. Also, it brings down the direct interaction with the underlying Fabric network, by providing necessary abstractions where the developer can define the different sets of participants, assets, transactions, permissions, business logic and queries.

Apart from local development, Composer also provides an online playground making it even faster for developers to test code and build queries. Hyperledger Composer serves requests using a Representational State Transfer (REST) server with which the decentralized application interacts.

In contrast to the smart contract development on Ethereum, development of smart contracts on Composer is completely modular. Though Ethereum offers modularity in usage of front-end libraries to build the complete decentralized application, Composer helps in modularizing the smart contract itself. Development of applications on Composer involves writing separate files which together form a business network archive which in turn interacts with the underlying fabric network.

A business network archive on Composer contains:

- Model file (.cto): This file is used to define all the assets, participants and transactions that would be a part of the application. Assets carry value and are traded on the network between participants using transactions.
- Script file (.js): This file contains the transaction functions that would be invoked upon submission of transactions by participants in the network. Updating assets is done in this phase, using transaction functions. Also, transfer of ownership happens with the help of these functions.
- Access Control (.acl): As we've mentioned earlier, Composer and Fabric are helpful in developing permissioned blockchains. This file contains the access

control rules which are used to assign permissions to different entities involved in the application. Typically, this file has fields such as description of the rule, participant to whom the rule applies to, operation allowed, resource to which the rule applies, and the actions permitted by the rule.

- Query file (.qry): This query file defines all the possible queries that a user is permitted to execute. These queries by design obey the access control rules, that is, a query cannot be run on a resource to which the user doesn't have access to.

All the above-mentioned files are bundled into a business network archive that can be deployed either on local Hyperledger Fabric environment or the online playground. The front-end interface for this implementation was developed using AngularJS whose code is auto-generated using an open-source client-side scaffolding tool for web applications named Yeoman.

In order to issue and store certificates, we use IPFS as the decentralized data store. We've made this choice because, though the state database supported by Fabric, which is LevelDB or CouchDB supports file storage, IPFS is a public decentralized data store and brings added transparency to the entire procedure. A major advantage of using IPFS is that the certificates are made available over an independent peer-to-peer system without having to query the blockchain for them.

Identities in Composer are issued as business network cards. Users interact with the network using these business network cards. A single identity is stored in a business network card. Historian registry is used to record all the successful transactions, including the participants and identities that submitted them. REST server generates Open API for the business network and supports Create, Retrieve, Update and Delete (CRUD) operations.

The smart contract written is such that the employer can verify the authenticity of the candidate's credentials and certifications even without having direct access to the certificate's contents. Also, we start on the assumption that the employers have pre-registered with the university and are then made a part of this private blockchain network.

4.2.1   Stepwise Process of Certificate Verification on Hyperledger Composer

i.   (Transaction 1—Request): Candidates request the certificate from the university by providing the necessary inputs such as his/her ID and the reason for the request. He does this by creating an asset called *request* which would contain the above-mentioned inputs along with the requestID. This asset is sent to the university using *sendRequest* transaction. The request asset has two other attributes such as transaction ID and the submit status. These attributes cannot be controlled directly by any user, and they get updated by the smart contract upon the submission of a request.

ii. (Transaction 2—Response): University reviews the candidate's request and publishes the encrypted digital certificate on a decentralized data store. This encryption is done using the public key of the candidate, which is available to the university either at the time of candidate enrolment or through interaction on the smart contract. The decentralized data store returns the hash of the encrypted certificate. The university creates a *response* asset that contains the hash of the published certificate as returned by the decentralized data store. This asset contains a field called responseID with which this response can be identified, when queried. Also, as a part of this asset, the university would specify the name of the degree or diploma awarded and the grade obtained. This asset is then sent by the university to the candidate in a *sendResponse* transaction. This *sendResponse* transaction contains a reference to the request to which the university is responding and another reference to the response generated. The response asset also has two other attributes such as transaction ID and the submit status. These attributes cannot be controlled by any user and they get updated by the smart contract upon the submission of a response.

iii.   Candidate decrypts the certificate using his private key, after downloading a copy of the encrypted certificate from the decentralized data store.

iv.   (Transaction 3—Validation): Candidate creates an asset called *proof* that would contain the requestID of the request sent by the candidate to the university and the responseID of the response sent from the university to the candidate. Also, the proof asset would contain the hash of the encrypted certificate which the candidate mentions. This proof contains a field called check hash status which cannot be set by the candidate. In addition to this the proof also contains two other fields, namely the transaction ID and the submit status. These attributes cannot be controlled by any user. They get updated by the smart contract after the submission of the *sendProof* transaction. Once the candidate submits this asset called proof as a part of the send proof transaction, the underlying smart contract checks the hash of the certificate as mentioned by the candidate against the hash of the certificate as mentioned in the response from the university. Check hash status is a Boolean field which would get set based on the result of this verification. Since this transaction is addressed to the employer, the employer can view the result of the verification. Also, after this transaction, the employer would have the permissions to view the request, response and the proof. This way, the employer can obtain the validation of the certificates of the candidate, without even viewing them.

Here again, there is no need for the university to explicitly be a part of the verification process. The credentials of the candidate can be viewed by the employer, who gains read access to the response asset after the send proof transaction. There is no need for the employer to actually see the contents of the certificate because the verification is done through a smart contract which all the parties trust. Also, the employer has access to the response asset which contains the credentials and the degree or diploma awarded to the candidate.

This four-step verification procedure is what is used in the implementation on Hyperledger Composer. As we can see from the steps mentioned, this way of verification eliminates the direct interaction between employer and university. This is a great benefit because it makes the entire verification process driven by the candidate directly and is a completely transparent process. In addition to this, there is no need for the employer to create any asset or submit any transaction. Also, the entire verification process is logged with the help of transactions recorded on the blockchain.

4.3   Results and Analysis

This section focuses on the results obtained after implementing our design on Ethereum and Hyperledger Fabric. Also, we'll be discussing the performance in terms of time taken per transaction on either of the platforms.

One must take note of the fact that we'd implemented our design on a public blockchain (Ethereum) and also a private blockchain (Hyperledger Composer). In the case of Ethereum, the smart contract used for our decentralized application has been written in Solidity programming language, which is a standard programming language for writing smart contracts. This smart contract has been deployed on the Ropsten Testnet of the Ethereum network. The details of the deployment of the contract are given below:

- Transaction hash: https://ropsten.etherscan.io/tx/0x7480e356dfd5f24e6f990b3a7 8c55ad434a833cf801434cda281b4271736f5ca
- Contract Address: https://ropsten.etherscan.io/address/0xe846f21b0321511061 21601b51eb7dc7517353d
- Block number: https://ropsten.etherscan.io/block/9619434
- Transaction fee: 0.00689404265216 Ether
- Gas limit: 3,249,191
- Gas used: 3,249,191 (100%)
- Gas price: 0.000000002121772051 Ether (2.121772051 Gwei)
- Confirmation time: about 43 s.

Using the above-mentioned transaction hash, one can view the transaction details on https://ropsten.etherscan.io/ the Ropsten Testnet of the Ethereum network. Since we've used a Testnet for deployment purposes, the ether used was fake ether. If we have to give a real currency value to the cost of deployment of the contract, it is $12.40 or 903.54 Indian rupees as of 04:26 pm on 10th February 2021. The confirmation time indicates the amount of time taken between triggering the transaction and the successful mining of the block in which the transaction is a part of.

In the following table we've logged the transaction details of 10 request transactions submitted on the Ropsten Testnet. We've done this to provide a comparison of the performance of our application on Ethereum and Fabric platforms. Though any transaction can be used for this comparison we've chosen the request transaction because it is the first transaction that would take place during the verification process (Table 1).

Table 1 Transaction details of 10 request transactions on Ethereum

Transaction hash	Transaction fee (Ether)	Confirmation time (seconds)
https://ropsten.etherscan.io/tx/0x9fefbd2 0b682344124839e0a92d27c82b1adc286 230f7e3def2814881b8c75d5	0.00957054486563	31.738
https://ropsten.etherscan.io/tx/0x671b 26254f7731148063962166c3faab85380 d2a0be9389c8660494bf5fc9e77	0.00957066400996	23.003
https://ropsten.etherscan.io/tx/0x2d63 59d9d423dd1eb0da5f455174dd51814a5 06a505dc44cc7296b106e830965	0.00957057890687	23.109
https://ropsten.etherscan.io/tx/0x4b66 bc82d2e3b04f13417006fe09bc67f9092 a0cd273cc7888183af1e12ee6a0	0.00957047678315	63.662
https://ropsten.etherscan.io/tx/0xa388 5fc1567c9bc16eb305acba10b824a8951 8a77cf2afb73f3c7118832d9269	0.01631792378315	22.940
https://ropsten.etherscan.io/tx/0xc194 752002da4f83ffc1836020321eb5705f2 ee7f3146a52b2deb0462f2a2fa1	0.01294426934501	25.114
https://ropsten.etherscan.io/tx/0xf51c 4449682b2e65d5af63b9b8c19505a023a 24017fee773040f48804ab7fa8a	0.01496840776254	45.230
https://ropsten.etherscan.io/tx/0x8c93 9b9d9e2f96ad4ca67b67c17478d5ef474 af7b9514d4547bbb9cd9c60f238	0.02306541180377	22.898
https://ropsten.etherscan.io/tx/0xf15c 019c35dde90b91271e44611b9d2acd918 e927d2c10d8e4e6a2796e0e1bf6	0.02306545282439	35.916
https://ropsten.etherscan.io/tx/0xb7b9 139b678a9defbc314ed6425084351ab19 e38ea789ace352033ad7d1b8447	0.00957047678315	62.222

We can use the transaction hash to obtain the details of the transaction from https:// ropsten.etherscan.io/. It can be observed from the above table that the confirmation time varies quite a bit, though all the transactions are of the same type (request transactions). This is because the confirmation time depends on a variety of factors. Once a transaction is triggered, it becomes a part of the pending transaction pool. Miners form blocks by selecting a set of transactions from the transaction pool. To get the maximum benefit out of mining a block, miners select those transactions from the pool that offer the maximum transaction fee. The node triggering the transaction sets the transaction fee by deciding on the gas price and the gas limit. The transaction fee is the product of the gas used and the gas price. Any excess ether is refunded to the node triggering the transaction. Since the amount gas is fixed for each operation on Ethereum, the gas price is what dictates the transaction fee offered. So, the confirmation time for a transaction depends on multiple factors such as the gas price offered, the availability of miners, difficulty of computing the Proof-of-Work and also on the gas price offered by the other pending transactions in the transaction pool.

We've varied the gas price for the transactions and observed changes in the time taken to confirm a transaction on the Ropsten Testnet. We observe that the *average confirmation time* for a transaction on the Ropsten Testnet for our request transaction in these iterations is *35.583 s*. Also, the *average transaction fee* paid is: *0.013821421 ether*. This when converted to real currency translates to $24.87 which is 1811.45 Indian rupees as of 04:27 pm on 10th of February 2021.

The smart contract written on composer is a combination of 4 files namely, the Model file (.cto), the Access Control file (.acl), the Transaction functions file (.js) and the Query file (.qry). These files together form the business network archive (.bna) file. This file can be deployed on any Hyperledger Fabric network.

Being a private-permissioned blockchain, the transaction details are stored in a local historian registry that is accessible only to the participants that are a part of the network. So there is no publicly available log to see the transaction details as we could do in the case of Ethereum with the help of a website by the name https://ropsten.etherscan.io/. The time taken for the **deployment of the smart contract** on Composer is: *4 s*.

We've used the request transaction as a benchmark to compute the transaction confirmation time of Composer. Recall that we've used the request transaction as a benchmark for performance on Ethereum implementation also. Being a private-permissioned blockchain, there is no transaction fee for confirmation of transactions on Hyperledger Fabric (Composer).

The average time taken for the **confirmation of a request transaction** on Composer has been observed to be: *0.86 s*.

4.3.1   Analysis

We've noticed that there was a significant difference in the speed of transaction processing on each of these platforms. On Ethereum we've seen that a user needs to wait for an average request transaction confirmation time of 38.2 s. On the other hand, average request transaction confirmation time on Hyperledger Composer is just 0.86 s.

Based on the above observation we **must not conclude** that Hyperledger Composer is a better platform for development of decentralized applications when compared to Ethereum. This is because Ethereum by design is a public permission less blockchain. So, the number of nodes involved is huge at any given point of time. Also, Ethereum implements the Proof-of-Work consensus mechanism. Owing to the computational intensity involved in this consensus mechanism, it takes a significant amount of time for the confirmation of a transaction. A major advantage of using a public blockchain such as Ethereum is that there is no dearth of transparency in the verification process. Moreover, the transaction data is always available for open audit.

On the other hand, Hyperledger Composer is a private-permissioned blockchain platform for development of decentralized applications. Due to this, the number of nodes participating in the network are significantly low compared to those present in a public blockchain. Also, the consensus mechanism involved is not as computationally intensive as the Proof-of-Work consensus used by Ethereum.

Another comparison of the results obtained on each of these platforms is relating to the cost of implementation. On Ethereum we had to pay 0.00689404265216 ether for the deployment of our contract. There is no need for any fee for the deployment of contracts on Composer. Also, the average transaction fee that we've paid on Ethereum was 0.013821421 ether. On Composer there is no need for any transaction fee. Again, this doesn't put the Hyperledger Composer platform on a higher pedestal compared to Ethereum. This is because, Ethereum being a public blockchain platform needs to incentivize miners who form the blocks on the blockchain. Also, having transaction fees on a public blockchain helps prevent the possibility of Denial of Service or Distributed Denial of Service attacks.

We must also bear in mind the different approaches employed by the two platforms in processing the transactions. Ethereum uses the order-execute approach. That is, the transactions are first pooled to form a block and then executed and validated. Because of the Proof-of-Work consensus mechanism used, block formation takes up a significant amount of time. On the other hand Hyperledger Fabric uses an execute-order-validate strategy. That is a major reason for transaction execution times being low on Hyperledger Fabric.

Though the transaction latency in either of the platforms is staggeringly different, the choice of each platform is justified, given the advantages of transparency provided by public blockchains such as Ethereum and the ease of defining permissions provided by Hyperledger Composer. Also, compared to the existing verification procedure we've brought down the verification speed significantly to near real time.

4.3.2　Comparison of ShareCert with Disciplina and Blockcerts

We'd compared the models of Disciplina and Blockcerts with ours across several parameters, the summary of which can be found from the below Table 2.

5　Conclusion

In this work, we've tried to address the problem of certificate and credential verification between the candidates, employers and universities. The existing methodology isn't transparent and involves placing trust on third parties. Due to the lack of transparency in the existing procedure, fraudulent verifications can take place which can prove risky for the employer and the candidate. Also, the existing process is time consuming and is an additional overhead to the university and the employer.

Table 2 Comparison with existing models

	Disciplina	Blockcerts	ShareCert
User Privacy preserving?	Yes	Yes	Yes
Certificates encrypted?	Not on the private chain	No	Yes
Any certificate format supported?	Yes	No. Openbadges standard adopted	Yes
Number of blockchains involved	Two. A private chain and a public chain	One	One
Storage of certificates	Private chain	In a wallet	On a decentralized data store after encryption
Supports implementation on both public and private blockchains?	No	No	Yes. On Ethereum and Hyperledger Composer
Additional entity required for verification?	Yes. Witness nodes are necessary	Not required	Not required
Transparency in verification?	Yes	Yes	Yes

As an attempt to address these existing issues, we've proposed a design to complete this verification process in a transparent and decentralized manner while also preserving the privacy of the entities involved in the verification. We've achieved this with the help of blockchain technology. Blockchain technology helps in bringing together decentralization, transparency and immutability of data while preserving user and data privacy. We note that these are the exact properties lacking in the existing methodology.

Both the implementations of this design on Ethereum and Hyperledger Composer shift the responsibility onto the candidate to initiate and complete the verification procedure. Also, there is no necessity for the university and the employer to interact during the verification process. All the verification is done as a part of the business logic embedded in the smart contract. This helps in reducing significant overhead for the university and the employer.

The reason for choosing Ethereum and Hyperledger Fabric is that while Ethereum is a public blockchain, Hyperledger Fabric is usually used to create private blockchains. So, the proposed design could be successfully implemented as decentralized applications on both the types of blockchains. Also, we've observed that the proposed design can complete the verification process in near real time as opposed to the existing procedure which takes about 30 days on average.

Acknowledgements Our work is dedicated to Bhagawan Sri Sathya Sai Baba, the Revered Founder Chancellor of the Sri Sathya Sai Institute of Higher Learning.

References

1. Arvind Narayanan, J.B.: Bitcoin and Cryptocurrency Technologies. Princeton University Press (2016)
2. Benet, J. IPFS-Content Addressed, Versioned, P2P File System (2014)
3. Cloudfare. Famous DDoS attacks|The largest DDoS attacks of all time. Retrieved from Cloudfare: https://www.cloudflare.com/learning/ddos/famous-ddos-attacks/(2020)
4. Hyperledger. An Introduction to Hyperledger (2018)
5. Khan, N.M Is blockchain the new database? Retrieved from Quora: https://www.quora.com/Is-blockchain-the-new-database(2019)
6. Kuvshinov, K.N. Disciplina: Blockchain for Education (2018)
7. MIT Media Lab Learning Initiative. Blockcerts—An Open Infrastructure for Academic Credentials on the Blockchain. Retrieved from Medium: https://medium.com/mit-media-lab/blockcerts-an-open-infrastructure-for-academic-credentials-on-the-blockchain-899a6b 880b2f. (2016)
8. Murali, A. Tamper-proof degree certificates to be India government's first blockchain project. Retrieved from Factor Daily: https://archive.factordaily.com/degree-certificates-india-blockchain-project/(2018)
9. Nakamoto, S. Bitcoin: A Peer-to-Peer Electronic Cash System (2008)
10. Swan, M. Blockchain: Blueprint for a New Economy. O'Reilly Media (2015)
11. Wood, G. Ethereum: A Secure Decentralised Generalised Transaction Ledgereip-150 Revision (2017)

Prospects of Semantic Web and Blockchain Technologies in Libraries

Faiza Bashir and Nosheen Fatima Warraich

Abstract Blockchain and the Semantic web are the recent revolutions that have already begun to transform the industry, social and political interactions, and any other kind of exchange of value. It's not just a trend; it's a growing phenomenon that's already in progress. For LI Centres, Semantic web and Blockchain's are emerging technologies producing new forms of records or new modalities of recordkeeping with which records and information professionals will need to en-gage. The adoption of these technologies in libraries has connected the support provided by libraries and the rapidly evolving and increasing interests of custom-ers. There are many benefits associated with these emerging technologies, i.e., Trust, Decentralization, Protects Privacy, Distributed, Permission-less Metadata, Interlibrary Loan, Resource Sharing and Voucher System, Intellectual Property, Digital Rights Management, and improve efficacy in many aspects. These tech-nologies have their limitations and challenges, i.e. legal challenges, technical is-sues, financial challenges, and social limitations and challenges. The LIS profes-sionals require more effort and technical skills to understand SW and BCT. Policy makers must prepare the LIS community for these forthcoming and inevitable technologies.

Abbreviations

AI	Artificial intelligence
BCT	Blockchain Technology
DRM	Digital Right Management
HTML	Hyper Text Markup Language
HTTP	Hypertext Transfer Protocol
IP	Intellectual Property

F. Bashir (✉)
Department of Library Science, Government Graduate College for Women, Township, Lahore, Pakistan

N. F. Warraich
Institute of Information Management, University of the Punjab, Lahore, Pakistan
e-mail: nosheen.im@pu.edu.pk

IPFS	Inter-Planetary File System
LAM	Library, Archives & Museums
LI Centres	Library & Information Centres
LIS	Library & Information Science
OWL	Web Ontology Language
RDF	Resource Description Framework
SKOS	Simple Knowledge Organization System
SW	Semantic Web
W3C	The World Wide Web Consortium
WWW	World Wide Web
XML	Extensible Markup Language

Learning Outcomes of the Chapter

The significance of this chapter will primarily be of interest to libraries, library stakeholders, policymakers, research communities, publishers, and third-party mediators. It is anticipated that this chapter fills the knowledge gap. Finding advancement within libraries and encouraging creative individuals involved in new technologies will help libraries prepare themselves for these potentially disruptive technologies. The learning outcomes of the chapter are:

- Introduction of semantic web and Blockchain technology
- Components of the semantic web and Blockchain technology; its framework, architecture, standards, languages, etc.
- Figure out the capabilities and potential benefits of the semantic web (SW) and Blockchain (BC) technologies in library setups.
- Core limitations and challenges associated with these technologies may act as hurdles for realizing these goals in a true sense and suggest ways for their effective uses in the future within the context of libraries.

Highlights

- Blockchain and the Semantic web are the recent revolutions that have already begun to transform the industry, social and political interactions, and any other kind of exchange of value. It's not just a trend; it's a growing phenomenon that's already in progress.
- For LI Centres, Semantic web and Blockchain's are emerging technologies producing new forms of records or new modalities of recordkeeping with which records and information professionals will need to engage.
- The adoption of these technologies in libraries has connected the support provided by libraries and the rapidly evolving and increasing interests of customers.
- There are many benefits associated with these emerging technologies, i.e., Trust, Decentralization, Protects Privacy, Distributed, Permission-less Metadata, Interlibrary Loan, Resource Sharing and Voucher System, Intellectual Property, Digital Rights Management, and improve efficacy in many aspects.

- These technologies have their limitations and challenges, ie. Legal challenges, Technical issues, financial challenges, and Social limitations and challenges.
- The LIS professional requires more effort and technical skills to understand SW and BCT. Policymakers must prepare the LIS community for these forthcoming and inevitable technologies.

1 Introduction

We live in an information-centric world. The rapid technological developments change the whole scenario of information. At present, people have become accustomed to digital information. The WWW has opened the doors to information with just a single click. The revolution brings new channels and offers more opportunities to access additional resources. The interlinking between similar concepts and presenting them through artificial intelligence technology makes the web world more dynamic.

Blockchain and the Semantic web are the recent revolutions that have already begun to transform the industry, social and political interactions, and any other kind of exchange of value. Libraries have a major opportunity to use these technologies to get the desired benefits and make the best use of these technologies. Both of these technologies are used to store data in a dispersed and tamper-resistant setting. This chapter aims to explain the potential of the semantic web and blockchain as a technology and platform to implement these technologies in libraries and information centers. These technologies are still emerging, and more technical changes are expected from them while implications for LI centers. The adoption of these technologies in libraries has connected the support provided by libraries and the rapidly evolving and increasing interests of customers. Libraries are becoming the information centers where resources are not limited inside the library only. These technologies have affected overall housekeeping works in libraries; especially the procurement of document types has extended to electronic and digitized materials. Managing these materials is also becoming more challenging day by day. These technologies are the perfect solution to those challenges [4]. MARC, ISBD, WebOPAC, Dublin Core, Union Catalogue, OAI-PMH, etc., are the common technologies libraries use for storing, managing, retrieval, and disseminating the various resources to serve their users [34, 36, 45].

1.1 What is the Semantic Web?

The term "semantics" comes from the Greek word "smantiká" (neuter plural of smantikós), which means "study of meaning." The Semantic Web is a platform for sharing and reusing data across apps, businesses, and communities. It's a collaborative effort led by W3C that includes a large number of academic and commercial partners. Its purpose is to convert all unstructured online documents into web data. It is built on the Resource Description Framework (RDF). According to Tim Berners-Lee, the Semantic Web is "a web of data that can be processed directly and indirectly by machines"[7]. Semantic Web standards are developed by the World Wide Web Consortium (W3C). In recent years, the semantic web (SW), an extension of the traditional web, has become a hot topic. SW is a highly valuable tool for people and computers to make information more relevant and intelligible. It establishes a semantic context of internet pages to enable human agents and artificial agents for learning content inside web applications [7]. As a result, Semantic Web offers a setting where software agents can interact through web resources and accomplish advanced tasks.

1.1.1   Components and Architecture of the Semantic Web

It is very important to understand SW's technical aspects before discussing its applications in different areas of society. The Semantic Web expands on the solution. It entails the use of data-specific languages such as Resource Description Framework (RDF), Web Ontology Language (OWL), and Extensible Markup Language (XML). HTML is a markup language that represents documents and the relationships between them (Fig. 1).

Fig. 1 *Source* Semantic web architecture layer

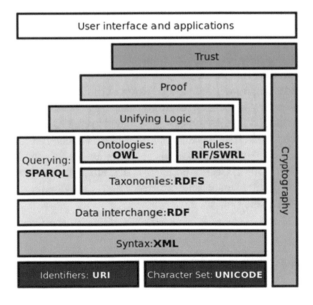

1.2   Blockchain

Blockchain is a peer-to-peer network that enables two parties to transact business, which removes the need for a third party to provide verification and trustworthiness of transactions [1, 11, 15–17]. The Bitcoin blockchain is a set of blocks strung together that record Bitcoin transaction. Each data block contains a batch of transaction information on the Bitcoin network [8, 9, 14]. The theory behind Bitcoin was first explained in 2008 white paper written under the pseudonym "Satoshi Nakamoto". Although the blockchain database system has been established as part of cryptocurrency technology, it has several other potential applications. Soon, everything from medical records to library checkouts could be linked to a blockchain database containing verifiable time stamped production and ownership records [23].

1.2.1   Framework and Architecture of the Blockchain Technology

Blockchain technology with its underlying applications is a new disruptive technology that has emerged recently: it has excited many people because of its 'potential to transform everything' [26, 27, 40]. BCT, which consists of a series of data blocks generated by cryptography, is one type of "distributed ledger" or "distributed ledger technology." It is the underlying technology of Bitcoin, a form of digital currency, which was first released in January 2009. Simply it's a chain of blocks that contain information (Fig. 2). It was originally introduced to timestamp digital documents and prevented the tempering of records [19, 22].

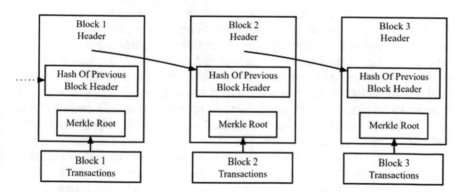

Fig. 2 *Source* Blockchain structure

2   Why Disruptive Technologies for LI Centres

Changes in the technologies used to create and maintain records are nothing new to information workers. Several changes took place in recent decades, resulting in new types of records and recordkeeping systems [24]. The semantic web and blockchains are emerging technologies resulting in new records and recordkeeping mechanisms that records and information workers will need to learn [5, 6, 10, 12, 29].

Libraries and information centers are confronted with insurmountable challenges and inconceivable transitions. Innovation has shifted from a concern to a necessity [21, 23, 31, 33, 45]. Information professionals' executives are under pressure to decide how technologies are adopted and implemented in their libraries in such a resource-constrained climate. Although the blockchain ledger system was developed as part of Bitcoin technology, it has many additional applications. Everything from medical data to library checkouts might be linked to a BC ledger soon, which would contain verifiable, time-stamped records of production and ownership [2, 3, 13, 32, 35, 46]. These systems could also transmit value between users, detect document changes, and protect data from tampering. Librarians and others in the healthcare field should consider their current systems and processes and see if moving them to the blockchain would be beneficial (23). LI centers currently have a lot of opportunities to make the most of this technology.

Librarians must understand SW and BCT capabilities, benefits, and hazards. We will examine potential applications of this groundbreaking technology and provide blockchain education to library users with this information. In the library environment, where interlibrary loans and reciprocal borrowing have shared a core virtue, some of these applications have much potential. But it's also vital to remember that blockchain isn't a "magic bullet" or a cure-all for data management issues [2, 18, 20, 21, 46]. Beyond simply planning for the future, LI centers must develop new technologies and determine how people will utilize them [20] Authors look at civic innovations that use blockchain technology and create a case for why the library would be a good place to start such efforts. They believe that libraries have a high level of community trust and that residents will connect the goals of these new developments to the purpose of libraries. They also suggested that BCT may be used to assist "badging" of abilities learned through training [35, 37, 43, 44]. Libraries could verify the content of personal skills portfolios. New services and platforms such as block stack, IPFS, and other media-driven decentralized technologies could be quite intriguing for libraries. Supporting inventive individuals with interest in developing technology and funding library experiments will help libraries prepare for this potentially disruptive technology [21, 23, 25, 28, 30].

LIS professionals have a significant chance to make the greatest use of modern technology and achieve the required results. The School of Information at San José State University has been working for many years to persuade and encourage people to pursue further education in this discipline. The project's website has a plethora of information concerning the application of blockchain technology in libraries.

This project provided a forum for blockchain experts, information workers, and those working in related industries to discuss practical ways libraries may use the technology to improve their services.

3 Capabilities and Potential Benefits of BC Technologies in LI Centres

SW & BCT is currently being used across several industries due to its unique features: tamper-proof records, authenticity, transparency, and elimination of third-party brokers. These technologies have great potentials, apart from the application in different fields of life, opening new opportunities for LI centers [19, 21, 24, 33, 46]. The characteristics of these technologies, such as decentralization, consensus mechanism, stable timing, and reliable data relations, can solve many problems. Let's take a quick look at some of these capabilities.

Digital Rights Management (DRM) tool is one of the most common potential uses of these technologies. For libraries, BCT could change how electronic resources are bought and paid for, including the maintenance of annual pricing plans. Libraries and vendors could use "smart contracts" to verify purchases and contract terms through these technologies. BC-based currencies for international financial transactions will be another potential benefit for LI centers.

Intellectual Property These technologies provide an e-book platform that would allow authors to self-publish and completely control the distribution of their books to readers, libraries, publishers, and more, which would change our perceptions about intellectual property, especially as it applies to digital objects. In addition, the unchangeable and traceable features of these technologies can effectively protect intellectual property as well. Another promising application for SW and BCT is coordinating, endorsing, and incentivizing research and scholarly publishing activities [21, 23, 38, 45].

Decentralization might be advantageous. These technologies can provide the trusted platform in a decentralized fashion and are resilient to adversaries, acknowledging that the current security solutions on the internet are vulnerable to attacks due to centralization and lack of immutable properties [39, 41, 42].

Establishing Trust in the validity of a record. In addition, the scourges of fake news, misinformation, and disinformation can be weakened by the potential applications of BCT and SW in the info sphere.

Library Verification of Credentials can be done through these technologies. They can rapidly promote the transformation of the library's intelligent service mode, create a credible data link for the circulation of various resources, build a platform for communication and learning between libraries and readers, and truly realize the function of Library Service centering on readers' reading needs.

Interlibrary Loan, Resource Sharing, and Voucher System for the existing library resource sharing service management system would break the barriers between libraries and provide another mode of operation.

Efficiency and security are areas where these technologies can benefit, such as increased security, privacy, or efficiency, and SW and BCT applications are currently being developed and tested in these areas outside of libraries. In LI centers, these technologies directly applicable to the peer-review process and for the chain of custody for digital repositories.

Integrity is the key strength of these technologies, and it helps to ensure the integrity of records through the way transactions are recorded and validated. It can also help in **Protects Privacy** in securing user records.

Building a Distributed, Permission-less Metadata Building a distributed, permissionless metadata archive has perhaps the most disruptive potential of these technologies. LI centers might use these technologies for the Inter-Planetary File System (IPFS). Digital preservation and tracking, corporate library records keeping, community-based collections to share objects, tools, services, and organizational data management are other potential benefits.

4   Core Limitations and Challenges Associated with These Technologies

The technical complexity of these technologies can make them difficult to understand. These technologies have their limitations and challenges, which can be divided into (1) Legal, (2) Technical, (3) Financial, and (4) Social limitations and challenges (Fig. 3).

4.1   Legal Concerns

Ownership of a decentralized ledger is a huge challenge when applying these technologies among LI centers. Zhang (2019) raises the question that who has the ownership of a decentralized ledger? [45].

4.2   Technical Concerns

Lack of unified standards for managing metadata and process is a major challenge for these technologies application in LI centers. **Speed** can be slow, and storage capacity may be insufficient. **Reliability and security** is another major concern as the rise of these technologies. **Data Removal** to fully remove the data, it would be required

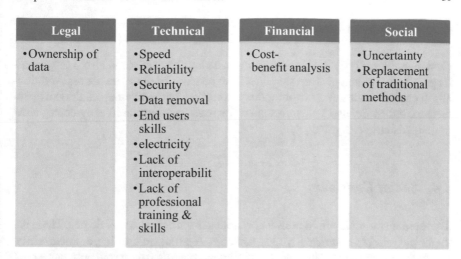

Legal	Technical	Financial	Social
• Ownership of data	• Speed • Reliability • Security • Data removal • End users skills • electricity • Lack of interoperabilit • Lack of professional training & skills	• Cost-benefit analysis	• Uncertainty • Replacement of traditional methods

Fig. 3 Challenges of SW and BCT

to delete it from every node running that blockchain. This would essentially make blockchain unusable for storing any data subject to these privacy laws.

Further, since legislation can change, it is important to consider how changes in the law could remove data from a blockchain system? Or even the complete shutdown of a blockchain itself. Transactions recorded on the blockchain are supposed to be immutable. What happens when that immutable data is wrong and needs to be corrected? Technology's inability to delete transactions, even inaccurate or illegal ones, has become a huge challenge for practitioners. **End Users Skills** It is not easy for people to take or manage their records in a complex setup. Hence there will be trouble in navigating and system by the general public. **Massive amounts of electricity required** the energy cost of a single Bitcoin transaction could power 1.5 American homes for a day [22]. Running a blockchain also requires a huge amount of computing power, which is a real environmental and economic concern. **Lack of interoperability** between different blockchain systems and the difficulty in scaling large amounts of data are other potential drawbacks. **Issues of reliability in transactions** with proper verification, using both proper security architecture and infrastructure management controls. **Lack of Training/Skills among Library Professionals** Libraries need staff with technical skills to keep things afloat. There is a real need for training information professionals on what these technologies are, how it works, and how it could be applied in libraries?. **Shutdown Problem** Blockchain is not yet the best approach to authentication for permanent or long-term records because there are significant risks posed if the blockchain has too little participation by working of users or nodes. Thus it raises questions about blockchain systems, such as what happens to the system where one eventually wants to stop its operations.

4.3 Financial Concerns

Cost–benefit analysis for implementing metadata blockchain is also a huge challenge. These technologies are significantly expensive for the average user. Additionally, there is also the key issue of funding: exorbitant internet prices, additional maintenance, hardware, and human resources increase pressure on already beleaguered or non-existent library budgets.

4.4 Social Concerns

There are many social limitations and challenges with these technologies. These are inefficient and environmentally unsustainable. **A great deal of hype, uncertainty**, and fraud surrounding it. In general and Bitcoin in particular, crypto currencies have an image problem and are regarded as something used for nefarious or criminal purposes. **Replacement of traditional methods** There will be difficulty moving libraries from traditional accounting practices to online cryptocurrency purchases. Libraries and their staff members are traditionally very conservative, which poses a challenge for introducing new technologies. As some communities are extremely cautious of change, such technology would be slow at best and non-existent at worst.

5 Summary

This chapter aims to explore the potential of the SW & BCT platform to implement in the LIS scenario. Furthermore, the study also investigated the core challenges and the limitations of these technologies. These technologies are still emerging, and more technical changes are expected for their implications for LI centers. This chapter aims to share its understanding of potential applications and core challenges/limitations of these technologies. We hope that this chapter can help to prepare LIS professionals for the future. In recent decades a lot of industries have been severely disrupted by technological innovation. These technologies have the potential to be the next major disruption. Blockchain technology, since its introduction in 2009, has sparked interest in every field of life. Every industry believes that blockchain technology has the potential to create efficiency. No doubt that SW and BCT can deliver many advantages over existing technologies. However, the two main characteristics, trust and decentralization. Experts are holding many discussions to explore whether these SW and BCT are suitable solutions to assist in resolving legacy workflows, correlating historical records, and improving scholarly communications to assist in disseminating scholarly works to the public. These technologies could change our perceptions about intellectual property, especially when applied to digital objects.

On the other hand, many factors determine the success or failure of any new technology. SW and BCT are still in their infancy. Like any other revolutionary technology, SW and BCT have advantages and disadvantages. Ownership of a decentralized ledger is one of the most highlighted issues associated with blockchain in libraries. Many researchers raised the question that who has the ownership of a decentralized ledger? There are many technical issues and challenges also linked with these technologies.

6 Concluding Comments and Recommendations

The rapid technological developments have affected every sapphire of life, including LI centers. These developments and innovation have changed the entire information landscape. Libraries have to face technological challenges that lessen the utility of technology. There is a need to eliminate the challenges associated with these technologies in the LIS domain to get the maximum benefits. The LIS professional requires more effort and technical skills to understand Semantic Web (SW) and Blockchain technology (BCT) to explore opportunities. To make the best use of these emerging technologies, all stakeholders need to collaborate. Therefore, policymakers should aware and take measure to prepare LIS community for these forthcoming and inevitable technologies. There is a need to design a curriculum to educate and inform practitioners about new emerging technologies and their potential uses, building a coalition of interested parties, and developing partnerships. Whether these SW and BCT are a usable solution for libraries, is still mostly theoretical until libraries are willing to start experimenting with applications for it. Nevertheless, librarians seem intrigued by the possibilities. More in-depth research to better understand what it can be applied to and overcome Legal, Technical, Financial, and Social challenges, along with regulatory oversight. SW and BCT might be too skeptical, but they demonstrate a divergence of opinion on the potential of these technologies. Therefore, further research is recommended.

7 Key Terms

- **OWL:** Web Ontology Language is a mechanism to process the content of web information. It provides more machine interpretability than XML, RDF, and RDF Schema by facilitating additional vocabulary with the existing semantics.
- **SKOS:** Simple Knowledge Organisation System is standard to support and organize knowledge in a systematic frame like thesauri, classification scheme, subject heading system, taxonomies, etc.
- **Markup Language:** It is the process of defining and presenting contents through the web in a syntactically distinguishable format. Hypertext Markup Language

(HTML) is the first and basic markup language, and then extensible Markup Language (XML) came into existence as an extended version.

- **W3C:** The World Wide Web Consortium is a web standard community that leads the web to its full potential and runs and maintains the web from a-z through standards.
- **RDF:** Resource Description Framework is a standard data model for exchanging data over the internet. It is a method of describing or modeling the concept of data. A set of classes having certain properties to provide basic elements to describe those data is known as RDF Schema. The set of classes are called vocabulary or ontology in the web language.
- **World Wide Web:** The World Wide Web or the web is a network of spaces available and accessible from any web-enabled device written in a particular language. The web is now transmitted to the semantic web from document-centric to data-centric.
- **Linked Data:** It is an interlink technology between two or more documents of related thoughts. Now the links become an open license for most use of thought contents.
- **HTTP:** Hypertext Transfer Protocol is an application-based protocol for transforming information through WWW. Through protocol can interchange information by text or media or can hyperlink to other resources/webpages.
- **Blockchain:** The longest path from the genesis block (the root of a tree) to the leaf is blockchain. The blockchain acts as a consistent transaction history on which all nodes eventually agree.
- **Block:** A block is a data structure used to communicate incremental changes to the local state. It consists of a list of transactions, a reference to a previous block, and a nonce.
- **Transaction (Bitcoin):** A transaction is a data structure that describes the transfer of bit coins from spenders to recipients.
- **Smart Contracts:** Nick Szabo coined the term Smart Contract to define a tool to automate human interactions. Only since the appearance of Bitcoin and not before, there exists a platform to program them as an algorithm that can self-execute, self-enforce, self verify, and self-constraint the performance of the contract.
- **Ontologies:** Ontology is a set of explicit formal specifications of the terms (classes or concepts) in a domain and relations (properties or roles) between them.
- **Semantic Blockchain:** Semantic Blockchain is the use of Semantic web standards on Blockchain-based systems. The standards promote common data formats and exchange protocols on the blockchain, using the Resource Description Framework (RDF).

References

1. Alaklabi, S., Kang, K. Factors influencing behavioural intention to adopt blockchain technology. In: Proceedings of the 32nd International Business Information Management Association Conference, IBIMA 2018-Vision 2020: Sustainable Economic Development and Application of Innovation Management from Regional expansion to Global Growth (2018)
2. Ayre, L., Craner, J.: Technology column: blockchain, linked data, and you. Public Libr. Q. **38**(1), 116–120 (2019). https://doi.org/10.1080/01616846.2018.1562317
3. Bambara, J.J. Allen, P.R. Blockchain: A Practical Guide to Developing Business, Law, and Technology Solutions, Xviii, 302p. McGraw-Hill Education, New York (2018)
4. Bashir, F., Warraich, N.F. (2020). Systematic literature review of Semantic Web for distance learning. Interact. Learn. Environ. 1–17 (2020)
5. Batista, D., Hofman, D., Joo, A., Lemieux, V. Blockchain technology and record-keeping. ARMA Mag. 14–17 (2019)
6. Begley, R. (2017). Information & Records Management and Blockchain Technology: Understanding its Potential. Masters thesis, Northumbria University.
7. Berners-Lee, T., Hendler, J., Lassila, O. (2001). The semantic web. Sci. Am. **284**(5), 34–43. https://doi.org/10.1038/scientificamerican0501-34
8. Bhatia, S., Wright de Hernandez, A.D. Blockchain is already here. what does that mean for records management and archives? J. Arch. Organ. 1–10 (2019)
9. Chen, G., Xu, B., Lu, M., Chen, N.S.: Exploring blockchain technology and its potential applications for education. Smart Learn. Environ. **5**(1), 1–10 (2018)
10. Coghill, J.G. Blockchain and its implications for libraries. J. Electr. Resour. Med. Libr. 1–5 (2018)
11. Davis, P. Bitcoin: a solution to publisher authentication and usage Accounting, The Scholarly Kitchen. Retrieved from https://scholarlykitchen.sspnet.org/2016/06/01/bitcoin-a-solution-to-publisherauthentication-and-usage-accounting/Enis (2016)
12. Dolan, L., Kavanaugh, B., Korinek, K., Sandler, B.: Off the chain: blockchain technology-an information organization system. Tech. Serv. Q. **36**(3), 281–295 (2019)
13. Ensign, D.: Copyright corner: blockchain and copyright. Ky. Libr. **82**(3), 4–5 (2018)
14. Figueroa, M., Griffey, J., Tomer, C., Howley, B., Garmer, A., Hess, R.R., Voto, A. Investigation of possible uses of blockchain technology by libraries-information centers to support city-community goals (Proposal Number. LG-98-17-0209) Investigation of Possible Uses of Blockchain Technology by Libraries-Information Centers to Support C (2018)
15. Findlay, C.: Participatory cultures, trust technologies and decentralisation: innovation opportunities for recordkeeping. Arch. Manus. **45**(3), 176–190 (2017)
16. Frederick, D.: Blockchain, libraries and the data deluge. Libr. Hi Tech News **36**(10), 1–7 (2019)
17. Ginsberg, D. Blockchain 3.0 or web 3.no? blockchain: what it is, how it's being used, and what it means for the future of law libraries. AALL Spectrum **22**(1), 36–39 (2017)
18. Griffery, J. Blockchain intellectual property. Retrieved from https://speakerd.s3.amazon aws.com/presentations/a6e4b49a7c9e4892b90041fb786b9bdc/Blockchain_for_Libraries_Int ernet_Librarian_2016.pdf (2016)
19. Hau, Y.S., Lee, J.M., Park, J., Chang, M.C. (2019). Attitudes toward blockchain technology in managing medical information: survey study. J. Med. Internet Res. **21**(12) (2019)
20. Herther, N.K.: Blockchain technology in the library. Online Search. **42**(5), 37–43 (2018)
21. Hirsh, S. Investigation of possible uses of blockchain technology by libraries-information centers to support city-community goals. In: 2017 National Leadership Grants for Libraries (2017)
22. Howley, B.: Blockchain, ledger legerdemain, and the public library. Inf. Today **33**(9), 14–15 (2016)
23. Hoy, M.B.: An introduction to the blockchain and its implications for libraries and medicine. Med. Ref. Serv. Q. **36**(3), 273–279 (2017)

24. Hughes, L., Dwivedi, Y.K., Misra, S.K., Rana, N.P., Raghavan, V., Akella, V.: Blockchain research, practice and policy: applications, benefits, limitations, emerging research themes and research agenda. Int. J. Inf. Manage. **49**, 114–129 (2019)
25. Huwe, T.K.: Blockchain and the library: beyond the numbers game. Comput. Libr. **39**(1), 8–10 (2019)
26. Ito, J., Narula, N., Ali, R. The blockchain will do to the financial system what the internet did to media. Harvard Bus. Rev. (2017). https://hbr.org/2017/03/the-blockchain-will-do-to-banks-and-law-firms-what-the-internet-did-to-media
27. Jo, J., Rathore, S., Loia, V., Park, J.: A blockchain-based trusted security zone architecture. Electron. Libr. **37**(5), 796–810 (2019)
28. Lemieux, V.: Blockchain recordkeeping: a SWOT analysis. Inf. Manag. J. **51**(6), 20–27 (2017)
29. Lemieux, V.L.: Trusting records: is blockchain technology the answer? Rec. Manag. J. **26**(2), 110–139 (2016)
30. Meth, M.: Blockchain in libraries. Libr. Technol. Rep. **55**(8), 1–24 (2019). https://doi.org/10.5860/ltr.55n8
31. Meth, M.: Understanding blockchain: opportunities for libraries. . Am. Libr. **51**(1/2), 65 (2020)
32. Nicholson, J.: The library as a facilitator: how bitcoin and block chain technology can aid developing nations. Ser. Libr. **73**(3–4), 357–364 (2017)
33. Ojala, M.: Blockchain for libraries. Online Search. **42**(1), 15 (2018)
34. Olnes, S., Ubacht, J., Janssen, M.: Blockchain in government: Benefits and implications of distributed ledger technology for information sharing. Gov. Inf. Q. **34**(3), 355–364 (2017)
35. Rubel, D. No need to ask: creating permission less blockchains of metadata records. Inf. Technol. Libr. (2019)
36. Sicilia, M.A., Visvizi, A.: Blockchain and OECD data repositories: opportunities and policymaking implications. Library Hi Tech (2018)
37. Singhal, B., Dhameja, G., Panda, P.S. Beginning Blockchain: A Beginner's Guide to Building Blockchain Solutions. Apress, India, Xv, 386p (2018). https://doi.org/10.1007/978-1-4842-3444-0
38. Smith, C. Blockchain reaction: How library professionals are approaching blockchain technology and its potential impact. (cover story). Am. Libr. **50**(3/4), 26–33 (2019)
39. Swan, M. Blockchain: Blueprint for a New Economy. O'Reilly Media, Inc, Beijing, Xvii, 128p (2015)
40. Tapscott, D. How the blockchain is changing money and business. TED Summit (2016)
41. Ugarte, H. A more pragmatic Web 3.0: linked blockchain data. Bonn, Germany (2017)
42. Van Rossum, J.: Blockchain for research. Digital Sci. Rep. (2017). https://doi.org/10.6084/m9.figshare.5607778.v1
43. Vazirani, A.A., O'Donoghue, O., Brindley, D., Meinert, E. Implementing blockchains for efficient health care: systematic review. J. Med. Internet Res. **21**(2) (2019)
44. Wang, L., Luo, X. (Robert), Lee, F. Unveiling the interplay between blockchain and loyalty program participation: a qualitative approach based on Bubichain. Int. J. Inf. Manage. **49**, 397-410 (2019)
45. Xidong, L. A smart book management system based on Blockchain platform. A smart book management system based on Blockchain platform (2019)
46. Zhang, L.: Blockchain: the new technology and its applications for libraries. J. Electron. Resour. Librariansh. **31**(4), 278–280 (2019)

Faiza Bashir is an energetic and ambitious professional in the field of Library and Information Science (LIS). Currently, she is working as Assistant Professor / Head of Department of Library Science at Government Graduate College for Women, Township Lahore, Pakistan. She has more than eighteen years of working experience in LIS as an academic librarian and a teacher. She is doing her PhD in Information Management from the Institute of Information Management, University of the Punjab, Lahore. She did her Masters in Library Science (2003) and

History (2006) from the University of the Punjab, Lahore. She has published her research work in well reputed international and national journals. She regularly presents her research at different National and International conferences. Faiza, working as journal editorial board member and reviewer for many national and international reputed journals.

Nosheen Fatima Warraich (Dr.) works as Director, Professor at the Institute of Information Management, University of the Punjab, Lahore, Pakistan. She received her Ph.D. from the University of the Punjab (2005–2011) and is honored to be the first Ph.D. of the department regular MPhil leading to the Ph.D. program. She is privileged to get Fulbright Fellowship to complete her Post Doctorate from State University of New York (SUNY) at Albany, the USA, in 2015–16. Dr. Warraich has more than 75 publications, including Impact Factor and HEC recognized national and international refereed journals, conference proceedings, books, and book chapters on her credit. She also supervised and has been supervising several research theses for Masters MPhil and Ph.D. students. She regularly presents her research at prestigious conferences in the field. She was awarded ASIS&T New Leader Award 2017–19 and SIG/III (International Information Issues) and currently serving the chair of IPC (International Paper Contest) 2021.

Scalability Analysis of Proof-of-Work Based Blockchain

Diksha Malhotra, Shilpa Dhiman, Poonam Saini, and Awadhesh Kumar Singh

Abstract Recent advancements in technologies and today's digital era demands a high level of security and privacy in all aspects. The concept of Blockchain has emerged as a promising technology. As any node in a peer-to-peer network does not fully trust the other nodes, the technology uses cryptographic hash functions (essentially the message digest) as digital signatures. The chapter focuses on discussing Blockchain basics, its working, types of Blockchain, its architecture along with various Blockchain applications. Also, we have provided a detailed discussion and comparison between various existing Blockchain consensus mechanisms viz. Proof-of-stake, Proof-of-Work, Proof-of-Activity, etc. The block size of proof-of-work based Blockchain, such as Bitcoin, is generally limited to 1 MB. As a result, when the number of transactions grow, the systems tend to become slower, expensive, and inefficient. Hence, it is necessary to perform scalability analysis on a network and incorporate necessary scalability solutions. The chapter discusses various system parameters involved with Blockchain scalability issues in detail. Further, we will present our series of experiments that have been performed for scalability analysis of Proof-of-work based Blockchain using a simulation tool. It provides details of the process along with the analysis results to examine the impact of varying parameters such as block-size, number of nodes, and transaction fee.

Keywords Applications · Blockchain · Consensus · Scalability analysis

1 Introduction

The concept of Blockchain came into existence when Satoshi Nakamoto published a seminal paper titled *Bitcoin: A Peer-To-Peer Electronic Cash System* [1] in the year 2008 and described the technology as a decentralized distributed ledger. The

D. Malhotra · S. Dhiman (✉) · P. Saini
Computer Science and Engineering Department, Punjab Engineering College (Deemed to be University), Chandigarh, India

A. K. Singh
Computer Engineering Department, National Institute of Technology, Kurukshetra, India

core technology behind the emerging Cryptocurrencies such as Bitcoin, Ethereum is Blockchain. Its birth brought about a movement which distorted the entire tech industry. Enthusiasts call this movement as Web 3.0 as the technology facilitated the decentralization of the World Wide Web. Blockchain can be defined as a consensus-oriented distributed ledger which stores data over a peer-to-peer network. These ledgers are distributed among various untrusted participating nodes which are updated based on global information. The decentralization of the ledgers resolves issues faced by the centralized databases such as need of a trusted third party, single point of failure, and lack of security.

Blockchain can be seen as a chain of blocks where each block contains the hash of the parent block in order to achieve immutability property. Every piece of information is stored in the form of transactions which collectively form blocks. Each block comprises a set of transactions, hash, nonce and hash of the parent block. Further, once any transaction or event is recorded on a block, it is not possible to modify the details being already shared among peers in the network. A set of peers called *miners* participate in a competition to validate the newly created block by presuming a random number or *nonce* to include the block with the help of signatures (Fig. 1). Some of the features of Blockchain technology are decentralization, distributed network, immutability, anonymity, open source and autonomous. Further, Hofmann et al. [2] distinguishes between centralized, decentralized and distributed systems by giving a pictorial representation of the same (Fig. 2, Table 1).

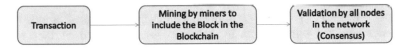

Fig. 1 Basic blockchain life cycle

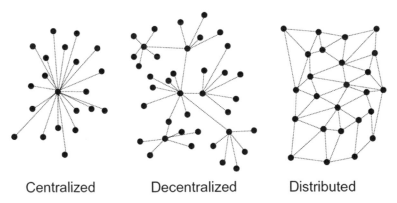

Fig. 2 Graphical representation of centralized, decentralized and distributed systems [2]

Table 1 Centralized versus decentralized versus distributed networks		Centralized	Decentralized	Distributed
	Failure	Single point of failure	Can tolerate multiple failures	No failure
	Scalability	Not scalable	Scalable	Scalable
	Points of coordination	Single point of coordination	Multiple points of coordination	No centralized coordination

Further, the block size of proof-of-work based Blockchain, such as Bitcoin, is generally limited to 1 MB. Hence, as the number of transactions grow, many scalability issues emerge like *increased response time, less throughput* and *more computational cost*. Thus, there is a possibility that current technology may not be able to sustain the growing demand. Blockchains start facing an increase in waiting time for completion of each transaction. To improve such issues, it is therefore necessary to perform scalability analysis on a network and incorporate necessary scalability solutions.

In the following sections, the chapter discusses Blockchain basics, its working, types and architecture along with various Blockchain applications. Further, it compares some of the consensus mechanisms like Proof-of-Stake (PoS), Proof-of-Work (PoW), Proof-of-Activity (PoA), etc. on features such as energy efficiency, 51% attack, double spending, block creation speed and pool mining. In addition, it performs a scalability analysis on simulated bitcoin Blockchain and analyses the impact of various parameters such as *block-size, number of nodes,* and *transaction fee.*

2 Blockchain Basics

The section discusses Blockchain architecture, its working, types, generations, and its applications in various domains along with related tools and interfaces.

2.1 Blockchain Architecture

Generally, a block contains data depending upon the kind of Blockchain, hash of previous/parent block, and its own hash. The blocks are sequentially added. The first block [1] of a Blockchain is known as a *genesis block* and every peer of the network is broadcasted with the same genesis block. The internal structure of a Blockchain consists of a header and block body.

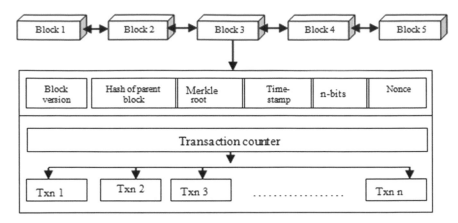

Fig. 3 Block diagram: block header and transaction counter

1. The block header contains:

   (a) *Block version*: the block validation rules to be followed.
   (b) *Merkle tree root hash*: the hash value of all block transactions.
   (c) *Timestamp*: the time at which block is generated.
   (d) *n-bits*: the target limit of valid hash of a block.
   (e) *Nonce*: a field which is initialized to 0 and keeps on increasing with every hash calculation.
   (f) *Parent block hash*: indicates a hash value of 256 bits which points to the previous block.

2. The block body [3] consists of transactions and a transaction counter as shown in Fig. 3. The maximum number of transactions that a block can hold depends upon the size of block as well as each transaction.

One of the fundamental components of Blockchain is a *Merkle tree*. It is a hash-based data structure, also known as a hash tree. Merkle trees are similar to binary trees in terms of branching factor i.e., each node can have up to 2 child nodes. Each leaf node contains hash of the transaction, and each non-leaf node contains a hash of its child nodes. A binary hash tree is created starting with the leaf nodes as shown in Fig. 4. Here, the hash tree, being binary in nature, requires an even number of leaf nodes. In case, there are an odd number of transactions, the last transactional hash is duplicated once to get an even number of leaf nodes [4]. All leaf nodes are grouped in pairs and the hash of grouped pairs creates a new parent block. Finally, all parent nodes are grouped in pairs and the same process is repeated until there is only one node left, termed as *root hash* or *Merkle root*. Merkle root contains all details related to transactional data. In order to check the presence of any transaction in a particular block, only Merkle root is required to easily scan through whole transactions. It also helps to maintain the data integrity. The Merkle root changes whenever there is even a slight change in transactional details or order. The Merkle tree requires very less

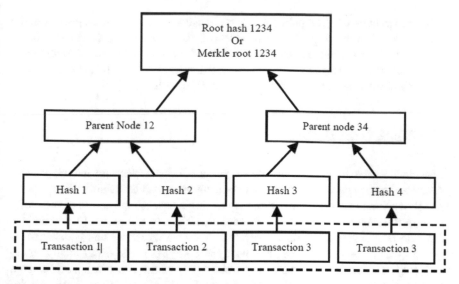

Fig. 4 Merkle tree

storage space and is computationally fast. It performs data verification along with data synchronization. Merkle trees are used in various existing Blockchains such as Bitcoin, Ethereum etc.

The main mechanism of Blockchain technology used for agreement between nodes of a network is called as a *consensus mechanism*. Since Blockchain is a distributed network, there is no chief entity to manage the Blockchain ledgers. Hence, consensus mechanism is accountable for ensuring Blockchain ledgers' validity. Once the block is validated, block gets affixed to the Blockchain. The following section describes several Blockchain based consensus algorithms to reach consensus in a network.

2.2 Blockchain Working

Following steps describe the working of a Blockchain network:

- Suppose a node k wants to send data to another node l in a Blockchain network. The sender node will store some transactions into a block and that block is broadcasted over the entire Blockchain network.
- Every other node in the network will receive the block and will attempt to validate it by using consensus algorithms like *proof-of-work*, *proof-of-activity*, etc. To validate transactions, an asymmetric cryptographic mechanism is used in the Blockchain. A digital signature is established on asymmetric cryptography that can be used in non-trusted environment. Each user has a pair of keys i.e., *public*

key and *private key*. The public key is advertised to all other nodes present in the network. The user keeps the private key with itself for signing the transaction.
- Once the block is authenticated i.e., validated with all correct transactions, it is attached to the Blockchain. The receiver can then retrieve the data from Blockchain and can be assured that the data is not tempered with.

2.3   Blockchain Types

Blockchain can be categorized into three types on a broader spectrum, namely, *public, private* and *consortium* [5] (Fig. 5). Wherein, anyone can enter a Public (Permissionless) Blockchain network, take part in the validation process and access the network. Existing Blockchain networks such as *Bitcoin, Ethereum* are some of the examples of public Blockchain network. However, a Private or Permissioned Blockchain is open only to member nodes i.e., any user can take part in the Blockchain mechanism only after being verified by the administrator of the Blockchain. Although a transaction can be read by anyone, only verified users can participate in the validation process. It can be seen as a centralized system up to some extent. Some of the examples of private Blockchain are Quorum, Blockstack and Multichain. Further, in a Consortium Blockchain such as Hyperledger and Ripple, anyone can interact with the network, but only a few predetermined nodes are permitted to participate in the

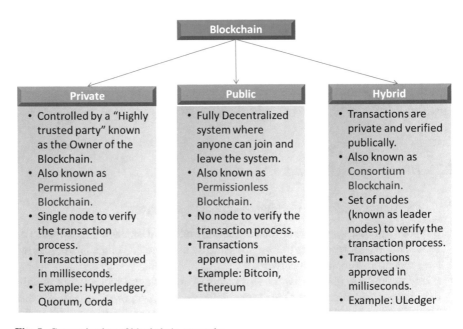

Fig. 5   Categorization of blockchain networks

validation process. The predetermined nodes can belong to different entities instead of one entity. Consortium Blockchain is hence, partially decentralized.

2.4 Blockchain Generations

With the Blockchain being coined as a term and core technology behind Bitcoin, various generations of Blockchain have been termed:

I. *Blockchain 1.0 (Cryptocurrency)*: Introduced in 2008 by pseudonymous developer of bitcoin, Satoshi Nakamoto. This earliest generation of Blockchain concentrated on implementing cryptocurrency/Digital payments using the Blockchain technology. The technology helps to maintain digital ledgers making it impossible to double spend the currency. Various examples of cryptocurrencies are Bitcoin, Litecoin, Ripple, Monero, etc.

II. *Blockchain 2.0 (Ethereum and Smart contracts)*: Introduced in 2015 by Vitalik Buterin. Ethereum allows the Blockchain developers to use Blockchain technology for applications beyond cryptocurrency using smart contracts. A smart contract is similar to a legal contract but a piece of code which can be deployed in order to perform transactions between parties.

III. *Blockchain 3.0 (Other applications)*: It includes other potential application areas such as health, government, science, etc.

2.5 Applications, Tools, and Interfaces

Blockchain technology can be used in diverse areas such as healthcare, finance, storage, Internet of Things (IoT) etc. [6, 7]. The subsection discusses various applications of Blockchain in different domains and discusses some of the tools and interfaces associated with Blockchain.

1. *Decentralized cryptocurrency*

Bitcoin is the paramount cryptocurrency based on Blockchain technology [1]. Here, the transactions are independent of any central entity and cryptocurrencies uses encryption techniques such as public key cryptography and hence, identities of users get anonymous. Each node, present in the Blockchain network, records and verifies each transaction that happens over the network (proof-of-work).

2. *Smart contracts*

A smart contract is a digital transaction protocol which defines a certain set of rules or conditions of a manual contract. In context of Blockchain, a smart contract is a form of script which is kept on the Blockchain. Each smart contract has a unique address and is triggered by addressing a transaction to it. Afterwards, smart contracts are executed independently on each node of the network in a pre-defined manner [8, 9].

3. *Internet of Things (IoT)*

IoT has a wider application range, for instance, smart grids, smart cities etc. However, increasing amount of data has given upsurge to various privacy and security issues [10], for example, lack of control, scalability, access management etc. To overcome such issues, Blockchain can be used in IoT that leads to a fully distributed system along with access control management [8, 9].

4. *Healthcare*

Blockchain is nowadays considered as a valuable tool for healthcare industry [11]. Various tasks such as tracking pharmaceuticals, decentralizing history of patient's health, and improving payment options, can be achieved using Blockchain technology. Hence, it will help revolutionize the healthcare industry worldwide. However, the healthcare records can't be stored on public Blockchain because of sensitivity of the data.

5. *Other applications*

There are many other use cases of Blockchain in numerous areas such as identity management [12], audit trails [13], supply chain, insurance, prediction market, storage etc.

Further, various tools and interfaces that can be used for developing Blockchain applications are mentioned in the Table 2 [14].

3 Consensus Mechanisms

In order to reach an agreement between the nodes of a network, a consensus mechanism [15] is generally employed in a Blockchain. The section overviews some of the widely used consensus mechanisms along with their variants and compares them on various parameters.

Table 2 Tools and interfaces

Application/use case	Tool/interface
Front-end	HTML, CSS
Back-end	Javascript, Python, C++, C#
Decentralized applications	Blockchain as a Service, Truffle, Embark, Blockchain Testnet, IOTA Tangle
Bitcoin applications	Coinbase's API, Tieron
Smart contracts	Ethereum (Solidity), Serpent, Mist, Geth
Smart contracts compiler	Remix, Solc, Ether contract
Blockchain emulator	Ganache CLI
Storage	Storj, BigchainDB, Zcash

3.1 Proof-of-Work (PoW)

The proof-of-work consensus mechanism was initially used in Bitcoin Blockchain. It includes a complex computational scheme for authentication and validation of a transactional block. In PoW, every miner is involved in calculating the hash of a block for validation. The consensus is reached when the calculated hash value is equal to or less than a predefined difficulty value. Firstly, a block which needs to be validated is hashed along with a nonce (*random number*) to get the preferred output i.e., hash value with certain number of zeroes prefixes (*n-Bits*). In case, miners do not achieve the desired output after one round of hash, the process of hashing is repeated with several nonces. Once the results are obtained by one of the miners, all other miners confirm the correctness for block validation. Lastly, the block is added to Blockchain while the miner who computed the desired hash is incentivized. *Example*: In Bitcoin, on successful block validation, miner gets 12.5 Bitcoins as an incentive. PoW uses SHA-256 hashing algorithm which makes it difficult for any user to guess the value of nonces. After every 2016 appended blocks, difficulty of PoW is adjusted to limit the addition of new block rate.

Further, in 2015, memory bound and scalable proof-of-work system, *Cuckoo Cycle*, was proposed [16]. This system allows miner nodes to successfully append the block. The authors have proposed a cuckoo hashtable [17] comprising of two hash tables with same size, though, different hash functions, thereby, resulting in two locations for each hash function. On insertion of a new key, if one of the positions is empty, the key is inserted. However, if both positions are occupied, new key is inserted in first location replacing the older one. Afterwards, the displaced key is inserted in another location and the same process is repeated until a vacant position is found, or maximum number of iterations is reached. The Cuckoo hash function can create a bipartite graph known as *Cuckoo graph*. In this graph, vertices contain displaced positions whereas edges contain displaced values of removed positions. The original proof-of-work is replaced by a new puzzle work, in which, miner needs to guess many nonces and guessed nonces are inserted in the hash functions until a graph is created with sufficient iterations. Lastly, miner broadcasts the block comprising of guessed nonces.

Besides, *Primecoin* [18] implements a proof-of-work mechanism which explores prime number chains. It is an open-source peer-to-peer cryptocurrency. The authors have proposed to search longest prime number chains instead of meaningless nonces. Primecoin uses *bi-twin chains* and *Cunningham chains* [19]. It has three requirements i.e., larger length than the given value, Cunningham chain form and origin value (midpoint of first two chain primes) divisible by a required number. Whenever a node broadcasts a block, other miners use an algorithm called the *Fermat Primality Test algorithm* to check the chain. As the chain length increases, the task of finding chain of prime numbers become exponentially difficult.

Further, in 2017, authors have proposed a generalized proof-of-work mechanism based on original PoW [20]. Here, instead of a single miner, a group of miners with certain stake or luck are required to append the block. Hence, the miners can easily find many possible nonces for a block. It consists of many appending rounds, and in every round, each miner finds possible nonces for their newly created block. If miner finds any nonce, it broadcasts the block along with nonce to all other miners who are responsible for validating the block and the validated block is stored in a temporary array. Afterwards, a random number is chosen for the round. If the block index of miner is equal to the random number, miner becomes a member of the group and adds the block to Blockchain. To control and check number of miners in a group, few miners are rotated after certain time within the group. The incentives are distributed equally among miners of the group after every round of appending block. However, the PoW has following tradeoffs:

- *Sybil attack resilience*: There is no possibility of malicious user attack to create fake nodes and suppress honest nodes in order to control Blockchain.
- *Incentivization*: The incentive amount is miner's hash rate dependent.
- *51% Attack vulnerability*: As soon as miners start mining in a pool and their control exceeds 50%, the pool can certainly control the complete Blockchain for its malicious demands.
- *Fork creation possibility*: A fork is created when two miners find the desired nonce at the same time. While one miner's nonce is broadcasted to some nodes, other nodes get nonce of another miner which may lead to creation of fork.
- *Computational time and efficiency*: The consensus mechanism takes longer computational time i.e., 10 min per block. Also, PoW is energy inefficient.

3.2   Proof-of-Stake (PoS)

As PoW is computationally difficult as well as time and energy consuming, an alternative mechanism called as *Proof-of-Stake* (PoS) has been proposed in literature. PoS is more energy efficient as the scheme does not require all miners to compute the hash value. In PoS, stake (share percentage) is used to decide the miner of the subsequent block i.e., a node with 1% of Bitcoin can only mine 1% of PoS blocks. In addition, PoS is more resilient to attacks. In order to attack, a malicious user needs maximum stake share i.e., over 50% stakes of the total.

Nxt [21], a cryptocurrency as well as a platform for payments, uses pure proof-of-stake consensus mechanism. In this, a miner has more chance to append a block if he has more stakes. Suppose a miner A has 'm' coins whereas all miners collectively have 'n' coins such that $m < n$, then miner A has a mining chance of m/n. After every 60 min, a miner is randomly selected based on miner's stakes. Once the miner is selected to mine a block, miner will verify all newly added transactions and store them into a block. Afterwards, miner broadcasts the newly created block and receives incentive in return. Further, in 2016, authors proposed a method where miners get a chance to append a block not only on the basis of stakes, but also on the basis of block state [22]. The authors also proposed *follow-the-Satoshi* method. The smallest unit of currency, *Satoshi* value, has been considered. A Satoshi index, between 0 and total Satoshis, is given as input to follow-the-Satoshi method. Afterwards, based on

Satoshi index, miner will find the Satoshi of the block. To find a new miner (last owner of the given Satoshi) who will append the next block, all the transactions along with given Satoshi are found. The hash function used takes three inputs to select Satoshi. The first input is a value returned by a function, the second input is total number of blocks in the Blockchain, and thirdly a random integer. For the selected Satoshi not being able to create a block, authors have proposed a solution. The proposed solution is that the selected Satoshi is blacklisted, if even after selecting for three consecutive times, Satoshi is not able to create a new block.

Later, in 2014, instead of using stakes for a chance to mine a new block, authors used stakes for voting called as *delegated proof of stake* [23]. Here, the nodes who possess stake will vote for witnesses or delegations, where the delegations are referred to as miners who are responsible for transactions verification. If a node has sufficiently large number of stakes, voting assigned to the delegation becomes stronger. Afterwards, a verification committee is made. The delegations within the committee verify the transactions and in addition, produce new blocks, valid or invalid. The role of delegation, essentially, a miner is kept rotating within the committee for fairness preservation. The proposed method creates a new block sequentially in 2s. A witness is removed from the committee if it is not able to produce a block. Otherwise, block is added into the chain and witness gets a reward in return. The PoS has following tradeoffs:

- Consumption of energy is low in contrast to PoW.
- Efficient in decision making i.e., takes less time for computation.
- 51% attack resilient as a miner requires 51% stakes of the Blockchain.
- Larger systems are 51% attack proof whereas smaller networks are 51% attack prone. This, in turn, results in centralization and complete control of the Blockchain by selected entities.

3.3 Proof-of Activity (PoA)

The PoA is a hybrid consensus mechanism [24], which combines both, PoW and PoS. It works as follows: firstly, an empty block i.e., a block header with former block hash without any transaction detail is generated by miners in consensus. Secondly, miner guesses the nonce of empty block. Upon finding a certain nonce, miner broadcasts the empty block to other remaining miners that are responsible for verification and validation of the received blocks. The above-mentioned steps follow proof-of-work which is further processed by proof-of-stake mechanism.

During validation and verification process, each miner checks itself for being stakeholder with higher number of stakes. The stakeholder i.e., Nth miner will generate a block consisting of an empty block along with numerous transactions and signature of N-1 miners along with his signature (Nth miner) to create a hash of the complete block. Thereafter, Nth miner broadcasts the block over the network and block gets attached to the Blockchain. In order to maintain fairness, equal incentives are given to the stakeholder along with N miners [24]. In addition, to reduce traffic,

Table 3 Comparison of various consensus mechanisms

Parameters	PoW	PoS	PoA
Energy efficiency	No	Yes	No
51% attack	Yes	Not in larger networks but possible for smaller networks	Yes
Fork creation	Yes	Very difficult	Possible
Double spending	Yes	Difficult	Yes (less impactful than PoW)
Block creation speed	Low	Fast	Low
Pool mining	Yes (easy to prevent)	Yes (difficult to prevent)	Yes
Example	Bitcoin [1]	Ethereum	Peer-to-peer coin [25]

transaction list is added into the empty block by the Nth miner. The PoA is more efficient in terms of double spending problem prevention.

Table 3 compares the three consensus mechanisms based on certain aspects. The Proof-of-Stake was introduced to address the drawbacks of Proof-of-Work algorithm such as 51% attacks, energy inefficiency etc. Mostly, Blockchain applications are based on PoW and PoS. The new consensus mechanisms e.g., Proof-of-Activity (PoA), are generally designed using both PoW and PoS [26].

4   Scalability Analysis of Blockchain

In recent years, cryptocurrencies and Blockchain have gained massive popularity, although, the initial scope was not wider. As the number of transactions grow, many issues emerge like *increased response time, less throughput* and *more computational cost*. Thus, there is a possibility that current technology may not be able to sustain the growing demand. Both Ethereum and Bitcoin use blocks to store and process transactions. However, the maximum size of the blocks (1 MB for Bitcoin) was limited initially. Although, size was restricted to make Bitcoin more protected, it led to scalability issues [27]. It is well understood that with each and every transaction in Blockchain, data is getting huge. Hence, with 1 MB for each block, only few payments can be processed at once, thereby, resulting in poor performance. Further, bitcoin has a throughput of three to four transactions per second.

However, in order to implement Blockchain in areas like *IoT*, *healthcare* etc., it becomes essential to process large number of transactions per second without any delay. Ethereum also has a similar problem i.e., the network has maximum capacity of 15 transactions per second. Here, the number of transactions per second can be increased by improving scalability. The main scalability concerns in cryptocurrencies are as under:

- *Latency* i.e., time taken to record transactions in a block.
- *Consensus time* i.e., time taken to reach final consensus.

This section covers scalability possibilities related to Blockchain by analysing thoroughly those factors which make it somewhat non-scalable. The various factors such as block-size, block-interval, throughput and transaction fee etc. affect Blockchain's performance along with its security. These are corelated to each other. Moreover, scalability limitations of various Blockchain technologies are discussed most frequently. The several system parameters which affect Blockchain scalability are discussed as follows:

1. *Block-interval*: It defines the latency i.e., time taken to record data in a block of a Blockchain. The block interval is inversely proportional to transaction confirmation time i.e., smaller block-interval will result in faster transaction confirmation. Hence, probability of producing stale blocks becomes higher. In PoW Blockchains, change in block-interval is directly related to the difficulty. As mentioned before, low difficulty results in large number of blocks within a network whereas high difficulty results in a smaller number of blocks for same time interval. Therefore, it is important to analyze how changes in difficulty will affect the security of PoW based Blockchains.

2. *Block-size*: It defines the number of transactions stored within a block. As mentioned before, block size of Bitcoin is fixed to 1 MB due to which a block can store up to 2400 transactions i.e. 7 transactions per second whereas payment platforms such as Visa have more than 6000 transactions per second. For PoW based Blockchains to match Visa like payment platforms, various factors such as *block-size* need to be analyzed. These numbers of transactions can be achieved by increasing the block-size. However, block-size increase in PoW based Blockchain has certain related tradeoffs:

   - Increase in block-size requires huge power consumption to store transactions in a block and mine that block.
   - Increase in blocksize leads to transmission of larger blocks over the previous network bandwidth. Hence, network propagation time increases and speed decreases.
   - Increase in block-size leads to higher probability of stale blocks.

   Hence, block-size controls the throughput of the network. Moreover, larger blocks lead to slow propagation speeds and therefore, stale block rate increases and security of Blockchain gets weaken.

3. *Transaction fees:* In PoW based Blockchains, whenever a user creates a transaction, it includes a transaction fee which is received by the miner who adds the block with that transaction into the Blockchain. There is no reward for miners to store a feeless transaction in a block. Since, block size in PoW Blockchains is fixed, it also restricts the number of transactions a block can hold. Hence, miners chose transactions with higher fee. The transaction fee directly affects the confirmation time. Therefore, the transactions with small fee have large confirmation time whereas higher transaction fee leads to less confirmation time. However,

few transactions might suffer starvation because of mid-range transaction fee though few others might get confirmed.

4. *Number of nodes*: The increase in number of nodes within a network has negligible impact on transaction rate i.c., *Transactions per second (tps)*, since each node processes the transactions independently. Rather than affecting the transaction rate, it has an impact on latency of the network because blocks with the transactions require more number of hops to propagate through the entire network.

5. *Number of miners*: It has negligible impact on transaction confirmation time. However, with the increase in the number of miners, stale block rate also increases gradually. Whenever a block is mined by various miners, everyone propagates the mined block through the network to make sure that other miners become aware of who has mined the block first to be added to the Blockchain. Since various miners have propagated the mined block, it is possible that nodes will receive different blocks, and this leads to creation of forks in the Blockchain. As PoW based Blockchains discard forks by always choosing the longest fork after certain time, this will lead to increase in abandoned blocks. Hence, increase in number of miners will increase the stale block rate and can lead to doubles pending attacks.

4.1 Simulation Details

This section discusses the process to analyze scalability issues by simulating a PoW based blockchain. The question arises here is that "*why can't we perform simulation on real Blockchain?*". The answer is due to huge utilization and higher computational cost, it is difficult to study and analyze the behaviour of a real PoW based Blockchain. Hence, simulation is the feasible option to analyze the performance of a Blockchain depending upon various parameters. This simulation is based on an existing framework [15] with few updations and addition of certain functionalies in order to support the scalability analysis of PoW based Blockchain. In our approach, we have discarded the network delay along with all networks. The various parameters for the analysis and hardware configuration are mentioned in Tables 4 and 5 respectively.

Here, we have analyzed the above-mentioned parameters such as block size, number of blocks etc. and studied their impact on transaction-confirmation time and block-confirmation time. Pseudocode 1 presents the pseudocode to generate transactions and adding them in a block based upon the transaction fee.

Table 4 Parameters for scalability analysis

Parameters	Description
Transaction fee	Fee given to miners in order to get the transaction confirmed
Block size	Block size in bytes
Transaction-confirmation time	Time taken by a transaction to be added into a confirmed block
Number of transactions	Number of transactions in a block depending upon block size
Number of blocks	Number of blocks mined in a certain time
Number of miners	Number of miners responsible for mining and if huge in numbers can
Block generation time	Time required to add a block to the blockchain

Table 5 Hardware configuration

Operating system	Ubuntu 18.04.2
Simulator	ns3.29
Processor	Intel(R) Xeon(R) W-2155 CPU @ 3.30 GHz 3.31 GHz
Memory	64.0 GB RAM

Pseudocode 1: Transactions (N, S)

```
Q = Ø                                    // Q is the priority queue
Total_size = N * S                       // N is the number of blocks and S is the block size
while Total_size > 0:
        Transaction t = new Transaction
        t.startTime = currentTime( )
        Q.enqueue(t,t.fee)
        Total_size = Total_size – t.size
//Add Block to Blockchain
while N > 0
        c = 0
        Block b = new Block;
        while c < S
                t = Q.dequeue
                b.add(t)
                c = c + t.size
        Mine block
        Add Block to Blockchain
        for each Transaction t in Block b:
                t.finishTime = currentTime( )
for each Transaction:
        Print Confirmation time
```

Here, Q maintains transactions according to the increasing order of transaction fee. The transaction fee is calculated implicitly by using Eq. 1.

$$\text{Total Transation fee} = (a * 180 + b * 34 + 10) * 1 \, \text{satoshi} \qquad (1)$$

Here, a and b are number of inputs and number of outputs respectively which are generated randomly within 1–5. For calculation of total fee, we have considered base fee as 1 satoshi which is equivalent to 0.0000001 BTC.

4.2   Simulation Results

As the Blockchain is simulated, hence, the simulation results are much faster than the real time Blockchain. The simulation is run for different parameters and then the conclusion is drawn based on the simulation results.

(1)   *Impact of the nodes*: Here, we study the impact of number of nodes on block confirmation time and block propagation time in proof-of-work Blockchains. For this, the simulator is run for different number of nodes ranging from 100 to 6000 nodes. Each simulation is run for 1000 consecutive blocks of 1 MB size. Firstly, it is observed that up to 1000 nodes block confirmation time is increasing rapidly and after 1000 nodes it is increasing linearly (Fig. 6). Secondly, it is observed that up to 700 nodes the block propagation time is increasing rapidly reaching peak and at 700th node there is sudden decrease in the time up to 1400 nodes. After that, time is decreasing slowly and after 2500th node the time is gradually getting constant (Fig. 7). Hence, it is observed from the simulation that with the increase in number of nodes, the block confirmation time is increasing linearly, and block propagation time is getting constant after increase and decrease simultaneously.

(2)   *Impact of block-size*: Here, impact of block-size on block confirmation time and number of transactions in Proof-of-work Blockchains is studied. For this, the simulator is run for different block sizes ranging from 1 KB to 8 MB. Firstly, it is observed from the simulation that as the block size increases from 1 KB to 1 MB, the number of transactions also increase linearly (Fig. 8). Secondly, it is observed that with increasing block size, the block confirmation time is almost constant up to 4 MB with very slow increase in the time. After 4 MB,

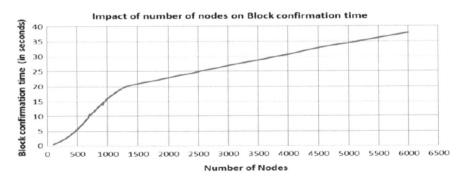

Fig. 6 Number of nodes versus average block confirmation time

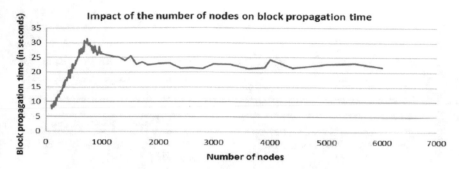

Fig. 7 Number of nodes versus average block propagation time

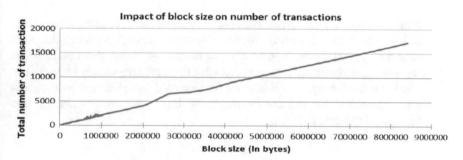

Fig. 8 Block size versus total number of transactions

time is increasing very slowly whereas after 8 MB block confirmation time is increasing rapidly (Fig. 9).

(3) *Impact of transaction fee*: Here, we study the impact of transaction fee on transaction confirmation time in proof-of-work Blockchains. For this, the simulation is run for 8 MB block size and depending upon their transactions the trend has

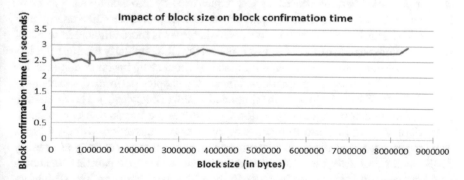

Fig. 9 Block size versus average block confirmation time

Fig. 10 Transaction fee versus transaction confirmation time

been followed. It is observed that lesser the transaction fee more is the transaction confirmation time (Fig. 10) and vice-versa. It is not important that each
time the transaction fee is inversely proportional to transaction confirmation
time. In the simulation study, we found that there are some cases when time is
less for lesser fee whereas it is more for increased fee. Hence, if a user gives
more transaction fee, the chances of getting the transaction confirmed in less
time are very high [28].

5 Conclusion

Blockchain is one of the fast-mounting technologies and is used in many applications like Healthcare, IoT, supply chain, insurance, etc. It provides decentralized
distributed peer-to-peer network in addition to trust less interaction among peers. The
chapter provided detailed discussion on Blockchain basics, its architecture, types,
generations, and various application areas. Further, for successful implementation
of Blockchain, efficient consensus algorithms and hybrid protocols are needed. The
chapter presented various consensus techniques used by Blockchain such as PoW,
PoS and PoA along with their comparison on various parameters like energy efficiency, 51% attacks etc. It has been observed that PoW is more efficient in terms of
security as compared to other consensus mechanisms. It is popularly used as Bitcoin's
consensus mechanism. In addition, it is noted that Blockchain becomes non-scalable
with increased number of transactions. Hence, the chapter performed a scalability
analysis for PoW based Blockchain (simulated Bitcoin Blockchain). The simulation
results indicate that more the transaction fee, lesser is the transaction-confirmation
time. Further, with increase in block-size, number of transactions within a block
also increases whereas block-confirmation time increases exponentially after every
8 MB block-size. Lastly, the results also indicate that there is an impact on Blockchain
performance with the increase in nodes in the Blockchain network. The scalability
issue may be addressed with different solutions such as *sharding, lightning networks*
etc. [29].

References

1. Nakamoto, S.: Bitcoin: a peer-to-peer electronic cash system. Decentralized Bus Rev 21260 (2008)
2. Hofmann, E., Strewe, U.M., Bosia, N.: Background III—what is blockchain technology? In: Supply Chain Finance and Blockchain Technology, pp. 35–49. Springer (2018)
3. Ali, O., Jaradat, A., Kulakli, A., Abuhalimeh, A.: A comparative study: blockchain technology utilization benefits, challenges and functionalities. IEEE Access **9**, 12730–12749 (2021)
4. Garewal, K.S.: Merkle trees. In: Practical Blockchains and Cryptocurrencies, pp. 137–148. Springer (2020)
5. Niranjanamurthy, M., Nithya, B., Jagannatha, S.: Analysis of Blockchain technology: pros, cons and SWOT. Clust Comput. **22**, 14743–14757 (2019)
6. Guo, Y.-M., Huang, Z.-L., Guo, J., et al.: A bibliometric analysis and visualization of blockchain. Futur. Gener. Comput. Syst. **116**, 316–332 (2021)
7. Malhotra, D., Saini, P., Singh, A.K.: How blockchain can automate KYC: systematic review. Wirel. Pers. Commun. 1–35 (2021)
8. Novo, O.: Blockchain meets IoT: an architecture for scalable access management in IoT. IEEE Internet Things J. **5**, 1184–1195 (2018)
9. Christidis, K., Devetsikiotis, M.: Blockchains and smart contracts for the internet of things. IEEE Access **4**, 2292–2303 (2016)
10. Dorri, A., Kanhere, S.S., Jurdak, R.: Towards an optimized blockchain for IoT. In: 2017 IEEE/ACM Second International Conference on Internet-of-Things Design and Implementation (IoTDI), pp. 173–178 (2017)
11. Mettler, M.: Blockchain technology in healthcare: the revolution starts here. In: 2016 IEEE 18th International Conference on e-Health Networking, Applications and Services (Healthcom), pp. 1–3 (2016)
12. Yasin, A., Liu, L.: An online identity and smart contract management system. In: 2016 IEEE 40th Annual Computer Software and Applications Conference (COMPSAC) pp. 192–198 (2016)
13. Malhotra, D., Srivastava, S., Saini, P., Singh, A.K.: Blockchain based audit trailing of XAI decisions: storing on IPFS and ethereum blockchain. In: 2021 International Conference on COMmunication Systems and NETworkS (COMSNETS), pp. 1–5. IEEE (2021)
14. Bahga, A., Madisetti, V.: Blockchain Applications: A Hands-On Approach (2017)
15. Rebello, G.A.F., Camilo, G.F., Guimarães, L.C., et al.: On the security and performance of proof-based consensus protocols. In: 2020 4th Conference on Cloud and Internet of Things (CIoT), pp. 67–74 (2020)
16. Tromp, J.: Cuckoo cycle: a memory-hard proof-of-work system. IACR Cryptol ePrint Arch **2014**, 59 (2014)
17. Pagh, R., Rodler, F.F.: Cuckoo Hashing. J. Algorithms **51**, 122–144 (2004)
18. King, S.: Primecoin: Cryptocurrency with Prime Number Proof-of-Work (2013)
19. Hardy, G., Littlewood, J.: Cunningham Chain (2020)
20. Tang, S., Liu, Z., Chow, S.S., et al.: Forking-free hybrid consensus with generalized proof-of-activity. IACR Cryptol ePrint Arch **2017**, 367 (2017)
21. Nxt wiki (2020). https://nxtwiki.org/wiki/Whitepaper:Nxt. Accessed 15 May 2021
22. Bentov, I., Gabizon, A., Mizrahi, A.: Cryptocurrencies without proof of work. In: International Conference on Financial Cryptography and Data Security, pp. 142–157. Springer (2016)
23. Yang, F., Zhou, W., Wu, Q., et al.: Delegated proof of stake with downgrade: a secure and efficient blockchain consensus algorithm with downgrade mechanism. IEEE Access **7**, 118541–118555 (2019)
24. Bentov, I., Lee, C., Mizrahi, A., Rosenfeld, M.: Proof of activity: extending bitcoin's proof of work via proof of stake. ACM SIGMETRICS Perform. Eval. Rev. **42**, 34–37 (2014)
25. King, S., Nadal, S.: PPCoin: Peer-to-Peer Crypto-Currency with Proof-of-Stake (2012)
26. Bamakan, S.M.H., Motavali, A., Bondarti, A.B.: A survey of blockchain consensus algorithms performance evaluation criteria. Expert Syst. Appl. 113385 (2020)

27. Chauhan, A., Malviya, O.P., Verma, M., Mor, T.S.: Blockchain and scalability. In: 2018 IEEE International Conference on Software Quality, Reliability and Security Companion (QRS-C), pp 122–128 (2018)
28. Kasahara S, Kawahara J (2016) Effect of Bitcoin fee on transaction-confirmation process. arXiv Prepr
29. Zhou, Q., Huang, H., Zheng, Z., Bian, J.: Solutions to scalability of blockchain: a survey. IEEE Access **8**, 16440–16455 (2020)

Smart Contracts: A Way to Modern Digital World

A. D. N. Sarma

Abstract This chapter presents an introduction that covers the background of the study includes traditional contract system and its limitations. In the past five four years, a large number of businesses, organizations, governments, and communities have been paying much attention to blockchain technology, as it opens up a new model of the legal contract known as smart contract. The smart contract is one of the applications of blockchain technology next to bitcoin, which can find a wide spectrum of potential applications in the digital economy including financial, insurance, health care, supply chain, trade, commerce, retail, and Internet of Things that transforms the world into a modern digital world. Blockchain technology enables us to execute, maintain, and monitor the processes of the contract system digitally from the beginning to the closure. Further, explained the importance of smart contracts in the modern digital world. Besides, briefly explain what a smart contract is and mention the elements of the smart contract. In addition, mention the evolution of various types of smart contract technology platforms. Moreover, present the life cycle of the smart contract and explain the functioning of each phase. In addition, briefly touch upon the major characteristics and benefits of a smart contract. Furthermore, mention three open challenges in developing smart contracts. Finally, discuss possible future trends and directions of smart contracts. Additional reading material is provided such as relevant books and papers that explore smart contracts in greater depth.

Keywords Blockchain · Blockchain platform · Blockchain types · Decentralized · Digital identity · Distributed ledger · Elements of smart contract · Hyperledger fabric · Off-chain and on-chain · Record keeping · Self-executing terms and conditions · Smart contract lifecycle · Types of smart contracts · Third party

Objectives

The objective of this chapter is to provide a holistic view of a smart contract using blockchain. Additionally, to provide a conceptual view of the existing legal contracts and its major limitations. In addition, the introduction to smart contracts,

A. D. N. Sarma (✉)
Centre for Good Governance, Hyderabad, India

© The Author(s), under exclusive license to Springer Nature Switzerland AG 2022
K. R. Ahmed and H. Hexmoor (eds.), *Blockchain and Deep Learning*,
Studies in Big Data 105, https://doi.org/10.1007/978-3-030-95419-2_4

the major elements of smart contract and their importance in the modern digital world. Besides, to provide the required background to understand the smart contract system that includes - what is a legal contract, how a legal contract will transform into a smart contract in the modern digital world. To provide the key elements that constitutes a smart contract framework and the life cycle of smart contract. To mention the key characteristics, benefits, and challenges of smart contracts. Furthermore, to mention the three challenges in developing blockchain based smart contracts and discuss future treads and directions.

1 Introduction

In the last five years, a large number of businesses, organizations, governments, and communities have been paying much attention to blockchain technology, as it opens up a new model of the legal contract. Blockchain technology enables us to execute, maintain, and monitor the processes of the contract system from the beginning to the closure. The term smart contract mainly refers to the automation of the legal contract [1]. More precisely, smart contract can refer as a digital version of the legal contract. Technically, the term smart contract means a piece of software code, which is executed on the top of the blockchain platform.

In general, traditional contracts take place between two or more parties in consultation with a legal advisor, and this is turned into a legal document, which is commonly referred to as a contract. This contract document is a legally enforceable agreement between the contracting parties. In business, the contact may often involve more than two parties. Primarily, any contract can include the following components: details of parties, term, value of goods or service, and terms and conditions. The contract gets executed as per the terms and conditions stated in the contract, and then the contract closes, whereas in other cases it may enter the maintenance and continue for several years. The terms and conditions of the contract to be monitored on a continual basis, during the currency of the contract, which means to monitor Service Level Agreement (SLAs) of the contract for delivery of goods or services.

In the maintenance phase of a contract, if the contract terms and conditions are properly fulfilled, then such a contract can be stated as compliance with formal requirements, otherwise the contract may become non-compliance. Due to any reason, one of the parties could not fulfil the contractual obligation, let us say the payment settlement between the parties gets delayed than stated, which means there is a deviation of contractual obligation between the parties. Moreover, when the terms and conditions of the contract are not fulfilled, then compensation maters between the parties will evolve such as penalties, interest charges shall come to force, which shall be taken up as a separate process for settlement. The following are few limitations in the traditional contract system: (i) manual process of contract, (ii) role of legal consultation during preparation of contract, (iii) lack of real time settlement (iv) a lot of friction may involve between the parties to complete the deal during non-compliances, (iv) need for third party involvement during legal settlements, (v) prolonged time delays during execution of settlements, and (vi) many times the spirt of the contract may become diluted, which may leads to several controversies between the parties.

The Hyperledger Fabric is one of the most popular permissioned blockchain platforms, which is an open-source platform that was developed by the Linux foundation. These platforms can enable us to build multiple applications, which range from small and medium to an enterprise scale. The Hyperledger [2, p. 37] describes the terms smart code as a chain code, this includes a set of rules that determines how the contracting parties can interact with each other. In an agreement (or a contract), whenever predefined rules are met, then the agreement enforces automatically, which is the purest form of decentralized automation. The smart contract code is responsible for facilitating, verifying, and enforcing the negotiation or performance of a transaction or an agreement. In this chapter, the author focuses on the approach of the smart contracts that is based on the Hyperledger fabric blockchain platform.

The phrase 'digital world' has numerous meanings in the literature. It is described as "a world of digital technology, which is made of digital systems and emerging technologies in the contemporary society". In [3], the phrase digital world is stated as the availability and use of digital tools to communicate on the Internet, digital devices, smart devices, and other technologies. In this paper, the term 'digital world' is used in the context of leveraging the use of emerging technologies for building a smart contract, which is based on the distributive ledger technology (broadly blockchain) for the benefit of users, organizations, and governments in the modern-day society. Blockchain-based smart contract systems provides a wide range of economic and social advantages in the modern digital world when compared to the traditional contract system. These are: (a) the contracting process is fully digitized, (b) the contract process can be made smart by efficient management of its lifecycle of activities; (c) improved transparency to do business [4], (d) speed and accuracy [5] (e) reduced transaction cost [6] and speed of execution, (f) reduced dependence on middlemen such as legal, (g) trust among the participants [5], and (h) to develop new business models [6] there by organizations ability to generate value to name few.

There is relatively little literature existing in regard to the study of the economic aspect of the blockchain technology and its applications like smart contracts. A recent study [6] presents how blockchain technology can shape innovation and competition in digital platforms. In addition, this study describes two costs that are affected by blockchain technology. These are (i) the cost of verification, and (ii) the cost of networking. The cost of verification relates to the ability to cheaply verify the state of the transaction, whereas the cost of networking refers to the ability to bootstrap and operate a marketplace without assigning the control to a centralized intermediary. These costs have implications for the design and efficiency of digital platforms, and open opportunities for new approaches to data ownership, privacy, and licensing; monetization of digital content; auctions and reputation systems. The reduction of the cost has come at the cost of increased market power and data access of the platform operators. According to the market report [7], forecasts suggest that global blockchain technology revenues will experience massive growth in the coming years, with the market expected to climb to over 39 billion U.S. dollars in size by 2025.

At the end of chapter, an additional reading material provided references to relevant standards, books and papers that explore smart contract in greater depth.

2 Background

In this section, briefly introduce background knowledge about blockchain. This includes explanation of various computing types such as centralized computing and Peer to Peer network. Further, briefly present what is a distributed ledger and mention various types of blockchain networks. Moreover, explain the working of on-chain and off-chain transactions in a blockchain. Furthermore, briefly outline transaction validation mechanism in a blockchain.

2.1 Centralized Computing System

In the early 1970s, the concept of the Internet leads to the away to the web technology, which has transformed the traditional client–server system landscape into a centralized architecture. In this environment, all computations are performed at a central location that is on a particular computer, which is known as dedicated server for processing data. In addition, the storage and network resources are also centrally located along with the server infrastructure, which is shown in Fig. 1. The

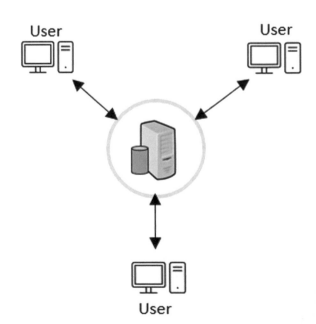

Fig. 1 A typical view of a centralized web application computing system

advantages of these systems are simple in design, easy to maintain, and highly scalable by increasing the computing resources as the number of the users increases to meet the workloads of the application. Most of the web applications are built using a centralized database (or ledger), which is more vulnerable to cyber-attacks and fraud, as it has a single point of failure. In a distributed environment, the computation is distributed to multiple computers, which join forces to solve the task, unlike a centralized system.

2.2 Peer to Peer Network

In a Peer to Peer (P2P) network architecture, the computers (or nodes) are connected to each other via the Internet, which is an example of distributed application architecture. Figure 2 shows a simple P2P network. In this network, each node is equally privileged, and workloads are distributed across nodes. The two main requirements to join a new node in a P2P network are: (i) an Internet connection and (ii) the P2P software. The common P2P software programs available in the market are Acquisition, BearShare, Kazaa, Limewire, and Morpheus to name a few. When all the nodes in a network are not consistent, such a network is called a disorganized P2P network. The P2P networks are more vulnerable to security attacks because nodes can exchange data. As a result, the participating nodes route traffic in the network, which opened up for security attacks.

Fig. 2 Organized Peer to Peer network

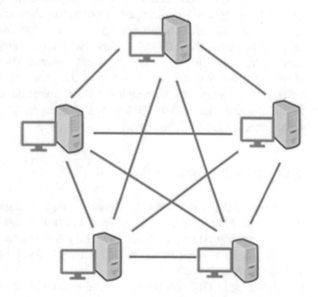

2.3   Distributed Ledger

The word ledger refers to a collection of files which is described as a set of records of transactional data. Most commonly, a ledger can be described as a database, a collection of files stored in a storage media, provides access, retrieval and use of data in a well-managed way. The distributive ledger is a digital record system of database which is most commonly known as a shared ledger or Distributive Ledger Technology (DLT). In this, the ledger is distributed, shared, and synchronized across the nodes in a network to the users for their access. All nodes on a blockchain network are connected to each other. Each computer maintains a copy of the ledger in a blockchain network, which synchronizes with ledgers located at other nodes. The Blockchain is one type of a distributed ledger [8]. Distributed ledgers use independent computers (referred to as nodes) to record, share and synchronize transactions in their respective electronic ledgers (instead of keeping data centralized as in a traditional ledger). Moreover, the data is organized in the blockchain in the form of blocks, which are linked and secured using a cryptographic hash. The Blockchain technology is often used by distributed ledger technology, which enables secure transactions between the parties without any presence of an intermediary, unlike a centralized system.

2.4   Blockchain

In recent years, blockchain is evolved as one of the emerging, disruptive technologies which has gained increasing attention worldwide in academic, business, industry, government, and research and development communities. It gave birth to Bitcoin, which is a decentralized digital currency without a central bank or single administrator, which uses the concept of the smart contract. Since its inception, the use of the smart contract is not only confined to the cryptocurrency applications but also extended its features for developing general applications. The applications cover in multiple domains such as the banking, finance, insurance, health care, trade and commerce, government, and more.

2.4.1   Definitions of Blockchain

According to [9] blockchains are immutable digital ledger systems implemented in a distributed fashion means that without a central repository and usually without a central authority. At its most basic level, they enable a community of users to record transactions in a ledger public to that community such that no transaction can be changed once published.

According to [10] blockchains are best understood as a new institutional technology that makes possible new types of contracts and organizations. Economizing on transaction costs leads to an efficient institutional structure of economic organization and governance.

According to [11] blockchain allows untrusting parties with common interests to co-create a permanent, unchangeable, and transparent record of exchange and processing without relying on a central authority.

2.4.2 Definition of Smart Contract

The following are a few definitions of smart contract by various practicing professionals, industry experts and academicians.

According to [5] Smart contracts are simply programs stored on a blockchain that run when predetermined conditions are met. They typically are used to automate the execution of an agreement so that all participants can be immediately certain of the outcome, without any intermediary's involvement or time loss. They can also automate a workflow, triggering the next action when conditions are met. At the most basic level, they are programs that run as they have been set up to run by the people who developed them.

According to [12] a smart contract defines the rules between different organizations in executable code. Applications invoke a smart contract to generate transactions that are recorded on the ledger. In sort, a smart contract defines the executable logic that generates new facts that are added to the ledger.

The terms smart contract defined [13] as computer protocols that are designed to automatically facilitate, verify, and enforce the negotiation and implementation of digital contracts without central authorities.

According to [14] a blockchain-based smart contract or a smart contract for short, is a computer program intended to digitally facilitate the negotiation or contractual terms directly between users when certain conditions are met.

2.4.3 Conceptual View of Blockchain System

A conceptual view of the blockchain network system is shown in Fig. 3. The important components of such a network system are nodes, ledger, miner, and consensus. It is essentially designed with a peer-to-peer (P2P) network of nodes. The node is essentially a computer system that maintains a copy of a decentralized ledger and a smart contract application known as *chaincode*, which acts as application software through with the services of smart contract available to the users. The user will initiate a transaction, which is the smallest building block of a blockchain system, such transactions are grouped in a predefined data structure known as block, and these blocks are distributed across all the nodes in the P2P network. Here a chain is a sequence of blocks that are arranged in a specific order, whereas miners validate new transactions and record them on a global ledger. The miner creates new blocks on the chain through a process called mining. Consensus is a mechanism that consists of a set of rules and arrangements to carry out blockchain operations.

Fig. 3 A conceptual view of blockchain system

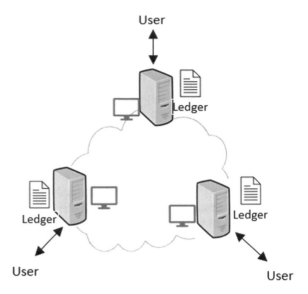

A blockchain can be described as a database, which is different from a traditional database in terms of storing data like decentralized and present information as well as past information, each participant will have a secured copy of the database, there is no administrator. The data being stored in blockchain is a decentralized manner that is in each node has a copy of data and all nodes will have similar data, which is validated by an internal mechanism known as consensus. The user will initiate a transaction which is endorsed, and such transactions will form into a block as per the predefined number of transactions per block, and these blocks are stored in a chain or a list. The data in each node in the ledger is cryptographically signed. Thus, each block consists of details like cryptographic hash of the previous block, time stamp, transaction date, and reproduces in the form of a Merkle tree. In a normal database, the storage model can be either a hierarchical, network, and object-oriented, whereas blockchain uses a data storage structure like linked lists (or a chain of blocks) in the form of a flat file. Directed Acyclic Graph (DAG) uses tree structure for a string of data. In DAG, there are no blocks of transactions, but the individual transactions are linked to multiple other transactions, which is like a tree. The examples of blockchain storage of databases are Bitcoin, Ethereum, Litecoin to name a few, whereas examples for DAG are Iota, Byteball, Nano (previously known as Railblocks), XDAG, NXT, and Orumesh. In addition, the other storage models are like Hashgraph, Hallochain and Tempo to name a few. The example of Tempo based blockchain is RadixDLT.

In a study [13] presented a framework for smart contract with reference to operational mechanism that comprise of six layers viz infrastructure, contracts, operational intelligence, manifestation, and application layers. The infrastructure layer deals the activities related to trusted development, execution, and trusted data feeds which is commonly known as Oracles. The contract layer deals the activities like static contract data, contract terms, scenario response rules, and interaction criteria. Besides, this layer covers the rules about contract invocation, execution, and communication, which acts as a static database of smart contract. The operational layer encapsulates the dynamic operations on the static contract that includes mechanism design, formal verification, security analysis and self-destruction. The intelligent layer encapsulates various intelligent algorithm, learning and decision making, the results will be fed back to the contract layer and operation layer for optimizing the contract design and operation in order to bring truly "smart" to the contract. The manifestation layer encapsulates various manifestation forms of smart contracts for potential applications including DApps, decentralized autonomous organizations, corporations, and societies. The application layer maintains all the smart contract application that are built upon the manifestation layer.

In a study [15] a taxonomy of blockchain-enabled smart contracts was presented focusing on smart contract improvement (i.e., addressing smart contract security, privacy, and performance issues) and smart contract usage (i.e., addressing domain-specific which is shown in Fig. 4.

In a study [16] outlines blockchain based service-oriented computing paradigm this study identifies analogies and difference between the Service-oriented Architecture (SoA) interpretation of smart contract and traditional web services technologies.

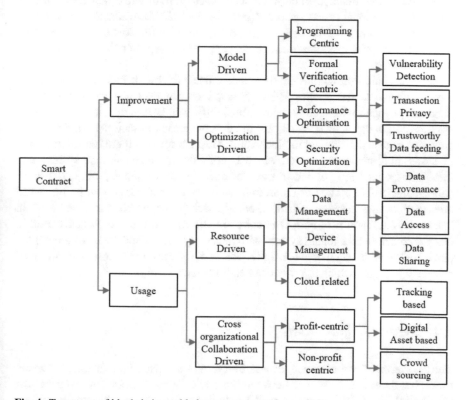

Fig. 4 Taxonomy of blockchain-enabled smart contract (*Source* [15])

This study mentions various challenges in enabling SoA in the Ethereum based smart contracts such as searching, discovery and reuse, cost awareness, interoperability and standardization, and performance.

2.5   On Chain and off Chain Transactions

The on-chain transactions that are performed within the blockchain platform are public, which means that the details of transactions are visible to the participants in the network. So, the users of the blockchain network can see all the transaction details. Thus, there is no privacy available for on-chain transactions of the smart contracts. Most of these transactions are structured data, such as the character, numeric formats. The amount of time taken to perform these transactions in a blockchain network depends on the volume of transactions, and number of nodes in a network. As the number of nodes in a network increase, then the transaction time increases because of the network congestion, these results in delay in completing the transaction. Moreover, the miner will take a longer time for completing the transaction in a public blockchain network. The on-chain transaction demands not only higher computational resources, but also time-oriented, this results in increasing the cost of the transaction drastically. In order to minimize the cost of the transaction and to provide privacy to the transactions [17], the concept of off-chain transaction was introduced.

The Off-chain transactions are happened outside the blockchain platform, which are not visible to the public. These transactions can take place in conventional databases, happen without any delay. Thus, Off-chain transactions are much faster as compared to on-chain transactions. In addition, these transactions cannot be irreversible because they are not in blockchain. Most commonly, these transactions will be stored in a third-party server, which is not a decentralized storage system, and these transactions are pointed to the blockchain with a pointer. This can also be called "oracle data". The Off-chine transactions are mostly unstructured like word document, images, objects, pdf files, emails, web pages, audio, and video files. In few cases, the features of both On-chain and Off-chain can be combinedly used to design a scalable blockchain system with necessary checks and balances among the parties of the blockchain network for reduction of the transaction cost, and improved privacy of the data which is known as a hybrid chain.

2.6   Types of Blockchains

The two terms distributive ledger and blockchain are used interchangeably many times, but there is a difference between these two terms. Blockchain is a subset of distributive ledger (DLT), whereas DLT is built on distributive database. According to end user's perspective blockchain can be classified into public and private whereas

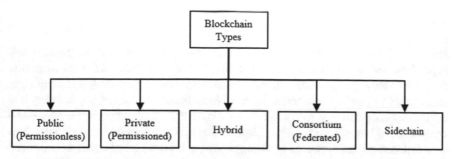

Fig. 5 Types of blockchains

based on network infrastructure it can be classified into three categories like permissioned, permissionless and hybrid. Figure 5 shows various types of blockchain network types. In a permission less blockchain everyone can become a validator for a blockchain, whereas in permissioned blockchains only specified (or preselected parties) can validate transactions.

2.6.1 Public Blockchains

A public blockchain has absolutely no access restrictions to the users. The participant user may be any individual whose identify does not require for performing transaction in the network. The user can perform transaction which may be either to buy or sell cryptocurrency and become a validator in the execution of a consensus protocol. These type of blockchain networks offer an economic incentive for those who secure them and utilize some type of a Proof of Stake or Proof of Work algorithm. Some of the largest, most known public blockchains are the bitcoin blockchain and the Ethereum blockchain. Few other examples of public blockchain are: Litecoin, Monero, Steemit, Bitcoin, Bitcoin Cash, Dash, Ethereum, IoTA, Litecoin, Monero, and Stellar to name a few. The main application of public blockchain is the issuance of a digital currency.

2.6.2 Private Blockchains

The private blockchains are owned by single organization that basically permissioned distributed ledger system. The read permissions on the blockchain network may be given to public or even restricted to the participants whereas write access is restricted to single individual of the organization. These private blockchains have some degree of external administration or control as compared to public blockchains.

2.6.3 Hybrid Blockchains

It is a combination of both public and private blockchain networks features. In this type of network is not open-up participation as public blockchain at the same time not fully restricted as private blockchain, but it has restricted and controlled participation. In contract to public blockchain, hybrid blockchains have improved security and scalability. Alternatively, hybrid blockchain is a combination of both centralized and decentralized blockchain network features. Thus, it consists of both centralized as well as decentralized networks, a certain portion of network is operated on centralized whereas as the remaining as decentralized that is based on the configuration of the network.

2.6.4 Consortium Blockchains

A consortium based blockchain is known as a permissioned blockchain, which is also known as a federated blockchain. In this type of blockchains, multiple organization are formed into a group (or consortium) for offering services to the members across different organizations in a group. The federated blockchain system can be best described as semi-private as well as semi-public, which is controlled by a user group, and works across multiple organizations. The consortium approach may be best fit to the following industries such as finance, banking, insurance, healthcare, trade, commerce, and logistic to name few. Thus, the influence of the single organization noticed in the private blockchain will eliminate at the same time it does not have as much as the degree of freedom of a public blockchain. It is a distributed system, unlike public blockchain, which is partially decentralized, and the control is given only to a few predetermined nodes. However, the control is given to a few members of a group as decided by the consortium. In addition, the consortium has to decide governing rules for the identification, setting-up, configuration of the blockchain platform, and the kind of services to be offered across the members of the organizations. There are several benefits of consortium systems over public and private blockchains. Firstly, infrastructure cost for the setup blockchain platform greatly reduces this acts as a backbone for the members of the consortium. Secondly, the members of the consortium can offer a gamut of services to their end users for inter-institutional and cross-cutting solutions. Thirdly, the transaction throughput of this type of blockchain is much faster and more scalable as compared to the public blockchain.

2.6.5 Sidechain Blockchains

A sidechain is a designation for a blockchain ledger that runs in parallel to a primary blockchain [18, 19]. Sidechain is an extended technology of blockchain, can provide a decentralized peer-to-peer platform to maintain the saved data while securely transferring key information between different systems [20]. A sidechain is a separate blockchain that is attached to its parent blockchain using a two-way peg. The two-way peg enables interchangeability of assets at a predetermined rate between the parent blockchain and the sidechain. Both mainchain and side blockchains need not fallow the basic structure of blockchain technology, including the block structure, The Proof of work PoW consensus mechanism, and new block generation procedure [21]. The main functions of sidechain are allowing the key information to transfer from one chain to others and reducing the burden of the mainchain [22], which help the system to gain both agility and freedom of using multiple networks.

2.6.6 Comparison of Various Types of Blockchains

See Table 1.

Table 1 A comparison of various types of blockchain types

Parameter	Public blockchain	Private blockchain	Consortium blockchain
Ownership	Nobody	Single organization	Group of organizations
Participants	Any individual without having their identity	Known identities	Known identities
Type of permission	Permissionless and Anonymous	Permissioned and Known identities	Permissioned and Known identities
Access	Anyone	Single organization	Group of organizations
Read	Anyone	Registered participants	Approved participants
Write	Nobody	Approved participants	Approved participants
Decentralization	Decentralized	Decentralized or centralized	Partially decentralized
Security	Low	High	Medium
Consensus	PoW and PoS	PBFT, RAFT Pre-approved participants Voting/multi-party consensus and Multi-signature	Pre-approved participants Voting / multi-party consensus
Speed of transaction	Slow	Fast	Fast
Scalability	Low	Medium	High
Examples	Bitcoin, Dash, Ethereum, Litecoin, Monero, Steemit, and Stellar	Ripple (XRP) and Hyperledger	Quorum, Hyperledger, Corda, R3 (banks), EWF (energy), Corda, and B3I

2.7 Transaction Validation

A method of authenticating and validating a transaction on a Blockchain, which is performed without the need to trust or rely on a central authority. It is also defined as the process of reaching an agreement between parties that do not trust each other. In other words, it means multiple servers agreeing on same information, something imperative to design fault tolerant distributed systems. Before a transaction is added to the blockchain it must be authenticated and authorised. The various consensus algorithms used for transaction validation are: Proof of work (PoW), Proof of Stake (PoS), Delegated Proof of Stake (DPoS), Byzantine Fault Tolerant (BFT), Proof of Importance (PoI), Proof of Authority (PoA), Proof of Capacity (PoC), Proof of Burn (PoB) to name a few.

2.8 Economic Aspects of Blockchain

There has been relatively little literature published on economic aspect of the blockchain discussed [6] on the two key costs that affects by the blockchain technology. These are (i) the cost of verification, and (ii) the cost of networking. The cost of verification relates to the ability to cheaply verify state of transaction whereas the cost of networking relates to the ability to bootstrap and operate a marketplace without assigning control to a centralized intermediary. These costs have implications for the design and efficiency of digital platforms, and open opportunities for new approaches to data ownership, privacy, and licensing; monetization of digital content; auctions and reputation systems.

2.9 Blockchain Platforms

In this section, we present various types of blockchain platforms and briefly describe their importance. In addition, the comparison between different blockchain platforms such as Bitcoin, Ethereum, Hyperledger, and R3 Corda, just to name a few, is presented.

2.10 List of Blockchain Platforms

Smart contract allows the development of blockchain-based applications and hosted on blockchain platforms. These applications can either be permissioned or permissionless or hybrid or consortium based. There are several players in the market providing blockchain platforms. The various blockchain platforms are Ethereum, Hyperledger, R3, Ripple, and EOS are a few names, a detail list is presented in Table 2.

Table 2 List of blockchain platforms

S.No	Name of the blockchain platform	Image/Logo of the platform	URL of the platform
1	Aeternity	æternity	http://aeternity.com
2	Ark	ARK.io	https://ark.io
3	BigChainDB	BIGCHAIN DB	https://www.bigchaindb.com
4	Bitcoin	bitcoin	https://bitcoin.org
5	Cardano	CARDANO	https://cardano.org
6	Chain	Chain	https://chain.com
7	EoS	eosio	https://eos.io
8	Ethereum	ethereum	https://ethereum.org/en
9	Heochain	HOLOCHAIN	https://holochain.org/
10	Hyperledger	HYPERLEDGER	https://www.hyperledger.org
11	IBM Bluemix	IBM.BlueMix	https://cloud.ibm.com/catalog/services/blockchain-platform
12	Lisk	Lisk	https://lisk.io
13	LTO Network	LTO NETWORK	https://www.ltonetwork.com
14	Microsoft Azure Blockchain	Microsoft Azure Blockchain as a Service	https://azure.microsoft.com/en-in/services/blockchain-service/
15	Monax	MONAX	https://monax.io
16	Monet	MONET	https://monet.network/

(continued)

Table 2 (continued)

S.No	Name of the blockchain platform	Image/Logo of the platform	URL of the platform
17	Nem	**nem**	https://nem.io
18	NEO	neo	https://neo.org
19	Openchain	openchain	https://www.openchain.org
20	Oracle Blockchain	ORACLE Blockchain	https://www.oracle.com/blockchain
21	Qtum	QTUM	https://qtum.org/en
22	Quorum	CONSENSYS Quorum	https://consensys.net/quorum
23	R3 Corda	r3.c·rda	https://www.r3.com
24	Ripple	ripple	https://ripple.com
25	RSK	rsk	https://www.rsk.co
26	Sovrin	sovrin	https://sovrin.org/
27	Stellar	Stellar	https://www.stellar.org
28	Stratis	Stratis	https://www.stratisplatform.com/
29	Swirlds	Swirlds	https://www.swirlds.com/
30	Tron	TRON	https://tron.network
31	Waves	**waves**	https://wavesplatform.com

2.11   Hyperledger Greenhouse

Hyperledger[1] is an open source focused on developing a suite of stable frameworks, tools, and libraries for enterprise-grade blockchain deployments which is started by the Linux Foundation. The objective of the Hyperledger project is to advance cross-industry collaboration by developing blockchains and distributed ledgers, with a particular focus on improving the performance and reliability of the systems. It has received contributions from IBM, Intel, and SAP Ariba, to support the collaborative development of blockchain-based distributed ledgers. This serves as a neutral home

---

[1] https://www.hyperledger.org.

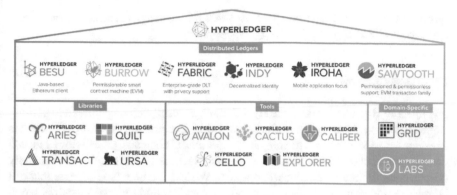

Fig. 6 The Hyperledger greenhouse (*Source* [23])

for various distributed ledger frameworks including Hyperledger Fabric, Sawtooth, Indy, as well as tools like Hyperledger Calliper and libraries like Hyperledger Ursa. It has frameworks like Fabric, Iroha and Sawtooth and tools like Caliper, Cello, Composer, Explorer, Quilt, Ursa. There are various business blockchain technologies being developed within the Hyperledger greenhouse that consists of DLT frameworks, tools, libraries, and domain specific applications are envisaged in Fig. 6.

2.11.1 Hyperledger Frameworks

All the frameworks, except Hyperledger Indy, are general purpose DLT, which can store whatever may be the type of asset. Hyperledger Indy is a framework that is focused on decentralized identity.

Hyperledger Besu

Hyperledger Besu[2] is an open-source Ethereum client developed under the Apache 2.0 license and written in Java. Its comprehensive permissioning schemes are designed specifically for use in a consortium environment. It runs on the Ethereum public network, private networks, and test networks such as Rinkeby, Ropsten, and Görli. Hyperledger Besu includes several consensus algorithms including PoW, and PoA (IBFT, IBFT 2.0, Etherhash, and Clique). Besu supports enterprise features including privacy and permissioning and can use to develop enterprise applications requiring secure, high-performance transaction processing in a private network. Besu includes a command line interface and JSON-RPC API for running, maintaining, debugging, and monitoring nodes in an Ethereum network. This API call can be made over HTTP or via WebSockets. This also supports Pub/Sub. The API supports

---

[2] https://www.hyperledger.org/use/besu.

typical Ethereum functionalities such as: Ether mining, smart contract development, and decentralized application (DApp) development.

Hyperledger Burrow

Hyperledger Burrow[3] is one of the Hyperledger projects, which operates as a permissioned Ethereum smart contract blockchain node. The primary function of Burrow is to execute the Ethereum smart contract programming code on a permissioned virtual machine. It supports both Ethereum Virtual Machine (EVM) and WebAssembly (abbreviated WASM) based smart contracts. WASM is a new type of code that can be run in a modern web browser, a standard WebAssembly virtual machine, leverages the features of low-level source languages such as C and C+ +. It uses BFT consensus via the Tendermint algorithm and it has a sophisticated event system and can maintain a relational database mapping of on-chain data. Governance and permissioning is built in and can be amended by on-chain proposal transactions. It is optimized for public permissioned proof-of-stake use cases but can also be used for private/consortium networks.

Hyperledger Fabric

Hyperledger Fabric [2, p. 8] is an open-source enterprise-grade permissioned distributed ledger technology (DLT) platform, established under the Linux Foundation. Fabric is the first distributed ledger platform to support smart contracts authored in general-purpose programming languages such as Java, Go and Node.js. Fabric has a highly modular and configurable architecture, enabling innovation, versatility, and optimization for a broad range of industry use cases including banking, finance, insurance, healthcare, human resources, supply chain and even digital music delivery. Typically, a fabric network comprises of "Peer nodes" and "Order nodes". The execute smart contract (or chaincode), access ledger data, endorse transactions and interface with applications whereas "Orderer nodes" ensure the consistency of the blockchain and deliver the endorsed transactions to the peers of the network. Authentication of member identity and their roles by the use of Membership Service Providers (MSPs) which is generally implemented as a Certificate Authority (CA).

The Hyperledger Fabric, which supports the use of one or more networks, each managing different Assets, Agreements and Transactions between different sets of Member nodes. The various types of nodes in Hyperledger can described as: client, peer and orderer. The client submits an actual transaction-invocation to the endorsers, and broadcasts transaction proposals to the ordering service, peer–commits transactions and maintains the state, and a copy of the ledger. In addition, peers can have a special endorser role. The orderer runs the communication service that implements a delivery guarantee, such as atomic or total order broadcast.

---

[3] https://www.hyperledger.org/use/hyperledger-burrow.

Hyperledger Indy

Hyperledger Indy[4] is a distributed ledger, which is built for decentralized identity. Hyperledger Indy provides tools, libraries, and reusable components for providing digital identities rooted on blockchains or other distributed ledgers so that they are interoperable across administrative domains, applications, and any other silo. Indy is interoperable with other blockchains or can be used standalone powering the decentralization of identity.

Hyperledger Iroha

Hyperledger Iroha[5] is a framework that supports distributed ledger system with emphasis on developing mobile applications. The features of Hyperledger Iroha are simple and easy to use, permissioned network, written in C + +, support of client libraries in Java, Python, JS, and Swift. ready-to-use set of commands and queries, and support for multi-signature transactions. Besides, it supports for a new BFT consensus algorithm known as YetAnotherConsensus (YAC). Hyperledger Iroha is being used in Cambodia to create a new payment system for retail payments alongside the National Bank of Cambodia, and in various other projects across healthcare, finance, and identity management.

Hyperledger Sawtooth

Hyperledger Sawtooth[6] is a modular blockchain suite that decouples transaction business logic from the consensus layer. So smart contracts can specify the business rules for applications without needing to know the underlying design of the core system. Hyperledger Sawtooth supports a variety of consensus algorithms, including Practical Byzantine Fault Tolerance (PBFT) and Proof of Elapsed Time (PoET).

2.11.2 Hyperledger Tools

Under Hyperledger greenhouse umbrella they are various types of tools developed in addition to the framework that can use to build variety of applications more effectively using Hyperledger blockchain stack. The tools such as Avalon, Cactu, Caliper, Cello, Composer, Explorer, and Quilt.

---

[4] https://www.hyperledger.org/use/hyperledger-indy.

[5] https://www.hyperledger.org/use/iroha.

[6] https://www.hyperledger.org/use/sawtooth.

Hyperledger Avalon

Hyperledger Avalon is a ledger independent implementation of the Trusted Compute Specifications published by the Enterprise Ethereum Alliance. It aims to enable the secure movement of blockchain processing off the main chain to dedicated computing resources. Avalon is designed to help developers gain the benefits of computational trust and mitigate its drawbacks.

Hyperledger Cactus

Cactus[7] is a blockchain integration tool designed to allow users to securely integrate different blockchains. With the use of Cactus, enterprises can able to confidently move forward with the blockchain platform that best meets their needs today, with the assurance they can integrate, communicate, operate, and transact with any other tech down the road.

Hyperledger Caliper

Hyperledger Caliper is a blockchain benchmark tool and one of the Hyperledger projects hosted by The Linux Foundation. Hyperledger Caliper allows users to measure the performance of a specific blockchain implementation with a set of predefined use cases. Hyperledger Caliper will produce reports containing a number of performance indicators, such as TPS (Transactions Per Second), transaction latency, resource utilization etc. The intent is for Caliper results to be used by other Hyperledger projects as they build out their frameworks, and as a reference in supporting the choice of a blockchain implementation suitable for a user's specific needs. This project was initially contributed by developers from Huawei, Hyperchain, Oracle, Bitwise, Soramitsu, IBM and the Budapest University of Technology and Economics.

Hyperledger Cello

Hyperledger Cello aims to bring the on-demand "as-a-service" deployment model to the blockchain ecosystem to reduce the effort required for creating, managing, and terminating blockchains. It provides a multi-tenant chain service efficiently and automatically on top of various infrastructures, e.g., baremetal, virtual machine, and more container platforms.

---

[7] https://www.hyperledger.org/use/cactus.

Hyperledger Composer

Hyperledger Composer is a set of collaboration tools for building blockchain business networks that make it simple and fast for business owners and developers to create smart contracts and blockchain applications to solve business problems. Built with JavaScript, leveraging modern tools including node.js, npm, CLI and popular editors, Composer offers business-centric abstractions as well as sample apps with easy to test DevOps processes to create robust blockchain solutions that drive alignment across business requirements with technical development.[8]

Hyperledger Composer provides a GUI user interface "Playground" for the creation of applications, and therefore represents an excellent starting point for Proof-of-Concept work. This runs on top of Hyperledger Fabric, which allows the easy management of Assets (data stored on the blockchain), Participants (identity management, or member services) and Transactions (Chaincode, a.k.a. Smart Contracts, which operate on Assets on the behalf of a Participant). The resulting application can be exported as a package (a BNA file) which may be executed on a Hyperledger Fabric instance, with the support of a Node.js application (based on the Loopback application framework) and provide a REST interface to external applications.

Hyperledger Explorer

Hyperledger Explorer is a blockchain module and one of the Hyperledger projects hosted by The Linux Foundation which was initially contributed by IBM, Intel and DTCC [16]. Designed to create a user-friendly Web application, Hyperledger Explorer can view, invoke, deploy or query blocks, transactions and associated data, network information (name, status, list of nodes), chain codes and transaction families, as well as any other relevant information stored in the ledger.

2.11.3 Hyperledger Libraries

Hyperledger Aries[9] provides a shared, reusable, interoperable tool kit designed for initiatives and solutions focused on creating, transmitting, and storing verifiable digital credentials. It is infrastructure for blockchain-rooted, peer-to-peer interactions. This project consumes the cryptographic support provided by Hyperledger Ursa, to provide secure secret management and decentralized key management functionality.

Hyperledger Quilt[10] is one of the blockchain library, Java implementation of the Interledger protocol and enabling payments across any payment network—fiat or crypto. Quilt provides an implementation of all core primitives required for sending

---

[8] https://www.hyperledger.org/.

[9] https://www.hyperledger.org/use/aries.

[10] https://www.hyperledger.org/category/hyperledger-quilt.

and receiving payments in a ledger-agnostic manner, allowing developers to write application payments logic once while gaining access to any other payment system that is Interledger-enabled. Hyperledger Quilt and Interledger have major developer backing from a variety of organizations, most importantly Xpring.

Hyperledger Transact

Hyperledger Transact[11] is a library form Hyperledger Greenhouse. The aim of the project is to make developing distributed ledger solutions easier by separating the execution of smart contracts from the distributed ledger implementation. Hyperledger Transact takes an extensible approach to implementing new smart contract languages called "smart contract engines," that implement a virtual machine or interpreter that processes smart contracts.

Hyperledger Ursha

Hyperledger Ursa[12] is a shared cryptographic library, it enables implementations to avoid duplicating other cryptographic work and hopefully increase security in the process. The library is an opt-in repository (for Hyperledger and non-Hyperledger projects) to place and use crypto. Hyperledger Ursa consists of sub-projects, which are cohesive implementations of cryptographic code or interfaces to cryptographic code.

2.11.4 Hyperledger Domain Specific Applications

Hyperledger Grid[13] is a platform for building supply chain solutions that include distributed ledger components. It intends to provide reference implementations of supply chain-centric data types, data models, and smart contract-based business logic – all anchored on existing, open standards and industry best practices. It showcases in authentic and practical ways how to combine components from the Hyperledger stack into a single, effective business solution.

Hyperledger Labs allows community to start new projects and allow for work to be done within a legal framework that eases transition to a project. Hyperledger Labs is not directly controlled by the Technical Streeting Committee. Labs are proposed and run by the community. They can be created by a simple request (done by submitting a Pull Request) to the Labs Stewards. The aim is that Hyperledger Labs encourages more developers to get involved and experiment in the community.

---

[11] https://www.hyperledger.org/use/transact.

[12] https://www.hyperledger.org/use/ursa.

[13] https://www.hyperledger.org/use/grid.

2.12 Ethereum

Ethereum [24] is the first open public blockchain platform, support to build decentralized application that run on blockchain technology, and the most advanced for coding and processing smart contracts. Ethereum is a programmable blockchain. The founders of Ethereum are founders Vitalik Buterin, Gavin Wood, and Jeffrey Wilcke. It serves as a platform for many different types of decentralized blockchain applications, including but not limited to cryptocurrencies. This has a suite of protocols that includes a peer-to-peer network protocol to define a platform for decentralized applications. At the heart of it is the Ethereum Virtual Machine ("EVM"), which can execute code of arbitrary algorithmic complexity. The various programming languages used for programming smart contract applications that run on the EVM like JavaScript and Python. Furthermore, it can code whatever the programming language individual wishes but would have to pay for computing power with "ETH" tokens. Nowadays, Ethereum is not only described as public blockchain-based distributed computing platform but also as a distributed operating system [25] because of its popularity of its architecture.

2.13 R3 Corda

The R3[14] is an enterprise software firm that is pioneering digital industry transformation and foundation in enterprise blockchain technology, which is most commonly used in blockchain solutions that deliver trust across the financial services industry and beyond. The code in Corda [26] is written using Kotlin, a programming language from JetBeans that targets the JVM and JavaScript and the major reason for choosing Kotlin is the high level of integration.

2.14 Quorum

The Quorum[15] is based on Ethereum and is a fork of Go-Ethereum, which is distributed ledger and smart contract platform designed for enterprise applications. This platform can be customized according to specific business needs and provides support for transaction-level privacy as well as network-wide transparency. The project was launched in 2016 and later this is acquired by Consensus, a blockchain platform from J.P. Morgan.

---

[14] https://www.r3.com/about/.

[15] https://consensys.net/quorum/.

2.15 Chain Core

The Chain Core [26] is one of the enterprises blockchain platforms and enabling institutions to issue and transfer financial assets. This solution is mainly designed for financial Institutions can launch and operate a blockchain network or connect to a growing list of other networks that are transforming how assets move around the world. The features of this platform are envisaged as follows - permissioned network access, immutable ledger, smart contract enabled, native digital assets, Federated consensus, and Transaction privacy. This company was acquired by Lightyear in 2018 and it would be renamed as Interstellar.

2.16 Stellar

The Stellar Development Foundation is a non-profit organization founded in 2014 to support the development and growth of the open-source Stellar network.[16] The stellar is an open network for storing and moving money and this is a blockchain-based distributed ledger network that connects banks, payments systems and people to facilitate low-cost, cross-asset transfers of value, including payments. Stellar is known as an established distributed ledger technology (DLT) platform that supports a cross-border transfer and payment system to connect financial institutions.

2.17 Ripple

Ripple[17] formerly OpenCoin, which offers global payments network deploying blockchain technology. The company develops RippleNet, an enterprise decentralized solution for global payments. Its solution connects banks, payment providers, digital asset exchanges, and corporates to send money globally using blockchain technology. The company's solutions include XRP, a digital asset for payments in form of crypto-currency; xCurrent, a software for banks for cross-border payments services; xRapid, for payment providers and other financial institutions to manage liquidity costs; and xVia, for corporates, payment providers, and banks to send payments across networks using standard interface. Ripple was formerly known as OpenCoin and changed its name to Ripple in September 2013. Ripple is a peer-to-peer digital payment network, similar to Bitcoin in many ways, but with a number of distinguishing features.

---

[16] https://www.stellar.org/foundation.

[17] https://ripple.com/.

2.18 Waves

The Waves is an enterprise hybrid blockchain[18] that combines the advantages of both private and public permissioned blockchain technologies. Depending on the customer's requirements, a solution can be realized either on a public permissioned or a private network. Leased Proof of Stake (LPoS) is an enhanced proof of stake consensus algorithm that allows you to lease your tokens to a Waves node, earning a percentage of the node pay-out, as a reward. Waves security is guaranteed by time-proven Proof-of-Stake consensus with more than 300 nodes all over the world. Ride language is built on a tech stack designed to avoid programming mistakes potentially leading to serious errors. Interaction with blockchain-enabled applications is seamless and safe. Ride is a straightforward, developer-friendly functional programming language for smart contracts and decentralized applications (DApps) on the Waves blockchain.

2.19 Cardano

The Cardano[19] is a groundbreaking proof-of-stake blockchain network, which is developed into a decentralized application (DApp) development platform with a multi-asset ledger and verifiable smart contracts. Built with the rigor of high-assurance formal development methods, Cardano aims to achieve the scalability, interoperability, and sustainability needed for real-world applications. This is one of blockchain system which is a Cryptocurrency Ecosystem. It is designed to be the platform of choice for the large-scale, mission critical DApps that will underpin the economy of the future.

2.20 A Comparison of Blockchain Technology Platforms

See Table 3.

---

[18] https://waves.tech.

[19] https://cardano.org/discover-cardano/.

Table 3 A comparison of various blockchain platforms

Blockchain Platform → / Parameter ↓	Bitcoin	Ethereum	Hyperledger	R3 Corda	Quorum EEA	Cardano	Stellar	Waves	IOTA	Ripple
Nature of blockchain	Cryptocurrency	Generic blockchain	Modular blockchain	Enterprise blockchain	Cryptocurrency	Cryptocurrency	A decentralized exchange for crypto, forex, and securities	A decentralized exchange for custom tokens, ICOs & crowd sales	Cryptocurrency	Cross-border transfer and payment system
Industry match	Fintech	Fintech	Cross Industry	Finance	Finance	Fintech	Fintech	Fintech	Fintech	Fintech
Location	Saint Kitts and Nevis	Zug, Switzerland	San Francisco, California	New York	New York	Zug, Switzerland	San Francisco	Incorporated in Switzerland but HQ in Russia	Berlin	San Francisco California
Original Author(s)/Founder(s)	Satoshi Nakamoto and Martti Malmi	Vitalik Buterin, Gavin Wood and Jeffrey Wilcke; and Charles Hoskinson	Tama Blummer and Christopher Ferris	David E Rutter	JP Morgan	Charles Hoskinson	Jed McCaleb and Joyce Kim	Sasha Ivanov	Sergey Ivancheglo, Serguei Popov, David Sønstebø, and Dominik Schiener	Chris Larsen and Jed McCaleb
Company	Saint Bits LLC	Ethereum Foundation	Hyperledger	R3 Corda Llp	Quorum Club Llp	Cardano Foundation	Stellar Development Foundation	Waves Tech	IOTA Foundation	Ripple Labs Inc
Initial release	2008	2014	January 2018	Sept. 2015	2016	2015	July 2014	2016	October 2015	2012
Written in	C++	Go, C++, Rust, Python and Solidity	Go, Java, JavaScript, and Python	Kotlin	Solidity	Haskell	C++, Go, Java, JavaScript, Ruby and Python	Scala	C++, Go, JavaScript, Java and Rust	C++, Node.js and JavaScript

(continued)

Table 3 (continued)

Blockchain Platform → / Parameter ↓	Bitcoin	Ethereum	Hyperledger	R3 Corda	Quorum EEA	Cardano	Stellar	Waves	IOTA	Ripple
License	Open source	Open-source Ether for transaction and computational services	Open source	Open source	Open source	Open source	Open source	Open source	Open source, IOTA Token	Open source
Distributed ledger	Yes	Yes	Yes	Yes	Yes	Yes	Yes	Yes	Yes, Tangle	Yes
Smart Contract	Yes	Yes	Yes	Yes	Yes	Yes	Yes	Yes	Yes	Yes
Decentralized/ Centralized	Decentralized	Decentralized	Decentralized	Not decentralized	Centralized	Decentralized	Less decentralized	Decentralized	Not decentralized	Decentralized
Type of blockchain	Public	Permissions, public, private	Permissioned, and private	Permissioned, and private	Public and private	Public	Public	Public and Private permissioned	Permissioned	Public and Private
Consensus	Proof of work	Mining based on PoW, Ledger level	Pluggable Framework	Specific (i.e. notary nodes) Transaction level	BFT, PBFT	Ouroboros (in house protocol) like PoS	Stellar Consensus	Leased PoS	Fast probabilistic consensus and cellular consensus	Ripple Protocol
Supported languages	Solidity	Solidity	Go, Java, and Python	Java and Kotin	Java, Kotin, Solidity	Plutus and Solidity	Not Available	Ride	C, JavaScript and Python	JavaScript
Currency	Bitcoin	Either	None	None	None	None	Lumen	None	None	None

3 Smart Contract Systems

3.1 Conceptual View of Smart Contract System

Most of the existing contracting models focus on a centralized approach and control over the contracting system excreted by a single individual. Thus, all the transactions happening in the system are stored in a single storage location, which is central to all the users of the system. A few examples for this type of systems such as banks, insurance companies, third-party financial services organizations, and most online platform companies. They are several advantages of this type of system like simple to design, deploy, operate, and maintain. Despite its advantages, the centralized system has the following limitations: (i) the parties are not known to each for exchange of goods and services thus there is less promise to hold the transaction, (ii) trusted authority for individuals to perform transaction (iii) difficulty in maintaining individual's data to perform transactions (iii) lack of trust of underlying platforms of centralized systems, and (iv) scalability of platforms because of single or limited users' authority.

The evolution of blockchain technology has opened to a new era of the decentralized application development that overcomes the limitations in the centralized system despite its own challenges in implementing the system. In this, the transactions are conducted in a Peer-to-Peer network. Further, each transaction is distributed into all peers in the network instead of a single copy of transaction in a centralized system. A conceptual view of a smart contract system is envisaged in Fig. 7, which consist of three major elements: (i) parties, (ii) smart contract, (iii) execution, and (iv) actions.

The party may be either individuals or organizations or things who enter the contract to perform a business transaction. Typically, business transaction may be a sale of goods or service. A smart contract is a self-executing contract with the terms and conditions of an agreement between the parties. It is written in the lines of code that can run directly in the given environment. A platform to run the line of code of

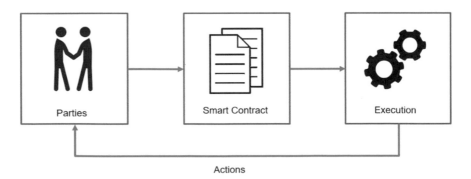

Fig. 7 Conceptual view of smart contract system

a contract is known as execution. An action is the results that is obtained from the execution of the contract that resulting to the parties this is shown in the figure with a feedback loop. For example, Party A and Party B enter into a smart contract on the weather betting in a specific city on a certain day. Based on the Party A, the weather of that day is a rainy day, whereas Party B is a sunny day. The weather on that day is rainy, so Party A shall pay $10 to Party B, which will take place automatically on the execution of the smart contract and this type of action in the smart contract is known as an event. Here, there is no third party for this contact and the system automatically executes the contract which is known as self-executing the contract. The very purpose of the smart contract is to eliminate the third party when establishing relations. This means that the contracting parties can transact directly with each other without any involvement of the third party.

The blockchain technology is a subset of the decentralized ledger technology, whereas the smart contract is a subset of the blockchain. A smart contract is a computer program which is stored on a blockchain that is used digitally to store terms and conditions of a contract. When the contract comes into execution then the actions of terms and conditions of that contract automatically as defined in the contract without any presence of third parties. There are several advantages of smart contracts over traditional contracts, which eliminates the limitations of traditional contracts like without an intermediary involvement, trust among participants in the transaction to name a few. A typical smart contract system contains as many conditions as required for the party's tasks to complete the contract satisfactorily. To establish the terms, participants to a blockchain platform must determine how transactions and their data are represented, agree on the rules that govern those transactions, explore all possible exceptions, and define a framework for resolving disputes. The entire process is usually iterative in nature that involves both developers and business stakeholders.

3.2 Framework for Building of Smart Contract

The process of building a smart contract is one of the most important activities in the overall smart contract ecosystem. The smart contract ecosystem will cover the entire life cycle of activities from the initiation to the closure of the smart contract. The process of building a smart contract can be divided into three-stage, as shown in Fig. 8. The first stage is the formation of the human form of contract, whereas in the second stage generation of a legal form of contract from the first. In the final stage, generates business logic from the legal form of a contract. The process at each stage is represented by an oval that consists of set of parameters like a_1, a_2, ..., a_i. These parameters will have user defined values, and these parameters to be mapped to the corresponding parameters of the next stage. This way, one form of contract can be transformed into other. In the final stage, the legal form of parameters is mapped to the corresponding code segments or blocks.

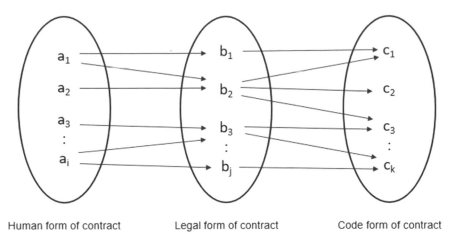

Fig. 8 Phases in building smart contract

Human form of contract: The contracting parties will discuss each other and come to some agreeable understanding to perform a business transaction through an agreement. The business transaction may be sale of goods or services. For example, Party A would like to sell an asset to Party B. A typical process involved in mapping of the parameters for building a smart contract is depicted in Fig. 9. In order to obtain human form of contract, the details of parties like name, age, location and identify are to be captured. In addition, parameters like timelines, value of an asset, and terms and conditions for payment to be specified.

Legal form of contract: In this stage, the parameters of human form of contract to be transformed into a standard legal form of contract that includes like date and time of contract, digital identity of parties, value of an asset as per the contract, terms and conditions, and SLAs if any. In addition, the details like closure of the contract to be specified.

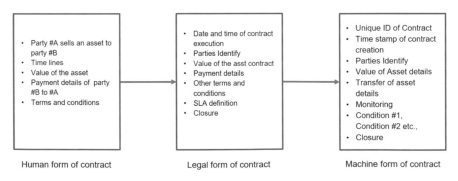

Fig. 9 A three step process of smart contract

Machine form of contract: In this stage, the legal form of contract is translated into a corresponding business logic, which in turn is converted into bits and pieces of code (or lines of code or chain code according to the Hyperledger terminology) logic which is implementable in a given smart contract platform for execution. The machine form of contract contains logical details such as unique id for each contract, time stamp of contract creation, party identity details, value of asset details, transfer of asset as per the terms of conditions mentioned in the legal contract.

The human form of contract parameters shall be mapped to a legal form contract, which is normally done by a legal advisor. The structure of the legal form of the document shall comply with the legal obligations. A traditional contract system has only the first two stages, whereas the smart contract system includes all the three stages. In the third stage, the primary focus is the generation of business logic. The generated business logic includes actions as per the as per the terms and conditions stated in the legal form of contract. Thus, the third stage is crucial in the entire process of building a smart contract. This is clearly shown from the above that in each phase of the contract be transformed into the other by a suitable mapping of the key parameters from one state to the other. The entire mapping process stated in the above can be performed by building a suitable model for an auto generation of a smart contract. The input to the system is a manual form of contract, whereas the output is source code (or business logic). Thus, there is a great need of tools for building smart contracts to ensure an error free and a standard form of smart contract code generation This entire process can be performed by means of an automated tool, which is known as smart contract generation system, as shown in Fig. 10.

The proposed framework is of a certain theoretical mode and practical value for researchers, consultants, and practitioners. This framework covers the key elements that are required for full lifecycle of smart contract. Further, the framework indicates the research direction and possible development trends.

In a study [28] outlines a novel methodology for autogenerating smart contracts that enforce the specifications of a transaction-focused system. This framework uses ontologies and semantic rules to encode domain-specific knowledge and then leverages the structure of abstract syntax trees to incorporate the required constraints. It first translates the smart contract template into an abstract syntax tree (AST). The AST can be walked through and manipulated to insert the instance-specific restrictions in the appropriate functions, which intrun produces source code file, thus generating a new smart contract which is the combination of our ontology-derived template and the dynamic values extracted from specific settings. This smart contract encapsulates the rules and processes governing all future transactions.

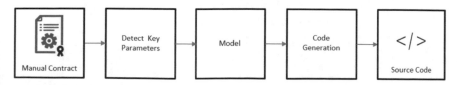

Fig. 10 Smart contract generation

3.3   Life Cycle of Smart Contract

The lifecycle management of smart Contract is the process of managing the activities of the entire lifecycle activities of a contract from beginning to end. The life cycle of smart contract can be described into the following five major phases namely (i) Negotiation, (ii) Creation, (iii) Execution, (iv) Governance, and (v) Closure. Figure 11 shows various phases in the life cycle of the smart contract. In the negotiation phase, contracting parties will discuss each other on a particular deal and come to an understanding to form a contract. The outcome of this phase is a human form of contract. During this phase, there is no legal binding on the parties to accept the deal until the parties do agree and sign on a contract. The creation phase mainly consists of the following activities: mapping of the human form of contract to legal form, generation of code, deployment of the code on the top of a blockchain platform, and user signing. In the execution phase, the contract is executed on a given blockchain platform. The activities related to the post execution of contract will fall under the governance phase. The governance phase includes to direct and control the functions of the contract like storing, signing, search, retrieval, settlement, compliance, auditing, renewal, amendments, renewal, monitoring, and obligation. The way out from the contract generally refers to the exit phase. The term contract closure refers to the process of completing all activities that are related to the terms and conditions of the contract. During closure phase, the exist criteria of the contract are properly verified. Finally, the contract is closed.

Fig. 11 Lifecycle management of smart contract

3.4 Smart Contract Issues and Indicative Solution Mapping

The author [29] has conducted a systematic mapping study on smart contracts to identify and tackling of smart contract issues. This study identifies four key issues namely codifying, security, privacy, and performance. The results of this study can be envisaged in the Table 4.

Table 4 A summary of smart contract issues and solution mapping

Smart contract issues		Proposed solution
Codifying issues	Difficulty of writing correct smart contracts	Semi-automation of smart contracts creation Use of formal verification methods Education (e.g., online tutorials)
	Inability to modify or terminate smart contracts	A set of standards for modifying/terminating smart contracts
	Lack of support to identify under optimized smart contracts	Use of 'GASPER' tool
	Complexity of programming languages	Use of logic-based languages
Security Issues	Transaction-ordering dependency vulnerability	Use of 'SendIfReceived' function Use of a guard condition Use of 'OYENTE' tool
	Timestamp dependency Vulnerability	Use of 'SendIfReceived' function Use of a guard condition Use of 'OYENTE' tool
	Mishandled exception vulnerability	Use block number as a random seed instead of using timestamp Use of 'OYENTE' tool
	Re-entrancy vulnerability	Check the returned value Use of 'OYENTE' tool
	Use of 'OYENTE' tool	Use of 'Town Crier (TC)' tool
Privacy Issues	Lack of transactional privacy	Use of 'Hawk' tool Use of encryption techniques
	Lack of data feeds privacy	Use of 'Town Crier (TC)' tool Use of encryption techniques
Performance issues	Sequential execution of smart contracts	Parallel execution of smart contracts

3.5   Use Cases

The following Fig. 12 shows real-world use cases of blockchain based on vertical specific such as government, energy, real estate, rail-ways, supply chain to name a few.

3.6   Characteristics, Benefits and Challenges of Smart Contracts

3.6.1   Characteristics of Smart Contracts

The key characteristics of smart contracts are envisaged as follows.

- **Digital Identity**—Each smart contract that is created in the blockchain must be uniquely identified for not only to access the details of the contract, but also helps in continuous monitoring of various service levels of the contract.
- **Record Keeping**—The smart contracts need to be stored data on a distributed ledger without compromising data security as well as performance of the application. Thus, it provides an electronic record keeping of the terms and conditions of the contract.
- **Trust among the parties**—The smart contract brings trust among the contracting parties.
- **Self-executing, self-verifying and tamper-resistant**—The other main characteristic of smart contract is the ability to execute and verify the details on its own. Moreover, the smart contract data is tamper-proof.

3.6.2   Benefits of Smart Contracts

The benefits of smart contracts are most apparent in the business.

- **Immutability**—The biggest benefit of blockchain is that once the record is created, it cannot be easy to alter, which is known as immutability of the system.
- **Decentralization**—The very function of blockchain is that record keeping is by means of a distributed ledger. Thus, data is stored in each node of the network.
- **Elimination of intermediaries**—Smart contract are self-executing and self-verifying on their own, create trust among the counter parties and built on a decentralized approach. All this set to elimination of intermediaries for approvals and trust services.

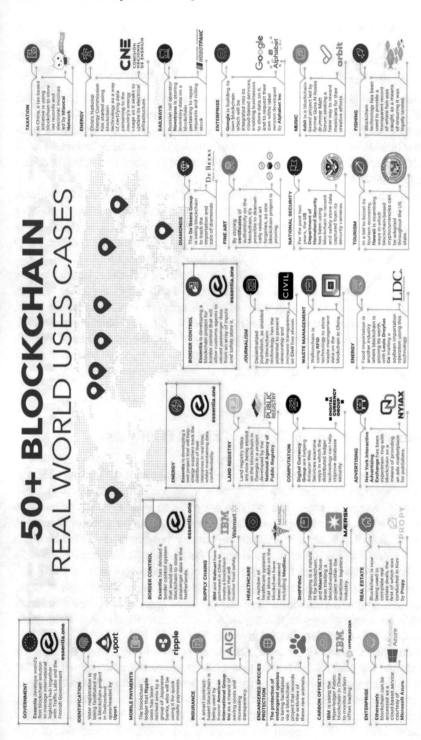

Fig. 12 Real word use cases of blockchain (*Source* [30])

3.6.3   Challenges of Smart Contracts

The following are a few challenges of the smart contracts: universal acceptance like a standard form of contract, interoperability, and custom trust models.

- **Standard form of contract**—Nowadays, organizations and communities are continuously evolving with new blockchain platforms. There is no common approach for the execution of the smart contract across the blockchain platforms. In addition, no two blockchain platforms will have similar features. This is resulting in reusability of smart contract code across the platforms which becomes a major challenge.
- **Interoperability**—The reference architecture of blockchain differs from platform to platform. In the general run of things, integrating multiple blockchains networks will increase the value chain across the organizations as well as the users of the system. As the reference architecture of blockchain platforms differs, it is becoming difficult to integrate multiple blockchain platforms.
- **Trust Models**—In the last 3 years, the importance of Off-chain transactions in the blockchain networks are steadily gaining, this will continue further because of its own degree of freedom in terms of private access compared to the On-chain transactions. For building a suitable trust model for a private blockchain network is interesting and becomes a challenging task. Thus, the model has to suit the need and requirements of the network organizations to support a balanced mix of Off-chain as well as On chain transactions.in blockchain networks.

4   Future Development Trends of Smart Contracts

This section presents the need for the future development trends in three main aspects of the smart contracts, namely the mapping tools, interoperability, and security model. The concept of smart contract can provide a support to build a variety of business applications which are totally new and further keep evolving on a continuous basis. The use of smart contracts will continue to increase in business and also continue to grow its demand for the next two decades because of the inherent features in the blockchain. Once a smart contract is made, making changes in it very difficult. So, there is a great need for development of tools to automate the process of smart contact activities such as building and validate of smart contract to name a few. These automated tools are not only improving the efficiency and effectiveness of the smart contracts, but also minimizes human errors.

A conceptual framework (or theoretical model) for building a smart contract has been presented in Sect. 3.2. The working of this framework is explained with the aid of a mapping process. So, looking into the process of building of smart contract, the major activities of the framework can be drawn as follows: (i) transformation of contract from one to another, for example translation of the human form contract to the legal form and (ii) generation of code. The scope for implementation of such mapping tools can be further examined and further research suggestions to come out of expanding the framework. The design of such model can be presented in several ways. The following are the few approaches of the model designing like template driven approach for transformation of the contract from one form to another, and the principles of Finite State Machine (FSM) for code generation from a legal form of the contract. We focused our approach for building a smart contract only with the aid of mapping tools, but it can open the following areas for further development of that framework like to design validating models for smart contract, and to explore how the interaction of code (generated from the tool) in runtime to keep variety of terms and conditions in the smart contract and provide a suitable trust model.

The first aspect of further direction of research is to explore the generated code with the help of tools to be secured, which means to protect the code without any tamper before deploying on a blockchain platform. It would be possible to include one or a number of these as a future research suggestion and make sure that any such suggestion before you take up to be justified either by your own findings or the literate.

The second aspect of further research should consider the direction of integration of blockchain platforms by identifying the required Application Programming Interfaces. This may be viewed in one of the two ways—first, integration between the On-chain and Off-chain in a single blockchain network, and the second is integration between different types of private blockchain networks.

The third aspect of further direction of research on smart contracts is to build a security model, which is according to the need and requirements of the participating organizations for building a large blockchain network, this includes a balance mix of the On-chain as well as Off-chain transactions and to enable the security to the Offline transactions in a private blockchain networks.

5 Conclusion

In this chapter, the functioning of the traditional contract system is discussed and mentioned its limitations. Additionally, explained a smart contract system and their importance in the modern digital world. Moreover, presented a list of major blockchain platforms and briefly explained a few. As well as, explained various frameworks, tools, and libraries available under the Hyperledger greenhouse. Moreover, presented a comparison between various blockchain platforms in a tabular form. Furthermore, presented a framework (or theoretical model) for building a smart contract and briefly explained the same. Finally, outline the future development trends of smart contracts from three main aspects such as mapping tools, interoperability, and security model.

References

1. Szabo, N. The idea of smart contracts. Nick Szabo's Papers and Concise Tutorials, vol 6 (1997)
2. Hyperledger-fabricdocs Documentation, Release master, 02 Sep 2020. https://readthedocs.org/projects/hyperl02edger-fabric/downloads/pdf/release-1.2/ (2020). Accessed 22 August 2021
3. Keenan, D. The 'digital world'. What does it mean? https://donnakeenan.wordpress.com/2015/03/24/the-digital-world-what-does-it-mean/ (2015)
4. Kukkuru, M.G. Smart Contracts: Introducing A Transparent Way To Do Business. https://www.infosys.com/insights/digital-future/smart-contracts.html (2021). Accessed 17 May 2021
5. Gopie, N. What are smart contracts on blockchain? https://www.ibm.com/blogs/blockchain/2018/07/what-are-smart-contracts-on-blockchain/ (2018)
6. Catalini, C., Gans, J.S.: Some simple economics of the blockchain. Commun. ACM **63**(7), 80–90 (2020)
7. Liu, S. Size of the blockchain technology market worldwide from 2018 to 2025. https://www.statista.com/statistics/647231/worldwide-blockchain-technology-market-size/ Accessed 1 Aug 2021
8. DLT. Blockchain & Distributed Ledger Technology (DLT).https://www.worldbank.org/en/topic/financialsector/brief/blockchain-dlt (2018)
9. Yaga, D., Mell, P., Roby, N., Scarfone, K. Blockchain Technology Overview, Draft NISTIR 8202, National Institute of Standards and Technology, January 2018, U.S. Department of Commerce (2018)
10. MacDonald, T.J., Allen, D.W.E., Potts, J. Blockchains and the boundaries of self-organized economies: predictions for the future of banking. In: Tasca, P., Aste, T., Pelizzon, L., Perony, N. (eds) Banking Beyond Banks and Money. New Economic Windows. Springer, Cham (2016). https://doi.org/10.1007/978-3-319-42448-4_14
11. Mulligan, C. Blockchain–a brief overview (p. 10), Imperial College Centre for Cryptocurrency Research and Engineering, Imperial College London (2021) https://na.eventscloud.com/file_uploads/b4d722450d854c8b9fdaf14823c49a0c_MULLIGAN_Blockchain-brief-overview.pdf. Accessed 18 Aug 2021
12. Smart Contracts and Chaincode. https://hyperledger-fabric.readthedocs.io/en/release-2.2/smartcontract/smartcontract.html. Accessed 25 Aug 2021
13. Wang, S., Ouyang, L., Yuan, Y., Ni, X., Han, X., Wang, F.: Blockchain-enabled smart contracts: architecture: applications, and future trends. IEEE Trans. Syst. Man Cybern. Syst. **49**(11), 2266–2277 (2019). https://doi.org/10.1109/TSMC.2019.2895123
14. Hu, Y. Liyanage, M., Manzoor, A., Thilakarathna, K., Jourjon, G., Seneviratne, A. Blockchain-based smart contracts, applications and challenges (2019), arXiv:1810.04699v2
15. Khan, S., Loukil, F., Ghedira, C., Benkhelifa, E., Bani-Hani, A.: Blockchain smart contracts: applications, challenges, and future trends. Peer-to-Peer Netw. Appl. **14**, 2901–2925 (2021). https://doi.org/10.1007/s12083-021-01127-0
16. Daniel, F., Guida, L. A service-oriented perspective on blockchain smart contracts. IEEE Internet Comput. **23**(1), 46–53 (2019). https://doi.org/10.1109/MIC.2018.2890624, https://ieeexplore.ieee.org/document/8598947
17. Idelberger F., Governatori G., Riveret R., Sartor G. Evaluation of logic-based smart contracts for blockchain systems. In: Alferes J., Bertossi L., Governatori G., Fodor P., Roman D. (Eds.) Rule Technologies. Research, Tools, and Applications, RuleML 2016. Lecture Notes in Computer Science, vol 9718. Springer, Cham (2016). https://doi.org/10.1007/978-3-319-42019-6_11
18. Raval, S. Decentralized Applications: Harnessing Bitcoin's Blockchain Technology. O'Reilly Media, Inc. p. 22 (2016). ISBN 978-1-4919-2452-5
19. Chowdhury, N. Inside Blockchain, Bitcoin, and Cryptocurrencies. CRC Press, p. 22 (2019). ISBN 978-1-00-050770-6
20. Singh, A., Click, K., Parizi, R.M., Zhang, Q., Dehghantanha, A., Choo, K.K.R. Sidechain technologies in blockchain networks: an examination and state-of-the-art review. J. Netw. Comput. Appl. **149**, 102471 (2020)

21. Li, M., Tang, H., Hussein, A.R., Wang, X. A sidechain-based decentralized authentication scheme via optimized two-way peg protocol for smart community (2020)
22. Casado-Vara, R., Chamoso, P., De Prieta La, F., Prieto, J., Corchado, J.M. Non-linear adaptive closed-loop control system for improved efficiency in IoT-blockchain management. Inf. Fusion **49**, 227–239 (2019). https://doi.org/10.1016/j.inffus.2018.12.007
23. Hyperledger landscape. https://www.hyperledger.org/use. Accessed 20 Aug 2021
24. What is Ethereum. https://ethdocs.org/en/latest/introduction/what-is-ethereum.html (2021)
25. Androulaki, E., Barger, A., Bortnikov, V., Cachin, C., Christidis, K., Caro, A., Enyeart, D., Ferris, C., Laventman, G., Manevich, Y., Muralidharan, S., Murthy, C., Nguyen, B., Sethi, M., Singh, G., Smith, K., Sorniotti, A., Stathakopoulou, C., Vukolic, M., Yellick, J. Hyperledger fabric: a distributed operating system for permissioned blockchains. In: Proceedings of the Thirteenth EuroSys Conference, pp. 1–15. EuroSys (2018) https://doi.org/10.1145/3190508.3190538
26. Corda. https://www.corda.net/blog/kotlin/#:~:text=When%20people%20start%20looking%20at,targets%20the%20JVM%20and%20Javascript (2021)
27. https://chain.com/docs/protocol/papers/whitepaper (2021)
28. Choudhury, O., Dhuliawala, M., Fay, N., Rudolph, N., Sylla, I; Fairoza, N., Gruen, D., Das, A. Auto-translation of regulatory documents into smart contracts. IEEE Blockchain Technical Briefs, IBM Research, Cambridge, MA (2018)
29. Alharby, M., van Moorsel, A. Blockchain-based smart contracts: a systematic mapping study, AIS, CSIT, IPPR, IPDCA-2017, pp 125–140, CS & IT-CSCP 2017. https://doi.org/10.5121/csit.2017.71011
30. 50+ Examples of How Blockchains are Taking Over the World. https://medium.com/@essentia1/50-examples-of-how-blockchains-are-taking-over-the-world-4276bf488a4b

Additional Reading

31. IEEE is the world's largest technical professional organization dedicated to advancing technology for the benefit of humanity. IEEE and its members inspire a global community to innovate for a better tomorrow through its highly cited publications, conferences, technology standards, and professional and educational activities. IEEE has recognized the importance of blockchain standards for the development and adoption of blockchain technologies for multiple industry sectors. The URL as follows: https://blockchain.ieee.org/standards
32. Beyond Standards, IEEE is dedicated to promoting technology standards and celebrating the contributions of the individuals and organizations across the globe who drive technology development. The content of this website focus includes but is not limited to the following: Showcasing new standards applications in the marketplace, Featuring new and emerging technologies, Highlight innovative new areas of standards development Celebrate innovators and disruptors who collaborate to advance standards and technology, Encourage participation in standards development, Events and educational opportunities etc., The URL as follows: https://beyondstandards.ieee.org/beyond-standards/
33. Blockchain—an opportunity for energy producers and consumers? https://www.pwc.com/gx/en/industries/assets/pwc-blockchain-opportunity-for-energy-producers-and-consumers.pdf
34. A Draft Discussion Paper on Blockchain: The India Strategy, Part 1, January 2020, NITI Aayog, Government of India. [Online] Available: https://niti.gov.in/sites/default/files/2020-01/Blockchain_The_India_Strategy_Part_I.pdf
35. A curated list of blockchain resources are maintained on the world's leading software development platform known as GitHub by a team, which covers book on blockchain, cryptography, white papers, Bitcoin, Network effects, Ethereum and Smart Contracts, Infographics and Talks, courses, blogs, articles, discussion groups, tutorials, You tube channels to name few. The URL of this repository of resources as follows: https://github.com/blockchainedindia/resources

36. Manav Gupta. Blockchain for dummies, IBM Limited Edition, [Online] Available: http://gun kelweb.com/coms465/texts/ibm_blockchain.pdf, Jhon Wiley & Sons, Inc, Hoboken, NJ (2017)
37. Making Blockchain Real for Business Explained by Esra Ufacki, The URL as follows: https://tekinvestor.s3-eu-west-1.amazonaws.com/original/2X/f/fe2c93811cda816267 c77aca180700dee341394b.pdf (2016)
38. Blockchain for Business, It covers enterprise adoption patterns, Use case examples from practice, Hyperledger Meetup, Frankfurt, 11. May 2017, https://www.hyperledger.org/wp-content/uploads/2017/05/HL_Meetup_Blockchain_IBM__Mai_v2a-1.pdf
39. Lin William Cong and Zhiguo He. Blockchain Disruption and Smart Contracts, National Bureau of Economic Research, 1050 Massachusetts Avenue, Cambridge, Working Paper 24399, http://www.nber.org/papers/w24399 (2018)
40. A white paper on Building Value with Blockchain Technology: How to Evaluate Blockchain's Benefits, July 2019, World Economic Forum. This white paper can help organizations to understand the state of the blockchain environment and the path to adoption. The analysis highlights the main advantages of the technology (broken down by industry), and the interviews shed light on the benefits and challenges of blockchain technology. And for organizations unsure where to begin or how to build a business case to assess the technology, the value framework shows what blockchain enables and where one can expect to realize value from it. The URL of this resource as follows: http://www3.weforum.org/docs/WEF_Building_Value_with_Bloc kchain.pdf

Decentralized Public Distribution System on the Ethereum Blockchain Interact with MyEtherWallet

C. Devi Parameswari and M. Ilayaraja

Abstract Blockchain technology is an innovative platform that provides distributed ledgers. With decentralized and transparent transmission mechanism across many businesses and businesses. In this work we proposed the future innovation of the public distribution system using the blockchain platform. We propose a decentralized public distribution system on the Ethereum network using solidity language, and interact with MyEtherWallet, which has the most features and functionality. Here, we create a blockchain-based application that is used to record all transactions between parties. This paper is required for all food transactions involving the central government, state governments, district level offices, retail ration stores, and consumers.

Keywords Blockchain · Public distribution system · Smart contracts · Solidity · Ethereum · MyEtherWallet

1 Introduction

Satoshi Nakamoto founded Bitcoin in 2009 [1], and its digital transaction (virtual currency) is considered the first example of financial exchanges. Blockchain are described as a value transfer protocol. Value can contain anything that can be specified as part of a record/amount of data. This blockchain-based value transfer is much safer, faster and cheaper than traditional systems. As a result, any two parties may use blockchain to efficiently and permanently record transactions. Since, blockchain

C. Devi Parameswari (✉)
Department of Computer Applications, Kalasalingam Academy of Research and Education, Krishnankoil, India
e-mail: deviparameswari@klu.ac.in

M. Ilayaraja
Department of Computer Science & Information Technology, Kalasalingam Academy of Research and Education, Krishnankoil, India
e-mail: ilayaraja.m@klu.ac.in

© The Author(s), under exclusive license to Springer Nature Switzerland AG 2022
K. R. Ahmed and H. Hexmoor (eds.), *Blockchain and Deep Learning*,
Studies in Big Data 105, https://doi.org/10.1007/978-3-030-95419-2_5

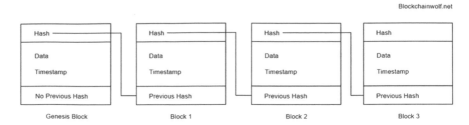

Fig. 1 Blockchain architecture

design is the safest way to transact blockchain [2]. This technology can be implemented and integrated in most application areas, for example events, medicinal information, identity organization and transaction processing [3–5].

In this Fig. 1 shows that architeture of blockchain. A blockchain is essentially a series of blocks linked together. Each block contains data about a transaction or contract. With its own signature and earlier volume. The only exception is the unsigned Genesis block of the previous volume [6]. This is because the Genesis block is the initial block of a blockchain. Transaction details are disseminated between multiple systems of a network without the requirement for a middle man.

Blockchain has the ability to change the corporate and inter-organizational business processes [7]. Technology can make it possible to track and exchange information in real time by providing a digital repository for transactions between stakeholders. In the chain of supply, the Consortium Chain, with its mostly decentralised or semi-decentralized governance structure, makes the participating organisations equally involved in consensus and decision-making processes, making it very attractive to supply chain applications [8].

1.1 Benefits of Blockchain Technology

Blockchain technology can be used in a number of real-world applications, most notably in the financial technology industry.

- The blockchain records and validates each transaction as a public, open ledger system, making it safe and accurate.
- All transactions are only checked by certified miners, rendering them permanent and unchangeable by someone else. This reduces the possibility of hacking.
- Peer-to-peer transactions can be achieved without the need for a middleman or a third party to handle them.
- Finally, there is technology decentralization.

In addition, smart contracts with this technology are able to automatically enforce agreements against special incidents in the supply chain. Since NITI Aayog correctly believes that blockchain is likely to solve many of the PDS problems already encountered, it calls for rigorous study and conception of how technology can be efficiently used in the complicated framework of the supply chain. PDS is the world's biggest distribution network.

2 Literature Review

Currently, the supply chain in the food industry will be briefly considered. For example [9], he once pointed out that there are many problems with China's farming food chain production being yet early. To address these issues, we need to capture the basis of agri-food chain [10]. The concept of agri-food chain management is the study of how to create the food chain of the agriculture products and logistics system, Apart from advanced agri-food chain technology, the market service system and management of quality also play an active role in government functioning.

Improving the distribution efficiency, obligations, anticorruption and decision-making will benefit from the implementation of information and communication technology (ICTs) as well as the Internet of Things (IoT), the supply chains with big data and artificial information (AIT). ICTs may revolutionize the entire supply chain, by making it possible to enhance the efficiency of existing supply chain operations through new business models. However, control over food traceability data, interoperability and immutability, guaranteeing records protection are some of the issues to be addressed. As an evolving and rapidly developing technology, Blockchain also has capability to provide solutions to those problems. Blockchain technology is also used in various non-financial fields. The health sector covers the application fields such as internet access to patients, public hospitals, medical accuracy and drug counterfeiting [11]. It is also used by applications like e-voting, citizenship and user support to enhance governance [12].

Blockchain Discovery provides us with instructions for managing IoT devices [13]. Expand a secure-based system that merge with smart devices to provide a secure communications platform in the smart city with the help of blockchain [14]. Blockchain can be implemented in a variety of agri-food chain structures to create a decentralized system that can provide transparency, security, neutrality and sustainability of all functions within the food chain network [15–17]. The agricultural chain plays an important role in supplying rice from farmers to each individual plate. Lack of communication and cooperation among persons associated with the agri-food chain is the main purpose of its ineffectiveness [18]. Therefore, the appropriate agri-food chain structure is crucial for effective manufacture, implementation, supply and transaction delivery, thereby meeting the consumer challenges and providing the best quality rice. It has the ability to develop, simplify and automate processes in business and industrial applications such as supply chain management and other applications such as education and integrity verification. In addition, various studies have also been presented in the field of blockchain-based supply chain applications such as improved monitoring and traceability, improved supply chain knowledge management, better inventory and performance management and intelligent transport systems.

3 PDS Design

In June 1947, India inaugurated the Public Distribution System (PDS). By ensuring that all activities are registered and safe, blockchain technology can be used to enforce these systems and fully eradicate unauthorized activity. The correct transactions made at lower levels of the supply chain (up to ration shops) cannot be monitored by the higher levels of the supply chain, especially by the government, under the current system. As a result, this flaw in the public delivery system provides an incentive for illegal or fraudulent activity. The proposed framework was created with solid language on the Ethereum blockchain platform, with blockchain technologies ensuring protection. Anyone can find transfer food products in the public distribution system, and it benefits authorities at all levels.

In this Fig. 2 made up of the Central Government, Food Corporation of India (FCI), State Governments, District Offices, Ration Stores, and customers/beneficiaries. The central government is the network's highest authority. Login is used to authenticate the elements mentioned above. Following efficient authentication, component-level exchanges can be carried out. Materials and state and district details was added to the blockchain network at the central government level. Every commodity's price, as well as the prices of registered ration shops, will skyrocket. Food Grains FCI distributes products to a variety of nations. Various districts receive food grains from the state government. The district office offers food grains to a variety of ration shops distributed in the district. Finally, rationing sells and distributes food grains to those

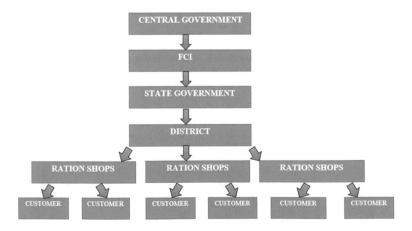

Fig. 2 Supply chain from the central government to the end user

who need them. Commodity transaction details are registered on the blockchain network at each stage. All food supply information will be monitored by the central government.

3.1 Ethereum and Smart Contracts

Ethereum, launched in 2015, is the world's programmable blockchain. This is a decentralized open source blockchain with smart contract functionality. Vitalik is the co-founder of Putter's Ethereum. It allows you to configure and execute smart contracts and distributed applications (Dapp) without the interference, duplication, control of third parties.

Smart contract is a contract among two people who use computer language. The details are stored in the blockchain database and the records cannot be altered. Transactions are sent automatically without a third person. You can write smart contracts using Solidarity, Serpent, Viper and LLC languages. Smart contracts run by using Ethereum Virtual Machine (EVM) in Ethereum platform and then used in the public Ethereum blockchain.

3.2 MyEtherWallet

- Taylor Monahan, who launched MyEtherWallet.com (MEW) in 2015 with Kosala "Kvhnuke" Hemachandra, declared a another organization and wallet service called MyCrypto.com.
- MyEtherWallet, otherwise called the MEW wallet, is one of the most well-known Ethereum wallets.
- MEW Wallet is a site that allows you to create an Ethereum wallet.
- You can utilize this wallet to store, send and secure ether.

4 Implementation

We present the system implementation of the public distribution system. We will use ganache-cli, Remix IDE and MytherWallet v3.40.0. Remix is a browser-based compiler and IDE that enables users to create Ethereum deals with Solidarity language and debug transactions. ganache-cli is a fast and customizable blockchain prototype. MyEtherWallet act as client interface. This implementation is done on the Intel® Pentium® 3558U @1.70 GHz processor with 4 GB RAM.

This framework clarifies how to implement different tools and use them.

We use our local operating system which is not accessible online is shows that Fig. 3, so you can download it from https://github.com/kvhnuke/etherwallet/releases and run the local host.

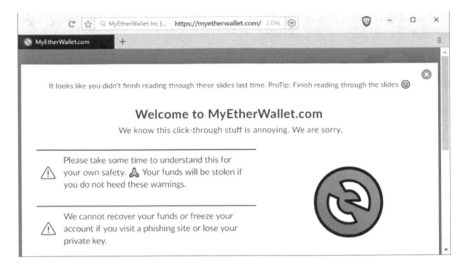

Fig. 3 MyEtherWallet v3.40.0

ganache-cli used Node.js package through npm and written using javascript. We used the command npm install-g ganache-cli. After installation, put the ganache-cli command and press Enter button. There are 10 default accounts with their private key, which is shows that Fig. 4.

In this Fig. 5 shows that, Open the Remix IDE and change the code above. Click Start to edit, and if it successfully compiles. Your code has been successfully compiled and you can now deploy then test it.

Fig. 4 ganache-cli

Fig. 5 Remix-IDE

4.1 Deploy Smart Contract and Test It

We use this for blockchain running MyStherWallet with ganache-cli. Open the cmd promt and capture the server URL link. Currently, open the index.html of MyEhter-Wallet. Select on the drop-down menu in MyEtherWallet and click Add Custom Network/Node.

Now go to the contacts in the menu and select Smart Contract, here you will want the byte code of particular contract. Go to Remix IDE, select on details and copy bytecode, which is shows that Fig. 6.

In this Fig. 7, We select the particular contract of an object, copy it, and paste it into the byte code in ordering the contract in MyEtherWallet. After pasting the byte code, the gas limit appears. To access your wallet, MyEtherWallet needs a private key. Open the ganache, copy a key from one of the wallets, then select Unblock button.

Open the command prompt, copy a private key from any account, and select Unblock button, which is shows that Fig. 8.

In this Fig. 9, after entering the byte code and accessing the wallet, press the Signature Transaction button, which will make the transaction verifiable. Now you can click deploy contract button, and click confirm the transaction button to send your smart contract to the blockchain network.

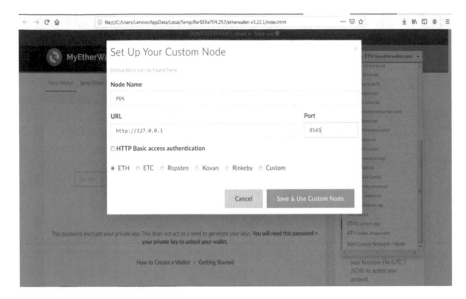

Fig. 6 Set up custom node

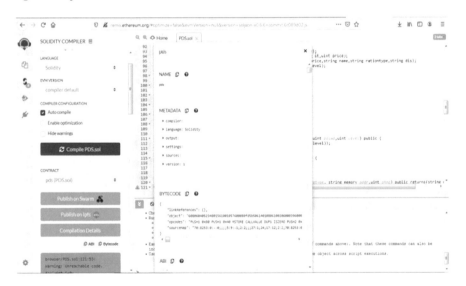

Fig. 7 Copy the byte code

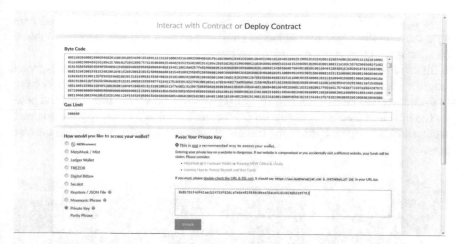

Fig. 8 Deploy a contract

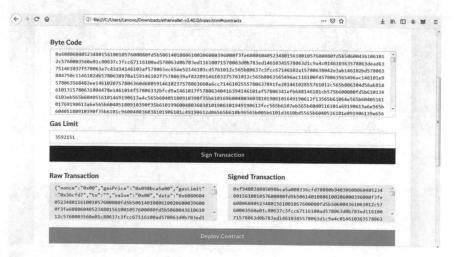

Fig. 9 Deploy contract

In this Fig. 10 shows that, you can check in ganache-cli whether it has been used successfully. Open the command window and here you can see that a contract creation transaction is being created in a block.

In this Fig. 11, Now let's test it, open the command prompt and copy the created contract address. Then Goto MyEtherWallet and paste the contract address. Next go to the Remix IDE afterwards, copy the ABI (Application Binary Interface) and paste it. The ABI interface holds details on the methods available in the specific contract. After selecting Access button, you can get all the functionality. We need to check it now.

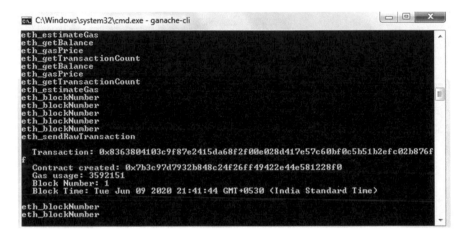

Fig. 10 Contract created

Fig. 11 Interact with contract

PDS Scheme-Implementation (Central Government)

This module contains the following:

- Districts Add state, county information and authorized ration shops in different districts
- Add the name of the goods in the blockchain network
- Buy grains from farmers
- Food grains are provided to each FCI network with a fixed price for each item
- Provide or delegate authorized users.

```
function AddCommodity (string memory name) public {

    commodityname.push(name); }

function  AddState(string memory state) public {

    stateinfo.push(state); }

function  AddDistrict(string memory District,uint rcode) public {

    districts.push(district(District,rcode)); }

function GetCommodityArray() public view returns (string[] memory){

    return commodityname; }

function Getstate() public view returns (string[] memory){

    return stateinfo; }
```

FCI (Food Corporation of India)

This module includes:

- Get the details of the products kept by the central government
- Assign the Various commodity to different states
- Transfer of Food Grain Details to Various State Governments.

```
function DistributetoFCI(uint name,uint pwd,uint x,uint qu) public returns(string memory){        for(uint
i=0; i<addauthorizedusers.length; i++){

if(addauthorizedusers[i].uname==name &&addauthorizedusers[i].pwd==pwd){

        string memory cname=comdits[x].cname;

        uint quan=comdits[x].quantity-qu;

        string memory measure=comdits[x].measurement;

        uint price=comdits[x].price;

        uint da=block.timestamp;

        distritofps.push(distofps(cname,quan,measure,da,price));

        emit addcom(cname,quan,measure,da,price);

        return "Authorized user,Commodity distribute to FCI"; }

    else{

        return "Unauthorized user"; }  }
function GetFPSstock() public view returns (distofps[] memory){

        return distritofps;}
```

State Government

This module contains the following:

- Receive details of items transferred from FCI
- Assign various commodities to different districts
- See details of food grains transferred to various district offices.

```
function statetodistrictt(uint name,uint pwd,uint x,string memory s,uint qu) public returns (string
memory){

    for(uint i=0; i<addauthorizedusers.length; i++){

      if(addauthorizedusers[i].uname==name &&addauthorizedusers[i].pwd==pwd){

    string memory cname=fpstostates[x].cname;

    uint quan=fpstostates[x].quantity-qu;

    string memory measure=fpstostates[x].measurement;

    uint da=block.timestamp;

    string memory dis=s;

    uint price=fpstostates[x].price;

    statetodistricts.push(statetodistrict(cname,quan,measure,da,dis,price));

    emit addstatecom(cname,quan,measure,da,dis,price);

    return "Authorized user,Distribute state to district";}

     else{ return "Unauthorized user"; } } }

function Getdistrictstock() public view returns (statetodistrict[] memory){return statetodistricts;}
```

District Office

- Transfer details of goods to State Government
- Assign items to various approved rations within the Within District
- See details of food grains transferred to ration shops.

```
function districttoration(uint name,uint pwd,uint x,string memory s,uint qu) public returns (string memory){

    for(uint i=0; i<addauthorizedusers.length; i++){

    if(addauthorizedusers[i].uname==name &&addauthorizedusers[i].pwd==pwd){

    string memory cname=statetodistricts[x].cname;

    uint quan=statetodistricts[x].quantity-qu;

    string memory measure=statetodistricts[x].measurement;

    uint da=block.timestamp;

    string memory ration=s;

    uint price=statetodistricts[x].price;

    districttorats.push(districttora(cname,quan,measure,da,ration,price));

    emit addstatecom(cname,quan,measure,da,ration,price);

    return "Authorized user,Distribute district to rationshop"; }

    else{ return "Unauthorized user"; } } }

    function Getrationstock() public view returns (districttora[] memory){

    return districttorats;}
```

Ration Stores

- Obtain details of goods transferred from the district office.
- Sell/distribute Grain cereals to beneficiaries
- See Grain Benefits Changed Food Grain Details.

```
function rationtopeople(uint x,uint ranum,string memory name,string memory rattype,string memory distri,uint qu) public{

    string memory cname=districttorats[x].cname;

    uint quan=districttorats[x].quantity-qu;

    string memory measure=districttorats[x].measurement;

    uint da=block.timestamp;

    rationtopeoples.push(ratopeop(cname,quan,measure,da,ranum,name,rattype,distri));

    emit rationtopeo(cname,quan,measure,da,ranum,name,rattype,distri);

    }

function Getpeoplestock() public view returns (ratopeop[] memory){

    return rationtopeoples; }
```

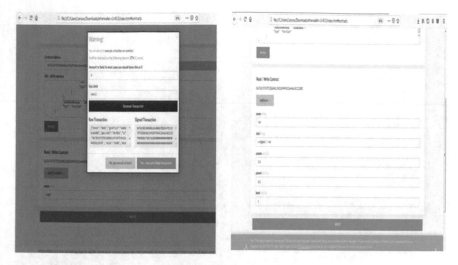

Fig. 12 Add commodity and authorized users

```
fpstostate[] fpstostates;

   addpeople[] addpeoples;

   district[] districts;

   statetodistrict[] statetodistricts;

   districttora[]districttorats;

   ratopeop[]rationtopeoples;

    event addcom(string cname,uint quan,string measu,uint da, uint price);

   event addstatecom(string cname,uint quan,string measu,uint da,string st,uint price);

   event rationtopeo(string cname,uint quan,string measu,uint da,uint price,string name,string
   rationtype,string dis);
```

Here, I add some pictures to access and test public distribution system functions.
In this Figs. 12, 13 and 14 shows that distribution activities between the components and block creation details in ganache–cli emulator.

5   Cost Estimation

In terms of integrating blockchain for PDS, it is important to quantify the cost associated with the implementation of smart PDS contracts. The ultimate aim is to introduce a scheme that can provide a viable PDS all the benefits of blockchain. All programmable calculations in Ethereum blockchain cost some fees to prevent network misuse and to solve other computer related problems. Therefore, to perform

Fig. 13 Distribute to FCI and FCI to State

Fig. 14 Distribute state to district and block details

all these activities, all operations, computations, message calls, smart contract creation/deployment and storage on EVM require gas.

The cost of deploying smart contracts for the public distribution framework has been compiled. There is an expense known as gas to run an application on the Ethereum blockchain. As the basic requirement to operate the company, all the transactions need 21,000 gas. If a consumer interacts with a smart contract with Ethereum,

21,000 gas is needed with additional gas associated with the running of that particular smart contract. The gas for medical smart contracts was compiled for contract deployment to communicate with the various contracts. The functions/operations involved in smart contracts consume more gas, resulting in more fees, which are more complicated. From the point of view of viability, it is clear from the findings that the cost of smart contract implementation for the public distribution system is very low. This cost is very economical in terms of the PDS system, and everybody would like to pay this small fee and preserve their lifetime PDS data.

6 Conclusion

In this work, we have proposed that the public distribution framework communicate securely with the MyEtherWallet interface. Promotion of Blockchain is a good innovation for government surveillance, monitoring, auditing the agri food chain, and helping manufacturers register with credibility. This technology not only benefits customers, manufacturers and supervisory sectors, but furthermore improves the productivity and functionality of the agri food chain. To prevent frauds and corruption in the current implementation of PDS programs, we can use blockchain technology to eliminate it and have a better system to help Indian citizens grow and benefit. Each transaction in the system is documented by the blockchain prototype and the history of the records is kept from beginning to present time.

Acknowledgements The first author expresses her gratitude to the management of Kalasalingam Academy of Research and Education for providing scholarship to carry out the research.

References

1. Nakamoto, S.: Bitcoin: a peer-to-peer electronic cash system (2008). https://www.Bitcoin.Org/Bitcoin.Pdf
2. Zheng, Z.: An overview of blockchain technology: architecture, consensus, and future trends. In: Big Data (BigData Congress), IEEE International Congress (2017)
3. Ølnes, S., Ubacht, J., Janssen, M.: Blockchain in Government: Benefits and Implications of Distributed Ledger Technology for Information Sharing (2017)
4. Sullivan, C., Burger, E.: E-residency and blockchain. Comput. Law Secur. Rep. **33**(4), 470–481 (2017)
5. Mansfield-Devine, S.: Beyond Bitcoin: using blockchain technology to provide assurance in the commercial world. Comput. Fraud Secur. **5**, 14–18 (2017)
6. Zhao, H.: Lightweight backup and efficient recovery scheme for health blockchain keys. In: IEEE 13th International Symposium on Autonomous Decentralized System (ISADS). IEEE (2017)
7. Al-Jaroodi, J., Mohamed, N.: Blockchain in industries: a survey. IEEE Access **7**, 36500–36515 (2019). https://doi.org/10.1109/ACCESS.2019.2903554
8. Mazumdar, S., Ruj, S.: Design of anonymous endorsement system in hyperledger fabric. IEEE Trans. Emerg. Top. Comput. **1** (2018). https://doi.org/10.1109/TETC.2019.2920719

9. Jin, X.U.: Supply Chain of Agricultural Products: Safeguard of Food Safety, China Logistics & Purchasing, pp. 68–68 (2005)
10. Hua, D.I.N.G.: On the Application of Supply Chain Theory in Enterprises Distributing Farm Produce, China Business and Market, pp. 17–21 (2004)
11. Farouk, A., Alahmadi, A., Ghos', S., Mashatan, A.: Blockchain platform for industrial healthcare vision and future opportunities. Comput. Commun. **154**, 223–235 (2020). https://doi.org/10.1016/j.comcom.2020.02.058
12. Warkentin, M., Orgero, C.: Using the security triad to assess blockchain technology in public sector applications. Int. J. Inf. Manag. **52** (2020). https://doi.org/10.1016/j.ijinfomgt.2020.102090
13. Huh, S., Cho, S.: Managing IoT devices using blockchain platform. In: ICACT2017, Korea, pp. 19–22 (2017)
14. Biswas, K., Muthukkumarasamy, V.: Securing smart cities using blockchain technology. In: IEEE 14th International Conference on Smart City, Sydney, Australia (2016)
15. Tian, F.: An agri-food supply chain traceability system for China based on RFID & blockchain technology. In: 13th International Conference on Service Systems and Service Management (ICSSSM), China, pp. 1–6, June24–26 (2016)
16. Bocek, T., Rodrigues, B.: Blockchains everywhere-a use-case of blockchains in the pharma supply-chain. In: IFIP/IEEE Symposium on Integrated Network and Service Management (IM2017), Portugal, pp. 772–777, May 8–12 (2017)
17. Devi Parameswari C., Mandadi V.: Public distribution system based on blockchain using solidity. In: Innovative Data Communication Technologies and Application. Lecture Notes on Data Engineering and Communications Technologies, vol. 59 (2021) https://doi.org/10.1007/978-981-15-9651-3_15
18. Somashekhar, I.C., Raju, J.K.: Agriculture supply chain management: a scenario in India. Res. J. Soc. Sci. Manag. RJSSM **04**(07), 89–99 (2014)

Applications

Adoption of Blockchain Integrated with Machine Learning in Intelligent Transportation System

Mahadev A. Gawas and Aishwarya R. Parab

Abstract An intelligent transportation System (ITS) is an unconventional application that aims to provide traffic congestion information and emergency information to other vehicles. In ITS, the compilation and distribution of traffic event information by vehicles are of the most extreme significance. To establish an effective data-sharing network in ITS, the vehicles need to share the data in a protected form in real-time. Conventional ITS system includes data streaming from On-Board Unit (OBU) sensors through Road Side Units (RSU) to centralized cloud servers. Nevertheless, it is still a challenge to distribute essential event information in a selected area under the dynamic ITS environment and in the existence of a malicious vehicle. The current fundamental issues which emerge include privacy concerns due to third party management of cloud servers, single point of failure which are a bottleneck in data flow among the vehicles.

In recent times, the blockchain technology which has been effectively applied in cryptocurrency has potentially emerged to be extremely secure and privacy-preserving technology for Internet of Vehicles (IoV) applications. Blockchain is an emerging decentralized and distributed computing paradigm that supports the bitcoin cryptocurrency which provides privacy and security in peer-to-peer networks. The adoption of blockchain on the Internet of Vehicles (IoV) can overcome a single point of failure and serve as an adequate means to securely and efficiently store and process Internet of Vehicle (IoV) data. However, introducing an efficient and scalable consensus mechanism in blockchain-based ITS is still an open research challenge given the circumstances of high mobility and resource constraints in vehicular networks. Because of the openness of communication networks, ITS security can be

M. A. Gawas (✉)
Department of Computer Science, Government College of Arts, Science and Commerce, Sanquelim, Goa, India
e-mail: gawas-dhe.goa@gov.in

A. R. Parab
Department of Computer Science, Government College of Arts, Science and Commerce, Quepem, Goa, India

compromised. We also focus on the capability of using Machine Learning in ITS, and how machine learning can incorporate and improve management and prediction tasks. The machine learning models can be used extensively in the prediction of traffic. It has been gaining popularity due to the applicable traffic scenarios in which the machine learning models are used. Arranging consistent and dependable applications in ITS requires effective protocols to attain confidentiality, data integrity, and authentication.

1 Introduction

The dynamic component and resource constraints of vehicular environments have presented huge difficulties to the design of an efficient vehicular IoT framework. The traditional cloud computing system requires the transmission of vehicle data to the cloud, resulting in high latency. To tackle this issue, mobile edge computing (MEC) innovations in vehicular networks have pulled in a serious level of consideration [1]. Future vehicular IoT frameworks should take into account the uncommon high unwavering quality and super low inactivity requirements to empower emerging applications, including community-oriented self-sufficient driving and intelligent control of traffic signals.

The two methods that have been introduced in order to exchange data among different entities in a vehicular network are Vehicle-to-vehicle (V2V) and vehicle-to-everything (V2X) Depending on these communications, the coordination between the entities, including base stations, vehicles, people, Road-side units (RSUs), etc., can be accomplished [2]. Nevertheless, it is hard to accomplish an effective coordinated effort among various entities belonging to different owners due to security and privacy issues. Because of the versatility of vehicles, stern application requirements, and restricted resources for communication, the regular centralized control is unable to give adequate nature of administration to associated vehicles, so a decentralized methodology is needed to fulfill the prerequisites of delay-sensitive and mission-critical applications [3].

A.   Blockchain Technology

Blockchain has been drawing in expanding interest from both scholastic and modern areas because of its promising features of giving a decentralized, open, and permanent framework [4]. Blockchain has proved to be a productive mode for storing data online. Furthermore, the decentralized network provides access to other users. Different users over the network store the transactional details in different ledgers. Blockchain can provide access to all these users over one single network. The transaction details can be stored and validated by the concerned users over the network. It is easy to get the transaction details as all the data is stored in the blockchain. Blockchain offers organizations to assertively identify the integrity of the data being generated. Consensus-driven time-stamping, proper audit trails and absolute entries will improve as blockchain begins turning out to be more standard.

Blockchain can also lead to new types of information adaptation given the following changes it offers:

- All organizations associated with a network will have access to the data as it is. This results in speeding up the processes like sharing, acquisition, and analytics of the data.
- The transactions will be enlisted and kept on a single file, giving a total outline of the transactions while removing the prerequisite for various powerful systems.
- Organizations will be able to manage and direct their data without the help of any third party.

A blockchain is a group of blocks, wherein all the blocks are connected together to form a chain using some cryptographic techniques. These blocks act as storage for the transactions, data, and scripts. The newly created blocks are consistently added to the chain in a digital ledger, and the ledger is kept up by all members of the network. Therefore, blockchain is also known as distributed ledger technology (DLT) [5]. Blockchain technology allows a framework to perform any secure transactions in an unsecured environment without any third-party security unit. Blockchain technology has been applied in varied domains. Figure 1 presents some of the applications of blockchain in the internet of things (IoT).

Blockchain is characterized as an "advanced, decentralized and distributed ledger wherein the transactions are logged and included sequential order to fully intent on making stable and carefully designed records". Blockchain depends on an open ledger that gives a stable and verified approach to handle transactions in a decentralized system, and it has motivated numerous scientists to study the utilization of blockchain in vehicular IoT.

The key advantages of blockchain are as below:

- Decentralization: A distributed network can be achieved in a blockchain without the use of a controller in the middle. Without the use of any third-party application for a global consensus, blockchain can establish a connection between two blocks by using cryptography technology.
- Irreversibility and traceability: A small change in a single block can cause irregularity among the other blocks as the connection between the adjacent blocks is certified by a hash function. This results in the irreversibility of the blockchain. This component empowers unchangeable records of the transactions. By revealing all the blocks to all the members in the network, blockchain can give traceability to all the arrangements made by the people.
- Fault resilience: Blockchain can accomplish a worldwide agreement without the requirement of a centralized entity. This empowers a better tolerance to unit faults, for example, when under a privacy attack or during a calamity. This component can be utilized to improve the security level of a framework or to plan a fault-tolerant system.

These features have worked with the quick development of the use of blockchain in different areas, including costs, funding, supply chain, medical services, insurance, resource management, etc. [6]. Numerous vehicular IoT frameworks expect vehicles

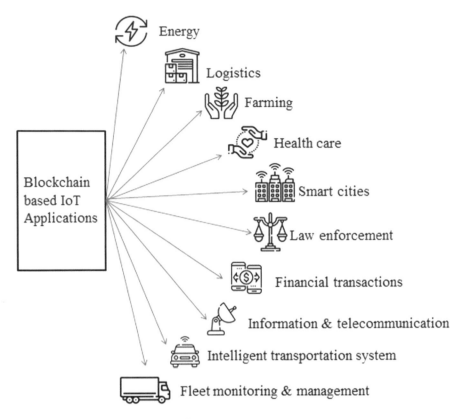

Fig. 1 Blockchain-based IoT applications

to direct quick and precise activities in decentralized conditions. Nevertheless, there are some significant issues that should be additionally researched to make blockchain relevant in vehicular networks:

- Decentralized agreement with faulty data: Blockchain empowers a distributed framework without using a centralized controller. A basic issue is the manner to accomplish an agreement in a complex vehicular network where every unit has restricted and faulty data.
- Vehicle mobility effect: The movement of vehicles presents difficulties to the administration of blockchain. A basic issue is a way to accomplish an agreement in a much of the time changing environment without giving up the reliability of a distributed system.
- Consensus delay effect: A majority of applications in vehicular IoT need a low latency, which represents a big test to the plan of a blockchain that causes a specific measure of time prior to arriving at an agreement. Hence, it turns out to be especially crucial to plan an agreement calculation that is solid and lightweight.

- Blocks distribution: In order to reach a consensus, blockchain distributes the blocks across the network. As the wireless resource in the vehicular environment is restricted, the effectiveness of the dispersal of blocks straightforwardly influences the performance of the framework. Thus, it is critical to address the issue of how to guarantee that the ledgers can be spread to all units in the network efficiently.

As shown in Fig. 2, the rise of blockchain innovation has brought another perspective about how to address the difficult points of the current IoT frameworks. Blockchain is an innovation that connects transaction blocks and a list of records that use cryptography to keep up transaction data in a decentralized way.

B. Intelligent Transportation System and Internet of Vehicles

The development of transmissions, detecting, and digital frameworks, has led to the replacement of traditional systems by high-level systems known as the Cyber-Physical System (CPS). This CPS is generally associated with the Internet advancements to form a link between the physical and the digital worlds. In recent years, CPS has gained popularity and has been extensively used in various domains of life [7]. An example of the CPS is the ITS which is referred to as the future of transportations. The main objective of ITS is to promote more agreeable, more secure,

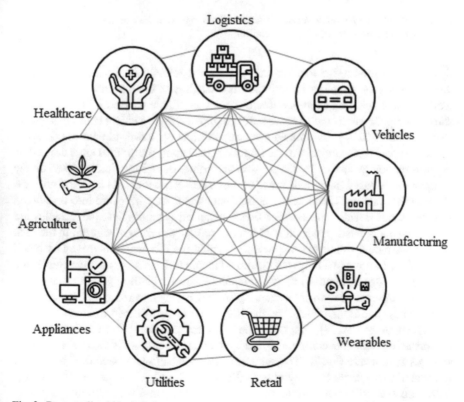

Fig. 2 Decentralized blockchain

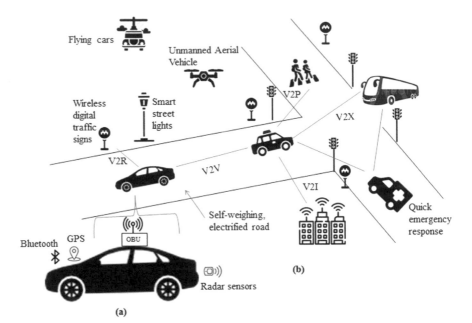

Fig. 3 **a** An outline of a Smart Vehicle; **b** A general description of the communication of an smart vehicle in an ITS that consists of various entities

dynamic, and productive transportation and good infrastructures. Simultaneously, the experience of traveling in a vehicle is changing as the automobile industry is working on developing smarter vehicles. In an intelligent transportation system, as shown in Fig. 3a, the vehicles are usually equipped with in-vehicle computational & storage units, OBUs, programmed systems, multiple sensors, and wireless devices.

In this specific situation, to make important decisions, the control units depend on the information generated by the sensors and cameras, and the link between all the components is based on various kinds of wired (CAN bus and LIN bus) and wireless (Bluetooth) systems. These trend-setting innovations may integrate into traditional vehicles soon. Besides, because of these advanced technologies, the vehicles are getting even more self-governing, which can possibly make a revolution in the ITS.

Another model called IoV is presented within the ITS which is driven by smart vehicles, the Internet of Things (IoT), and AI methods. In this model, the vehicles are associated with one another, individuals, and frameworks through communication systems so the vehicles can drive securely and smartly by observing and detecting the surrounding conditions. Therefore, the IoV system is considered to be an extended component of the CPS [8]. Other than the smart vehicles, the IoV environment in ITS comprises of vehicular communication systems, road-side units (RSU), and cloud storages as shown in Fig. 3b. The vehicular communication framework is frequently referred to as a vehicle to everything (V2X) which is only a communication model. V2X empowers communications from a vehicle to any unit (i.e., vehicle, structure, grid, pedestrian, and so on). Alternatively, the RSUs generally behave as access points and base stations to help VCS.

C.   The Challenges Associated with IoV

The growth in the IoV field is quicker than any time in recent years, because of the progressions of the Internet revolution and the tools used in a vehicle. Essentially, this promising IoV field is imagined to enhance the vehicular resources, transport structure, and the way of living, and to upgrade drivers' wellbeing in the coming years. A huge amount of information will be produced and moved to the cloud and edge storage from the vehicles and their services which will be used for IoV [9]. The vehicles in the coming years will likewise have excellent computational and storage measures. To provide a wide scope of benefits to the users, this information and resources will be imparted to one another. Especially, for Artificial intelligence (AI) based systems, the vehicles depend on cloud and edge computing units to assign the tasks.

Nonetheless, combining the present Internet systems to support IoV model opens up numerous difficulties [10–12], these include privacy, integrity, availability, and execution. The difficulties related to it will be expanded with the development of IoV network. In such a manner, numerous issues are interrelated with ITS. In reality, the IoV environment represents various attributes, and all the more explicitly, among these, some are one of a kind as compared with other IoT applications. Subsequently, the IoV environment may bring various new difficulties. Such exceptional features of IoV are described below.

High Mobility: In an IoV environment, both autonomous and semi-autonomous vehicles are referred to as moving units running on roads, unlike other smart devices. The speed of the vehicles may differ from each other which present different mobility especially for vehicles drive in manual mode. However, vehicles have adequate energy to send computational and communication resources, when the vehicles link with other units, the challenge will be to keep a steady communication because of the different and high mobility.

Intricacy in Wireless Networks: The IoV system depends on heterogeneous wireless communication networks, wherein other wireless technologies also exist. In this network, all vehicles are connected with each other. Technologies like Bluetooth, mmWave, and Dedicated Short Range Communication (DSRC) that use various wireless network services.

Latency Critical Applications: In an IoV network, more effective protocols are needed to transfer data among other vehicles. In reality, those applications are latency-sensitive and usually have short to medium transmission distances. Therefore, the data transmission delay caused between the source and the destination should be minimal [13]. For instance, in applications related to safety and emergency, the transmission and communication must be done within a specified time limit to avoid any fatalities. Due to these limitations, the potential Internet-based advancements to be included in this IoV model should avoid any delay in communication of transmission.

Scalability and Heterogeneity: Vehicles that visit larger areas have the advantage of obtaining scalability through RSU edge computing units, wireless networks, and vehicular ad hoc networks. Besides, the IoV tools, that display heterogeneous devices, look forward to a seamless integration of the present data and communication technology by considering the heterogeneity.

Artificial Intelligence (AI): The use of AI methods will be an integral part of the vehicular networks for the IoV application situations to adapt to the use of a wide range of vehicular information. The entire methodology from data collection to deployment is basically subject to various AI algorithms to guarantee self-learning capability [14]. However, since many AI algorithms basically depend on the data generated by the vehicle to rain the model; the dynamic and heterogeneous qualities of IoV situations further add challenges to execute ML methods.

D. Motivations of Using Blockchain in IoV

Blockchain has the capability to offer a significant number of inventive solutions to a large number of IoV applications [15]. A majority of IoV applications are mobile and real-time and they generate and transmit a huge amount of data. Many conventional techniques are not likely to be effective in IoV applications. In addition, the expanding network in such situations may allow the attackers to attack the weaker nodes. On the other hand, the incorporation of blockchain into IoV not only improves the security, privacy, and dependence but additionally upgrades the automation and the performance of the system [16, 17]. Therefore, to oblige adaptability and handle huge information, blockchain-like solid innovative technology must be utilized. Some major motivating reasons for the use of blockchain in IoV have been mentioned below.

Firstly, as we know one of the main highlights of blockchain is decentralization. Blockchain permits making decentralized IoV networks that contain distributed units such as RSUs, vehicles, and individuals. Simultaneously, these distributed unit substances can deal with their tasks autonomously. The functioning of the current IoV network which is principally dependent on a single decision will be moved into the decentralized model and become streamlined. Eventually, the decentralization will upgrade the vehicle's user experience. Secondly, blockchain does not depend on cloud-like frameworks for information storage and execution nor does it make use of any third-party entities. The units in the blockchain network can carry out the transactions and vehicular services on their own thus, reducing the operational costs. Thirdly, adopting blockchain for IoV can address security attacks like accessibility, interrupts, and single-point-of failure. This is because of the organization and replication of blockchain among all the units associated with the network. So even if one or more units are compromised, the services can still function smoothly. Blockchain technology is based on modern cryptographic techniques to guarantee security and privacy to the systems. Therefore the security and privacy of IoV networks in the blockchain are achieved by cryptography.

Another feature of blockchain is invariability for IoV networks. The blocks in the blockchain maintain a link with each other by using the hash values of each block. This invariability feature of blockchain prevents any alteration of data and also helps in precise auditing. Moreover, it empowers the organization and implementation of any pre-defined rules or scripts by making use of the smart contract. Blockchain offers distributed trading such as peer-to-peer (p2p) exchanging, sharing, and interaction between two entities [18]. In a p2p network, direct communication is set up between the entities that request and provide the services. This feature of p2p is used in IoV networks to exchange data between vehicles and RSUs safely. Since the elements don't have to interact with any mediator in the p2p network, it eventually brings about low latency applications and services. The IoV associates various units that do not trust one another. Since blockchain is based on the novel consensus mechanism, it is able to build a solid trust even among non-trusted entities. Other than the consensus mechanism, blockchain can also make use of smart contract to achieve trust between the entities. Lastly, the public blockchain does not require any authorization and can provide access to all the data stored in the blockchain thus, providing transparency in the IoV system.

2 Machine Learning for ITS

To accomplish an improved prediction of traffic flow, numerous prediction strategies have been proposed, like numerical demonstrating techniques, parametric and non-parametric strategies. In the non-parametric strategies, the most commonly used technique is based on Machine Learning (ML) [19]. In this segment, we focus on the capability of using Machine Learning in ITS, and how machine learning can incorporate and improve management and prediction tasks. Machine learning is a part of software engineering, which underlines the knowledge of machines in performing human-like jobs. The conventional machine learning approaches include Supervised Learning (SL) [20], Unsupervised Learning (UL) [21], Reinforcement Learning (RL) [22], and Deep Learning (DL) [23]. Using these machine learning approaches researchers have implemented various models to overcome the issues in the intelligent transportation system.

In recent years, several researchers throughout the world have evaluated the condition of advanced traffic prediction strategies. In [24], the researchers not just surveyed the prediction models classified in naive models, parametric models, and non-parametric models, they additionally had a profound audit of various sorts of open-source datasets. The dataset is characterized by the versatility of the sensor, and the researchers looked at the pros and cons of these datasets. Furthermore, the researchers looked at various sorts of data models utilized to predict traffic in different situations.

In [25], the authors have made use of models based on neural networks (NN) to predict the traffic. The basic ideas utilized in NN model prediction have been discussed. The NN models classified by their neural construction, layer design, and activation function were also reviewed. The results were based on the predictions made by NN networks in different situations.

With the advancement of the Deep Learning (DL) structure in the NN model, some studies are also grounded on deep learning-based traffic prediction models. In [26], the authors surveyed the various DL models and the connected work processes used to develop a usable DL prediction model. They also studied the DL models by applying them to various types of traffic state predictions and utilizing the equivalent dataset, which can give the reader a more instinctive perspective on the benefits and drawbacks of various DL expectation models.

According to the studies, machine learning-based prediction models have been gaining popularity in the past few years. The motivation behind this is limited knowledge of the past of the traffic patterns to build a model, constraints on prediction tasks, and improved non-linear features [27].

The machine learning models have been used extensively in the prediction of traffic. It has been gaining popularity due to the applicable traffic scenarios in which the machine learning models are used. In this paper, we will zero in on the ML models utilized in rush hour gridlock forecast undertakings. In the interim, we likewise focus on the application situations the ML model has been applied to. It is significant that we think about various kinds of models on their correctness, along with features like the capacity to manage some particular issues, and the effectiveness of the model, the equipment, and the information reliance of the model. As we probably are aware, ML is an extremely vast subject, and there are numerous sorts of classification models for ML models dependent on alternate points of view. In this paper, the strategies we discussed are sorted by various kinds of ML calculation, as per [28–31], and are additionally explained as follows:

- Regression model: By contemplating the connection between the reliable variable and the autonomous variable, the regression model attempts to utilize a curve or a line to fit the dataset.
- Example-based model: This model works on the prediction task by finding the differences and resemblances among the input data and previous data samples and uses the result to make the final prediction model. One such example is the k-Nearest Neighbors (KNN) model.
- Kernel-based model: In this type of model, a kernel function is used to map the input function into a high-request vector space, where the prediction tasks are not difficult to tackle. Some examples of this model are the Support Vector Machine (SVM) and Radial Basis Function (RBF) model.
- Neural Network model: Neural Network model is a sort of model developed by reproducing the manner in which data goes through neurons in the human brain. The data is passed through various organizational structures in different kinds of Neural Network models. In this type of network, the data is changed into an activation signal by using the activation function. Eventually, the prediction is made depending on the actuation signal. Examples of the neural network model are Feed Forward Neural Network (FFNN), Recurrent Neural Network (RNN), and Convolutional Neural Network (CNN).
- Hybrid model: In this type of model, the final prediction is made by consolidating at least two expectation results from various prediction models.

3 Applications of Blockchain-Enabled IoV

Blockchain innovation has been drawing in immense interest because of its ability to accomplish a decentralized, straightforward, and tamper-resistant framework [32]. Numerous studies are focusing on the utilization of blockchain in managing transactions and data in vehicles (Fig. 4).

In this section we will see how blockchain innovation can be used in IoV application situations, we confirm that blockchain is expected to adapt to various IoV applications and administrations including vehicular information security, managing the vehicle, and on-request transportation administrations. The following is a survey of incorporating blockchain technology in such IoV application situations.

(a) Data Protection and Management

With IoV applications gaining more attention, a lot of information will be generated by the vehicles in IoV network to improve the security and services of the vehicles. The vehicles will now depend on RSUs for sharing, storing, managing and using the data. Nevertheless, the computing nodes are positioned along the roads therefore they can be prone to security and privacy breaches. However, these nodes can be difficult to trust as they are operated by different service providers. In fact, the vehicles over the network may not share the data due to these conditions.

In the recent years, the implementation of blockchain in IoV applications has been increasing not only because of its privacy and security but also due to the trust among the nodes in the network. In [33], the authors make use of consortium

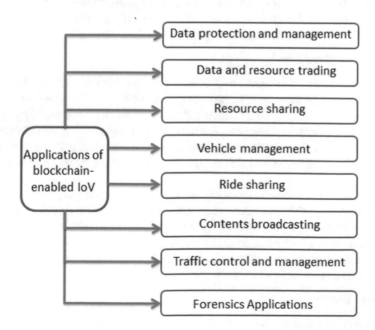

Fig. 4 Applications of blockchain-enabled IoV

blockchain, where they present a secure blockchain-based information system in the vehicular networks. The benefits of using smart contract in this proposed framework are double. The smart contract is firstly used to accomplish sharing of information securely. Furthermore, the smart contract guarantees that the information can't be shared without approval.

(b)   Data and Resource Trading

Businesses for vehicles and other related units in an IoV framework have been empowered by data and resource trading. This trading may prompt security issues, for example, access control authorization and protection of users. Moreover, thinking about the versatility of the vehicles and remote connections, the exchanging situations could be powerless against DoS attacks. These attacks could also be unintended. Eliminating any possible disturbance due to these attacks and to guarantee reliable trading among the entities is also a worry.

Recently, studies have proved blockchain to be a potential solution for the above mentioned concerns. In particular, blockchain could offer secure, shared, and decentralized resolutions in an IoV framework for data and resource trading.

(c)   Resource Sharing

IoV carries a potential to develop a resource sharing platform for all vehicles. The close entities can share their computational data using this platform. The platform can administer services that are latency sensitive and also AI applications. Though it has all advantages, it still has two concerns. Firstly it is hard to keep up to this platform as the vehicles may not trust one another. Nevertheless, using blockchain can help overcome this issue to some extent. Secondly a suitable methodology is required so that the vehicles can contribute in resource sharing.

(d)   Vehicle Management

Smart Parking and vehicle platooning are some famous examples of vehicle management. In a smart parking system the users know about the free space to park and book the spot beforehand. Thus reducing the time of the users and avoiding congestion in traffic. In vehicle platooning, a number of vehicles with common interest are combined together. Essentially, in a platoon one vehicle which is called as the platoon head, leads the other vehicles which are the members with a little space between each other.

A large number of vehicles, parking lots and roadside units have led to a complicated management of the IoV framework. This can further lead to security and privacy concerns of these units. The use of blockchain can essentially address these concerns. Blockchain can particularly address the concern of a centralized parking system that requires displaying the data to the users for booking of sots and using a framework that prevents and attacks and leakage of data.

(e)   Ride Sharing

The services provided by Uber are an example of ride sharing which have gained a lot of interest due to its convenience. These services are used by people that wish.to go to a same destination in the same vehicle thus reducing traffic congestion on the roads. In the near future ride sharing services will become a part of an intelligent transportation system due to the on-going demand.

Despite of the undeniable advantages, the current cloud-helped ride-sharing administrations have two inexplicable difficulties. In the first place, due to the concentrated cloud association, the clients may experience delays in communication. Also, privacy of the users is at risk due to the exposure of the users data. Hence, to encourage both vehicles and users to share their data, it is important to create productive and secure ride-sharing services.

As of late, blockchain has been now utilized in such applications. In [34], the challenges of carrying out the ride-sharing framework for self-ruling vehicles along with a conventional centralized architecture are addressed. The authors have developed a blockchain-based decentralized ride-sharing system grown with the goal that the autonomous vehicles will want to offer themselves in the system.

(f)   Contents Broadcasting

Broadcasting the data among the vehicles connected over the internet targets to improve the in-vehicle as well as safety services. These administrations can offer various applications like publications, advertisements, internet shopping, endorsing business items, entertainment, and emergency data declarations. Moreover, the edge computing units like RSUs can play a significant part to help vehicles by storing important data, transferring data to the nearby vehicles and reduce latency to improve user experience.

Nonetheless, blockchain can possibly empower p2p content sharing and broadcasting among the vehicles through V2V and V2I communications. In fact blockchain-based p2p model can guarantee a proficient, economical, and reliable stage for the IoT. Moreover, to speed up the transfer of data, blockchain empowers to develop innovative ideas.

(g)   Traffic Control and Management

The data produced by the vehicles like, traffic information, mishap occurrences, and sharing of data other vehicles will help in managing the traffic and other such services easier. The usage of the data can bring various opportunities and advancements for vehicles in ITS framework such as, managing the traffic in real time, and easing of traffic congestion. In such scenarios security is an important issue. The transfer of data between the vehicles and the roadside units can be vulnerable to security attacks. Another significant challenge is to develop an autonomous traffic control management system.

(h)   Forensics Application

One of the potential applications of blockchain is digital forensics. This application in IoV incorporates investigation of car crash, especially for autonomous vehicles. In such scenarios blockchain can provide assistance in calculating damage cost after an accident by using smart contract and collecting data by communicating with the other vehicles.

An extensive vehicular system called as Block4Forensic utilizing permissioned blockchain and vehicular public key foundation (VPKI) is introduced in [35]. The reason for this Block4Forensic system is to offer forensic assistance to analyse accidents of vehicles. In Table 1, a summary of the above mentioned blockchain applications are presented.

Table 1 Summary of blockchain application scenarios in the IoV

Applications	Description	Opportunities in blockchain
Data Protection and Management	– A lot of information is generated by the vehicles in IoV network to improve the security and services of the vehicles. This data is shared among the vehicles over the network – The vehicles depend on RSUs for sharing, storing, managing, and using the data	– Blockchain offers decentralization of the data generated by the vehicles and does not depend on any centralized management – Blockchain is used in the protection of vehicle data because of its immutability and decentralization features to overcome data integrity and accessibility issues – It offers an efficient vehicular data management system as compared to conventional database management system while guaranteeing security requirements through modern cryptographic techniques
Data and Resource Trading	– Businesses for vehicles and other related units in an IoV framework have been empowered by data and resource trading for communication and computing	– Blockchain can help to establish peer-to-peer (p2p) data as well as resource trading among the vehicles in a decentralized ledger network that consists of vehicles as users – This autonomous trading has facilitated in reducing the transaction costs significantly
Resource Sharing	– The consensus resource sharing among the vehicles possibly develops the latency-aware and AI- related services – It is difficult to achieve this due to the trust and mobility issues	– To accelerate the resource sharing among the vehicles, blockchain has excluded the dependence of trusted third parties, and intermediaries – Blockchain provides direct sharing of resources between a service provider and a user which results in lower latency services and a rise in the complete performance of services

(continued)

Table 1 (continued)

Applications	Description	Opportunities in blockchain
Vehicle Management	– Due to the diversity and a large number of entities involved in the network, it is challenging to preserve the privacy of users as well as implement the access control in a smart parking system that is under centralized controlled – Therefore it is difficult to manage the IoV network	– Blockchain has improved the smart parking and vehicle platooning systems by its decentralization feature – Vehicle management over the decentralized network has adopted the authentication of the service requests and influences through the smart contract
Ride Sharing	– The integration of vehicles and users into ride sharing services over the Internet is typically based on a cloud platform – Subcontracting the data of users to the cloud can affect the privacy of users	– Blockchain provides less delay in communication and helps in faster data retrieval as compared to cloud-based services – The ride sharing services are made easier by blockchain as it finds the users and vehicles using RSU rather than cloud platforms
Content Broadcasting	– Content broadcasting is the ability of IoV system to distribute the non-safety and safety contents of vehicles and RSUs which may be partially trusted, non-trusted, and vulnerable to attacks	– Blockchain can ensure the reliability of the contents by storing it into multiple nodes that can also help in fast retrieval – The invariability feature of blockchain prevents any data integrity attack
Traffic Control and Management	– The vehicle traffic management system relies on the data produced by the vehicles as well as the RSUs – This system may be vulnerable to security and privacy attacks	– As compared to the conventional system, – Blockchain provides decentralized management and control using consensus mechanisms – The strong invariability and multiple storage features of blockchain has enhanced the accessibility yet guaranteed better security protection against DoS attacks – The Smart contract is useful in traffic management and control in terms of authentication and authorization without depending on trusted key generation centers

(continued)

Table 1 (continued)

Applications	Description	Opportunities in blockchain
Forensic Application	– This application in IoV incorporates investigation of a car crash, especially for autonomous vehicles	– Blockchain can assist in calculating damage cost after an accident by using smart contract and collecting data by communicating with the other vehicles – Using blockchain with modern cryptographic techniques in forensic application systems enhances traceability and transparency

4   Designing an Efficient Blockchain System to Satisfy the Application Requirements in Vehicular IoT

- Blockchain for developing vehicular IoT applications: Every vehicular IoT application can vary in features and requirements. For instance, the systems for collision avoidance need very low latency, whereas the data analytics systems in vehicles strict protocols. It is essential to plan a productive blockchain system keeping in mind the qualities of every application. It is essential to plan a productive blockchain system thinking about the qualities of every application. A lot of research in the field of productive blockchain systems for arising vehicular IoT applications is still required.
- Privacy-mindful blockchain: The basis of attaining consensus in a blockchain system is the scattering of all previous transactions in the entire network. This brings about a security concern since all transaction information is displayed to the authorized entity, bringing about the chance of uncovering the genuine nature of an entity. In this period of big data, combining data from various entities can result in a threat to the privacy and security of the systems. Hence, it is the need of the hour to debate over the security concerns of the transactions in blockchain considering the vehicular IoT network
- Privacy and Security concerns: Even though blockchain has a higher level of security than the traditional centralized systems, there are still some security concerns. Although the security level of blockchain is higher than the ordinary brought together framework, there is additionally a worry of the 51% assault [36]. Because of the mobility in vehicles, some authentic vehicles may not add to the blockchain framework on time, causing a security issue. Therefore, avoiding privacy attacks in a dynamic network environment needs more research.
- Blockchain for intelligent collaboration: Vehicular IoT consists of a large number of devices for sensing communicating and computing. To accomplish greater computational capacity and lower latency, the vehicular IoT devices in the future can consider combining multiple entities as the resources at each vehicle are limited. As of late, federated learning has drawn incredible importance in using

the information from numerous devices to improve system intelligence. The traditional techniques in federated learning depend on a single server to collect feedback from various entities. Nevertheless, in some vehicular IoT systems, the centralized control server doesn't exist. Blockchain enables the communication between one or more entities without depending on any centralized server. The issue of utilizing blockchain to enhance the decentralized framework could be an area of research.

5 Improving Vehicular IoT Protocols to Support Blockchain

- Blockchain network performance: In a blockchain framework, an enormous amount of information is sent on the network prior to accomplishing consensus among the systems. Researchers still can't seem to address the huge measures required to maintain the blockchain which is particularly a huge issue in vehicular systems as the transfer speed is restricted. Although blockchain gives a decentralized methodology, it causes a higher overhead than the centralized methodology as it needs to transfer the information to all the entities over the network. Hence, it is critical to think about the balance between decentralization and networking systems. The networking performance of blockchain in vehicular IoT environments is still a topic that needs to be explored.
- Methodologies for communication, storing, and computing blockchain: The blockchain framework is influenced by the combined effort of communication, storing, and computing blockchain. As there is no centralized control server in a blockchain framework, combined allotment turns out to be more intricate. Vehicular IoT conditions possess new difficulties to assign resources as it makes use of the dynamic framework. To start with, the restricted communication resources can be a challenge in empowering a decentralized framework. Secondly, vehicles may perform computation collaboratively as some vehicles may not have the capability to perform such complex computations. Lastly, the system performance can be improved by efficiently caching the data in some nodes.
- Blockchain for vehicular conditions: The blockchain framework should take into account the qualities of vehicular conditions such as the mobility of the vehicle, densities, communication network, and other applications. For instance, studies show the use of RSUs as a medium to add the blocks in a blockchain framework. As continuous communication may happen between vehicle-to-roadside (V2R) units, vehicles associating with a similar RSU change after a certain period. With time and location, the vehicle density also differs. These features need to be considered when planning a consensus algorithm.

6 Conclusion

The advancement of the Intelligent Transportation System (ITS) in the blockchain is still unclear. Keeping in mind sustainability, there are very few studies that have explored the topic of blockchain in ITS. With the development of communication and computing systems, it is essential to analyze the effective utilization of communication and computing resources to help blockchain in vehicular conditions. ITS is a field of innovative work of quickly developing technologies wrapped into various kinds of stages for a group of advanced applications. Blockchain in vehicular networks has the potential of developing many unique applications. However, the dynamicity of vehicular conditions presents many difficulties in designing an efficient blockchain system. For the effectiveness of the running applications, the management and services of a large amount of data is a fundamental requirement. As a result, the advancement in Machine Learning has a significant impact on the technologies that drive a revolution in ITS.

References

1. Zhang, X., Li, R., Cui, B. A security architecture of VANET based on blockchain and mobile edge computing. In: 2018 1st IEEE International Conference on Hot Information-Centric Networking (HotICN), pp. 258–259 (2018)
2. Peng, C.; Wu, C., Gao, L., Zhang, J., Alvin Yau, K.L, Ji, Y. Blockchain for vehicular internet of things: recent advances and open issues. Sensors **20**(18), 5079 (2020)
3. Mollah, M.B., Zhao, J., Niyato, D., Guan, Y.L., Yuen, C., Sun, S., Lam, K.Y., Koh, L. Blockchain for the internet of vehicles towards intelligent transportation systems: a survey (2020)
4. Nguyen, Q.K., Dang, Q.V. Blockchain technology for the advancement of the future. In: 2018 4th International Conference on Green Technology and Sustainable Development (GTSD), pp. 483–486 (2018)
5. Živić, N. Distributed ledger technology for automotive production 4.0. In: 2020 28th Telecommunications Forum (TELFOR), pp. 1–3 (2020)
6. Wan, L., Eyers, D., Zhang, H. Evaluating the impact of network latency on the safety of blockchain transactions. In: IEEE International Conference on Blockchain (Blockchain), pp. 194–201 (2019)
7. Tuo, M., Zhou, X., Yang, G., Fu, N. An approach for safety analysis of cyber-physical system based on model transformation. In: 2016 IEEE International Conference on Internet of Things (iThings) and IEEE Green Computing and Communications (GreenCom) and IEEE Cyber, Physical and Social Computing (CPSCom) and IEEE Smart Data (SmartData), pp. 636–639 (2016)
8. Gupta, R., Tanwar, S., Al-Turjman, F., Italiya, P., Nauman, A., Kim, S.W.: Smart contract privacy protection using ai in cyber-physical systems: tools, techniques and challenges. IEEE Access **8**, pp. 24 746–24 772 (2020)
9. Hassan, N.U., Yuen, C., Niyato, D.: Blockchain technologies for smart energy systems: Fundamentals, challenges, and solutions. IEEE Ind. Electron. Mag. **13**(4), 106–118 (2019)
10. Chattopadhyay, A., Lam, K.Y., Tavva, Y. Autonomous vehicle: Security by design. IEEE Trans. Intell. Transport. Syst. (2020)
11. Hahn, D.A., Munir, A., Behzadan, V. Security and privacy issues in intelligent transportation systems: classification and challenges. IEEE Intell. Transport. Syst. Mag. (2019)

12. Li, W., Song, H.: Art: An attack-resistant trust management scheme for securing vehicular ad hoc networks. IEEE Trans. Intell. Transp. Syst. **17**(4), 960–969 (2015)
13. Baee, M.A.R., Simpson, L., Foo, E., Pieprzyk, J.: Broadcast authentication in latency-critical applications: on the efficiency of IEEE 1609.2. IEEE Trans. Veh. Technol. **68**(12), 11577–11587 (2019)
14. Yuan, T., da Rocha Neto, W.B., Rothenberg, C., Obraczka, K., Barakat, C. et al. Machine learning for next-generation intelligent transportation systems: a survey (2020)
15. Du, X., Gao, Y., Wu, C.H., Wang, R., Bi, D., Wu, W. Blockchain-based intelligent transportation: a sustainable GCU application system. J. Adv. Transp. **2020**, 1–14 (2020)
16. Ren, Q., Man, K.L., Li, M., Gao, B., Ma, J. Intelligent design and implementation of blockchain and internet of things–based traffic system. Int. J. Distrib. Sensor Netw. (2019)
17. Yuan, Y., Wang, F. Towards blockchain-based intelligent transportation systems. In: 2016 IEEE 19th International Conference on Intelligent Transportation Systems (ITSC), pp. 2663–2668 (2016)
18. Wu, X., Wu, H. Trust management of mobile users P2P nodes based on blockchain. In: 2020 IEEE 6th International Conference on Computer and Communications (ICCC), pp. 507–511 (2020)
19. Boukerche, A., Wang, J. Machine learning-based traffic prediction models for intelligent transportation systems. Comput. Netw. **181**, 107530 (2020). ISSN 1389-1286
20. Garcia, S., Luengo, J., S´aez, J.A., Lopez, V., Herrera, F. A survey of discretization techniques: Taxonomy and empirical analysis in supervised learning. IEEE Trans. Knowl. Data Eng. **25**(4), 734–750 (2013)
21. Hastie, T., Tibshirani, R., Friedman, J.: Unsupervised learning. In: The Elements of Statistical Learning, pp. 485–585. Springer (2009)
22. Kaelbling, L.P., Littman, M.L., Moore, A.W.: Reinforcement learning: a survey. J. Artif. Intell. Res. **4**, 237–285 (1996)
23. LeCun, Y., Bengio, Y., Hinton, G. Deep learning. Nature **521**(7553), 436 (2015)
24. Nagy, A.M., Simon, V.: Survey on traffic prediction in smart cities. Pervasive Mob. Comput. **50**, 148–163 (2018)
25. Do, L.N., Taherifar, N., Vu, H.L. Survey of neural network-based models for short-term traffic state prediction, Wiley Interdiscip. Rev. Data Min. Knowl. Discovery **9**(1), e1285 (2019)
26. Aqib, M., Mehmood, R., Alzahrani, A., Katib, I., Albeshri, A., Altowaijri, S.M.: Smarter traffic prediction using big data, in-memory computing, deep learning and GPUs. Sensors **19**(9), 2206 (2019)
27. Arrieta, A.B., Díaz, N., Del Ser, J., Bennetot, A., Tabik, S., Barbado, A., García, S., Gil-López, S., Molina, D., Benjamins, R., et al.: Explainable artificial intelligence (XAI): Concepts, taxonomies, opportunities and challenges toward responsible. AI. Inf. Fusion **58**, 82–115 (2020)
28. Goodfellow, I., Bengio, Y., Courville, A. Deep Learning. MIT Press (2016)
29. Shalev-Shwartz, S., Ben-David, S. Understanding Machine Learning. Cambridge University Press (2014)
30. Wang, J., Boukerche, A. The scalability analysis of machine learning based models in road traffic flow prediction. In: 2020 IEEE International Conference on Communications, pp. 1–6. IEEE (2020)
31. Sun, P., Aljeri, N., Boukerche, A.: Machine learning-based models for realtime Traffic flow prediction in vehicular networks. IEEE Netw. **34**(3), 178–185 (2020)
32. Liu, Y., Yu, F.R., Li, X., Ji, H., Leung, V.C.: Blockchain and machine learning for communications and networking systems. IEEE Commun. Surveys Tutor. **22**(2), 1392–1431 (2020)
33. Kang, J., Yu, R., Huang, X., Wu, M., Maharjan, S., Xie, S., Zhang, Y. Blockchain for secure and efficient data sharing in vehicular edge computing and networks. IEEE Internet Things J. (2018)
34. Shivers, R.M. Toward a secure and decentralized blockchain-based ride-hailing platform for autonomous vehicles. Ph.D. dissertation, Tennessee Technological University (2019)

35. Cebe, M., Erdin, E., Akkaya, K., Aksu, H., Uluagac, S.: Block4forensic: An integrated lightweight blockchain framework for forensics applications of connected vehicles. IEEE Commun. Mag. **56**(10), 50–57 (2018)
36. Viriyasitavat, W., Xu, L.D., Bi, Z., Hoonsopon, D.: Blockchain technology for applications in internet of things—mapping from system design perspective. IEEE Internet Things J. **6**, 8155–8168 (2019)

Empowerment of Internet of Things Through the Integration of Blockchain Technology

C. Muralidharan, Y. Mohamed Sirajudeen, and Y. Rajkumar

Abstract Due to the convergence of advanced technologies, the emergence of electronic things tends to increase tremendously. The Internet of Things (IoT) is the technology where enormous things such as sensors, softwares etc. are embedded together for enabling the connection as well as communication of data. With the evolution of Industrial IoT, the need for monitoring the things that are used for automation becomes one of the thrust among the industries. There is a necessity that the monitoring can be empowered through the integration of blockchain technology. Not limited to monitoring, it also enriches the scalability, interoperability, durability, device reliability, privacy, security, silo mentality, standardization of the devices that are used for the internet of things. The blockchain technology is the distributed ledger with the list of growing records or blocks which are linked using the cryptography. The one of the main advantage of blockchain technology is that all the transactions are captured with its respective timestamps and the data that are recorded in the blocks cannot be altered retroactively. This enforces timely updation of information which cannot be altered. The Industrial IoT plays a major roles in the IoT platform as there exists smart connected operations that covers plenty of assets which obviously needs capability such as high level connectivity, analytics on big data and the development of applications. To resolve these challenges of IIoT, several requirement has to be considered such as asset visibility, Bigdata handling, technology integration and security. Also it is felt that most of the existing facilities of the industries such as ad hoc networks, smart grids etc. could not connect to the IoT system as it is not with built in intelligence. These needs an interface for invoking the communication. Since the blockchain technology has the decentralized nature, the information of the communication that happens between the two devices such as state of the devices, interactions and data digest can be acquired all in a point. This might reduce the risk with the business processes that the industry faces. Also, in the field of logistics

C. Muralidharan (✉)
SRM Institute of Science and Technology, Kattankulathur 603203, India

Y. Mohamed Sirajudeen
Vardhaman College of Engineering, Hyderabad 501218, India

Y. Rajkumar
Vellore Institute of Technology, Vellore 632014, India

which is one of the most important field, the information about the goods will be kept out of the illegal persons. Hence, it is considered that the blockchain technology will hold a safer environment for the individual as well as the enterprises. Thus the integration of blockchain with the Internet of Things will regulate the process of IoT through decentralized control which is kept highly safe than other technology implications. This chapter discuss about the different components of block chain enabled IoT system and the application of implementing the blockchain in IoT.

Keywords Blockchain · Internet of things · Industrial internet of things · Blockchain and IoT

1 Introduction

After the evolution of portable internet services and lightweight computing device, the connectivity has increased exponentially. One of the example of lightweight computing devices is Internet of Things (IoT). It has a huge potential to provide excellent services across many areas from like manufacturing, intelligent transportation, military applications and utility organizations, social media, healthcare and smart cities. Internet of Things (IoT) is a group of heterogeneous devices with low computing power, low storage and limited battery connected to the internet. The sensors connected to the IoT microprocessor senses the information from the environment and transfers it to the cloud server through the internet [1, 2]. Collected information are later processed and analyzed in the cloud storages to obtain knowledge. The analyzed results helps in improving the processes and making better business decisions. Some of the advantages of IoT technology are, saves human effort, saves time, enhance the performance of data collection, collects huge number of data and helps in attaining best knowledge.

Though IoT has several advantages, it also has some issues to be addressed. According to the report of International Data Corporation (IDC), the number of IoT devices connected to the internet will reach 41.6 billion by the year 2025, and also it is predicted that the total size of the data generated by these IoT devices will cross 79.4 ZettaBytes (ZB). Managing and protecting such a large quantity of IoT generated data in cloud storage remains challenging to Cloud Service Providers (CSP). The "Global Risk Report–2019" published by the World Economic Forum (WEF) mentions that cyber-security attacks and data breaches are the fourth biggest threats. According to the statistical report of HIPPA Journal, a total of 3054 data breaches happened during 2009–2019. Nearly, 230,954,141 data were stolen, leaked, and impermissibly disclosed. It also mentions that the number of exposed records has tripled during 2018–2019.

IoT based data collection are used in healthcare and military applications. Losing those confidential information might lead to a great danger. The cloud architecture is created based on the centralize structure. The data collected by the IoT devices are stored in the centralized cloud storages maintained by the Cloud Service Providers.

The issue with this model is that the central server might act maliciously and can hijack and make fraudulent use of the user's confidential data. As far as cloud storages are concerned, the security and privacy remains as an open challenge yet. By converting the centralized storage setup into a decentralized form can improve the security and privacy of the user data. Decentralized IoT storage system allows the user's to securely store their confidential data in the nearby decentralized servers and gives the users the full authority over their data. Song et al. [3], proposed the idea of using Blockchain technology in the Internet of Things (IoT) network. It supports decentralized fault-resistant management and data immutability [3]. Later, Mocnej et al. [4], proposed the concept of decentralized IoT storage architecture. It proposes a five-key features of a decentralized IoT platforms such as, multi-network approach, scalable and interoperable implementation, low power consumption, Intuitive data and device management, artificial intelligence at edge nodes [4].

The blockchain (BC) technology is decentralized paradigm which provides privacy and security to the confidential data. The beauty of blockchain technology is, it omits the need of centralized authority in managing the confidential data [1, 5]. Also, it removes the centralized server and works with multiple number of decentralized service. The data transactions made by an IoT devices to a decentralized server are verified by other decentralized servers connected to the network. It builds the trust between the entities involved in the network without a centralized authorities [2]. Disruptive technologies such as big data and cloud computing have been leveraged by IoT to overcome its limitations since its conception, and we think blockchain will be one of the next ones. This paper focuses on this relationship, investigates challenges in blockchain IoT applications, and surveys the most relevant work in order to analyze how blockchain could potentially improve the IoT.

In blockchain technology, the entire data or the ledger is shared over all the nodes connected to the blockchain network based on the consensus algorithm. The security and privacy of the user's data is ensured in the blockchain network as the same copy of the ledger is stored in multiple nodes. The advantages of blockchain technology are transparency, immutability, auditability, public verifiability, security and privacy [5, 6]. Blockchain is predominantly used in crypto-currencies. However, the advantage of blockchain technology has extended its usage in intelligent transportation system, supply chain management, industry, agriculture, and many more. The use of decentralized servers, contiguous algorithms and self-executing smart contracts have make it as a promising technology for securing the IoT generated data.

The rest of the chapter is divided into the following sections. In Sect. 2, presents a brief background of IoT and its security and privacy issues. In Sect. 3, discusses the blockchain technology and its components, Sect. 4 discusses about the integration of blockchain and IoT, Sect. 5 discuss about the scenarios in integration of blockchain and IoT and Sect. 6 concludes the chapter.

2   Introduction to Internet of Things (IoT)

The idea of adding sensors and intelligence to basic objects was discussed throughout the 1980s and 1990s (and there are arguably some much earlier ancestors), but apart from some early projects including an internet-connected vending machine progress was slow simply because the technology wasn't ready [6–8]. Chips were too big and bulky and there was no way for objects to communicate effectively. Processors that were cheap and power-frugal enough to be all but disposable were needed before it finally became cost-effective to connect up billions of devices.

Currently, billions of embedded IoT physical devices around the world are connected to the internet. Those devices are attached with various sensors to sense the information from the environment. Microprocessor in the IoT devices are collects the information and sends it to the centralized cloud servers. IoT devices are super-cheap and has an ability to connect with the wireless communication model to transfer the sensed information. The Internet of Things is making the fabric of the world around us smarter and more responsive, merging the digital and physical universes.

2.1   Growth of Internet of Things (IoT)

The number of IoT devices and the volume of the data generated by them are increasing exponentially. According to the reports of Global IoT market 2018, the total number of IoT devices that are connected to the internet has reached 23.1 billion. Also, it is expected to reach 75.4 billion in 2025. The data generated by the IoT devices are valuable resources that can be used to optimize manufacturing and business operation. The global IoT market is expected to reach a value of USD 1,386.06 billion by 2026 from USD 761.4 billion in 2020 at a CAGR of 10.53%, during the forecast period (2021–2026).

2.2   Industrial IoT

The industrial Internet of Things (IIoT) enables the connection between the products as well as the processes for empowering the digital transformations. The enhancement of IoT to Industrial IoT is to support the industries for monitoring, analyzing and acting automatically [9]. One of the important reason for the implementation of IIoT is to optimize the operational efficiency and to monitor and maintain automatically. This obviously reduces the expenditure and maintenance cost. This also increases the efficiency of managing the consumer's demand by providing the timely service.

2.3 *Applications of IoT*

Below are the few IoT enabled applications,

(a) **Smart Wearables**

Wearables such as watches, fitness bands etc. are enabled with IoT which is used to track the calories burnt, walking steps, heart beats etc. This avoids the usage of particular health equipment separately as all these are provided at a single device.

(b) **E-Health**

The health of the patients can be monitored by the doctors with the IoT enabled wearables. These wearables are connected to different sensors that collects the real time data from the patients [10]. This supports the doctors to manage the patients in a timely manner.

(c) **Traffic Monitoring**

Vehicle traffic is one of the important problem throughout the world. Hence implementation of IoT will be useful in managing the traffic. Every part of world tend to develop smart cities which enables the connected things. The updation of traffic details can be observed with the help of smart devices through the usage of maps thereby avoiding the traffics. This also support the users in identifying the alternate route for their travel.

(d) **Agriculture**

The smart farming is the target of every part of world. Each and every data about the changes in the agricultural field can be observed with the use of IoT equipment. This support the farmers in observing the changes timely and do the needy things. This also supports the farmers in updating the knowledge about the crops.

3 Blockchain Technology

Blockchain is a digital form of information tracking system that tracks the information of all the transactions timely that cannot be modified or hacked. It is a distributed ledger that tracks the information of all the participants which are stored and distributed across the network of systems [11]. Each and every block includes the transactions and for every new transactions that occurs will be updated to the particular participant's ledger. It is usually called as Distributed Ledger Technology as it is managed by a number of participants [12]. They are immutable in nature as the data that are stored in the chain will not be modified without proper permission from the participants and will be updated with proper timestamp.

3.1   Components of Blockchain Technology

To understand better about the blockchain, there are necessary components that needs to understand well. They are,

- Application node
- Shared ledger
- Consensus Property
- Virtual Machine.

(a)   **Application node**

Application specific nodes has to be installed and used for proper implementation. In the Bankchai like applications all the participants will be restricted and each can be joined with the special permission [13]. The Bankchain like application are allowed only for the banks for implementing the nodes whereas in the Bitcoin ecosystem any participant can download and install the application.

(b)   **Shared Leger**

The blockchain follows distributed data structure where the data will be distributed over the networks. This data can be viewed by the participants once they run the node application [13]. E.g. If the participant is running Ethereum client, then they can see the Ethereum ecosystem ledger.

(c)   **Consensus Property**

This seems to be a logical component where the consensus algorithm will be implemented with the node applications. It provides set of rules for the arriving data at the ledger. As per the property, the single ecosystem can arrive at a single time. If one of the ecosystem happens ata second and the other will happen at a sub second and so on.

(d)   **Virtual Machine**

It is the last component of the node application which is implements by all the participants. The virtual machine is a type of representation of machine that are made with the help of set of instructions in a specific language. It is an abstract of the physical machine that is used for efficiently using the physical resources in the manner of virtual form.

4   Blockchain and Internet of Things

One of the fast growing technology that attracts multiple domains is Blockchain technology which is a form of Distributed Ledger Technology. Cryptocurrency is one of the root form of blockchain since 2014. Likewise Internet of Things is another domain that is been used in every part of the world to make everything connected. Connecting these two domains will empower the cybersecurity, logistics, finance etc. The IoT is a technology where number of sensors collect the information for automating the process whereas the blockchain is a ledger technology where the information or data will be stored in a distributed manner. Only the common thing needed for both the technology is Internet [14, 15]. When both the domains are connected with each other than the data that arc collected from the things will be stored in a distributed ledger with the use of blockchain technology.

IoT covers wide range of applications such as smart grids, management of health and smart cities. Experts says it is increasing invisibly which are dense that collects and process the data. But the serious hit that needs to be considered in the IoT is the security and privacy concerns of the data as it handles sensitive data [15]. Few of the privacy and security challenges includes the resources are heterogeneous, no central control, issues in scalability and so on. Hence security and the privacy are the major concern that needs to be concentrated in the field of IoT. This can be overcome by implementing the blockchain with the IoT.

All the research firms says that the blockchain is growing rapidly and are not only for the financial purposes. Initially it was developed for empowering the cryptocurrencies but later it is been suggested for different purposes such as empowerment of cybersecurity, distributed data management etc. Many of the companies are expecting the integration of these technologies through which solutions can be found for many of the services [16]. The integration of blockchain with the IoT support us to find how the devices are communicating with each other thereby finding the flaws that happens in the process.

4.1 Advantages of Blockchain and IoT Integration

The advantages of integration of Blockchain with the IoT is discussed below:

(a) Decentralization

The blockchain is decentralized in nature which reduces the need for third party in IoT network thereby solving the single point failures. If any disruption occurs in the blockchain it will not affect the IoT network because the data will be stored in multiple nodes and the whole system will be highly resilient for the technological failures [16, 17]. Thus the blockchain handles the network with proper validation with its peer to peer architecture.

(b) Transparency

All the transactions that happens over the blockchain is transparent and are accessible to all the network legal participants. Since the blockchain follows distributed model any document can be shared to any of the participants in the network without the need of individual copies. The shared documents can be modified with a proper consensus i.e. approval from all the users.

(c) **Security Enhancement**

Blockchain seems to be more secure and reliable in managing the records. While enabling the service itself all the participants will agree for being documenting all the transactions. All the transactions are encrypted and are connected to each other. The data or information is stored in multiple networks rather than storing it in the single server thereby preventing it from the hackers. Usually asymmetrical cryptography will be used in blockchain this enables the generation of keys randomly and preventing the hackers from finding the private key. When it is integrated with the IoT network the users are provided with trusted access control which authorizes all the IoT operations.

(d) **Traceability**

Any transactions are easily traceable with the integration of blockchain, this supports the supply chain in tracing any goods from its origin. The authenticity of any goods can be checked at any time as the blockchain holds history of data.

(e) **Data Privacy**

All the data are efficiently managed and are immutable in nature which protects the data from alteration in the IoT network. With the use of digital signature and the hashing chain the integrity of the data is preserved which enlights the privacy of the user's data.

(f) **Cost Reduction**

One of the major factor for every business is cost. Since the implementation of Blochchain in IoT avoids the third party and the deployment of infrastructure the cost is reduced much.

(g) **Immutability**

All the transaction that happens are recorded with timestamps hence the data is immutable. Every transactions will be validated by the blockchain before it is recorded, hence no data can be deleted or altered after the validation.

4.2 Challenges in Integration of Blockchain and IoT

1. High power is required for computing the blockchain algorithms which creates a trade off on consumption of power and performance of the algorithms.
2. The IoT devices generate enormous data continuously hence the streaming of data continuously will increases the concurrency rate.
3. The data is streamed continuously which requires the connection of higher storage and computing resources.

4. Big data has to be handled because for each and every participants local copy of ledger will be maintained and a copy will be broadcasted to all the block to make it append to its relative block.
5. Though the blockchain provides transparency it is risky in handling the health-care data that are collected from the IoT devices where higher confidentiality is required for the user's health data.
6. The immutability of data implies that the data will not be deleted from the distributed ledger when it is published.
7. The governance of data has to be increased as there is no filtration is used before the publication of data in the distributed ledger.

5   Use Cases of Blockchain and IoT Integration

Here are the use cases that have significant impact over the integration of blockchain and IoT:

(a)   Automotive Industry

Due to the digitization, the need for automation in the automotive industry increases drastically. The IoT enables sensors are been used in the automotives for making it autonomous. With the enhancement of decentralized network in the Industrial Internet of Things enables easy information exchange. Upon implementing blockchain in IoT will support the automotive industry to make changes in several aspects such as production of autonomous cars, autonomous traffic control, auto fuel payment and smart parking of vehicles. NetObjex is one of the industry that have implemented blockchain in IoT for enabling smart parking. This supports the system to find the empty space for parking the vehicle and auto payment for the parking with the use of crytowallet.

(b)   Logistics and Supply Chain

Usually the supply chain includes several participants such as material provider, broker, investors etc. which complicated the process of end to end visibility due to the different participants. This obviously increases the delivery time of the goods. Hence companies are towards investing on blockchain integration for easier tracking. Crisp is one of the industry which implements this integration for easier shipment. They uses motion sensors, temperature sensors, GPS, connected devices and other sensors for enriching the shipments. All the information that are collected from the sensors will be stored in the blockchain which makes the transaction distributed.

(c)   Smart Homes

Another important aspect that is expected by all the people is smart homes. This supports us to manage all the IoT devices. When the blockchain is integrated with the IoT network then the users can control the home security systems from the remote location and each and every changes can be recorded. Telstra is a telecommunication company that provides solution for the smart homes. They have connected all the smart devices with the blockchain and the biometric equipment which prevents the collected data from unwanted manipulation.

(d) **Pharmaceutical Industry**

Major problem resides in the pharmaceutical industry is the production of counterfeit pharma products. They are responsible for all the developing, manufacturing and distributing drugs. Hence the complete history of each and every pharma products has to be traced. This can be possible with the integration of blockchain and IoT. Mediledger is a blockchain integrated IoT module that tracks all the transactions of the medicines. This tracked information can be accessed by all the participants right from the manufacturer to the end consumer.

(e) **Agriculture**

Most expecting field for implementation of blockchain with IoT is agriculture where many IoT devices are used. Many of the sensors are used in the farms through which continuous data will be admired. When these devices are connected to the blockchain then all the information are traced which have its impact over the food industry. Pavo implements the blockchain with IoT. The data is collected from the Pavo devices and are stored in the blockchain. This supports the farmers in enhancing the farming techniques with the traced information. This also supports the distributors and end users in better knowledge about the particular crop or the food.

6   Conclusion

Thus the chapter elaborates the importance of Internet of Things and its emergence. The concept of blockchain is also discussed with the needed components. Further, the integration of blockchain with the Internet of Things is discussed in detail where this integration regulate the process of IoT through decentralized control which is kept highly safe than other technology implications. This chapter discuss about the different components of block chain enabled IoT system and the application of implementing the blockchain in IoT.

References

1. Panarello, A., Tapas, N., Merlino, G., Longo, F., Puliafito, A.: Blockchain and IoT integration: a systematic survey. Sensors **18**(8), 1–37 (2019)
2. Dwivedi, A.D., Srivastava, G., Dhar, S., Sigh, R.: A decentralized privacy-preserving healthcare blockchain for IoT. Sensors **19**(2), 1–17 (2019). https://doi.org/10.3390/s19020326

3. Song, J.C., Demir, M.A., Prevost, J.J, Rad, P.: Blockchain design for trusted decentralized IoT networks. In: 2018 13th Annual Conference on System of Systems Engineering (SoSE), pp. 169–174 (2018)
4. Mocnej, J., Pekar, A., Seah, W., Kajáti, E., Zolotová, I.: Internet of Things unified prtocol stack. Acta Electrotechnica et Informatica **19**, 24–32 (2019)
5. Atlam, H.F., Wills, G.B.: Technical aspects of blockchain and IoT. In: Advances in Computers, vol. 115, pp. 1–39 (2019)
6. Dorri, A., Kanhere, S.S., Jurdak, R., Gauravaram, P.: LSB: a lightweight scalable blockchain for IoT security and anonymity. J. Parallel Distrib. Comput. **134**, 180–197 (2019)
7. Buccafurri, F., Lax, G., Nicolazzo, S., Nocera, A.: Overcoming limits of blockchain for IoT applications, In: Proceedings of the 12th International Conference on Availability, Reliability and Security, pp. 1–6 (2017)
8. Lin, J., Shen, Z., Zhang, A., Chai, Y.: Blockchain and IoT based food traceability for smart agriculture. In: Proceedings of the 3rd International Conference on Crowd Science and Engineering, pp. 1–6 (2019)
9. Zhou, L., Wang, L., Sun, Y., Lv, P.: Beekeeper: a blockchain-based IoT system with secure storage and homomorphic computation. IEEE Access **6**, 43472–43488 (2018)
10. Novo, O.: Blockchain meets IoT: an architecture for scalable access management in IoT. IEEE Internet Things J. **5**(2), 1184–1195 (2018)
11. Dedeoglu, V., Jurdak, R., Putra, G.D., Dorri, A., Kanhere, S.S.: A trust architecture for blockchain in IoT. In: Proceedings of the 16th EAI International Conference on Mobile and Ubiquitous Systems: Computing, Networking and Services, pp. 190–199 (2019)
12. Liao, C.F., Bao, S.W., Cheng, C.J., Chen, K.: On design issues and architectural styles for blockchain-driven IoT services. In: 2017 IEEE International Conference on Consumer Electronics-Taiwan (ICCE-TW), pp. 351–352 (2017)
13. Roy, S., Ashaduzzaman, M., Hassan, M., Chowdhury, A.R.: Blockchain for IoT security and management: current prospects, challenges and future directions. In: 2018 5th International Conference on Networking, Systems and Security (NSysS), pp. 1–9 (2018)
14. Cui, P., Guin, U., Skjellum, A., Umphress, D.: Blockchain in IoT: current trends, challenges, and future roadmap. J. Hardw. Syst. Secur. **3**(4), 338–364 (2019)
15. Pavithran, D., Shaalan, K., Al-Karaki, J.N., Gawanmeh, A.: Towards building a blockchain framework for IoT. Cluster Comput. **23**(3), 2089–2103 (2020)
16. Alkurdi, F., Elgendi, I., Munasinghe, K.S., Sharma, D., Jamalipour, A.: Blockchain in IoT security: a survey. In: 28th International Telecommunication Networks and Applications Conference (ITNAC), pp. 1–4 (2018)
17. Reyna, A., Martín, C., Chen, J., Soler, E., Díaz, M.: On blockchain and its integration with IoT. Challenges and opportunities. Future Generation Comput. Syst. **88**, 173–190 (2018)

Application of Machine Learning Algorithms to Disordered Speech

Seedahmed S. Mahmoud, Qiang Fang, Musleh Alsulami, and Akshay Kumar

Abstract Aphasia is an acquired neurogenic language/speech disorder that can be assessed with one of the well-known assessment standards, such as the Chinese Rehabilitation Research Center Aphasia Examination (CRRCAE, for Chinese-dialects speaking patients), Aachen Aphasia Test (AAT, for German-speaking patients) and Boston Diagnostic Aphasia Examination (BDAE, for English-speaking patients). These standards are utilized by a skilled speech-language pathologist (SLP) to assess patients with aphasia (PWA). Generally, there are three aphasia assessment tasks where an SLP performs a comprehensive examination of the patient's communication abilities, including speaking, expressing ideas, understanding language, and reading and writing. These tasks are the discrimination between healthy and aphasic speech, the degree of severity of impairment for aphasic patients' assessment and the classification of aphasia syndromes (such as Global aphasia, Broca's aphasia, Wernicke's aphasia, and amnesic aphasia). Conventional methods of aphasia assessment and rehabilitation are resource-intensive processes that require the presence of an SLP. Due to the vast number of PWAs, the big data involves in the aphasia assessment and the financial difficulties, it is complex to fulfil this requirement manually. Therefore, automation of the aphasia assessment process is essential to attract researchers' attention. There are several research efforts in automating the aphasia and speech disorder assessment process using machine learning (ML) algorithms. This chapter presents the application of classical machine learning (CML) and deep neural networks (DNN) to aphasic speech assessment. Moreover, the challenges and limitations facing ML algorithms over disordered speech datasets are presented. Furthermore, the chapter provides recommendations on the suitable ML framework for aphasia assessment based on comparative performance results between various CML's classifiers and CNN algorithm over aphasic speech. This chapter also presents two practical examples of the ML application to aphasia assessment.

S. S. Mahmoud (✉) · Q. Fang · A. Kumar
Department of Biomedical Engineering, Shantou University, Shantou, China
e-mail: mahmoud@stu.edu.cn

M. Alsulami
Department of Information Systems, Umm Al-Qura University, Mecca, Kingdom of Saudi Arabia
e-mail: mhsulami@uqu.edu.sa

© The Author(s), under exclusive license to Springer Nature Switzerland AG 2022
K. R. Ahmed and H. Hexmoor (eds.), *Blockchain and Deep Learning*,
Studies in Big Data 105, https://doi.org/10.1007/978-3-030-95419-2_8

Keywords Aphasia assessment · Deep neural network · Machine learning framework · Mandarin · Speech impairment

1 Introduction

Aphasia is a condition in which individual's ability to comprehend and produce speech is impaired. It affects reading and writing skills as well [1]. The two main classes of aphasia are fluent speech if the patient can speak with ease but with grammatical errors or non-fluent speech if the patient speaks slowly and has difficulty in uttering words [2]. The following are the types of aphasia; expressive aphasia, conduction, receptive and amnesic aphasia [2, 3]. It is generally caused by stroke, brain injury, brain tumor, and progressive neurodegenerative disorders [2, 3]. However primary progressive aphasia (PPA) is associated with neurodegenerative disorders such as Parkinson's disease [4]. PPA is characterized by symptoms such as forgetting the meaning of some words, pronouncing words wrongly or speaking difficulty, and failure to recall some words [5]. Aphasia patients generally lead a poor quality of life (QoL) because they are faced with several challenges, i.e., social isolation due to inability to communicate well, frustration, and loss of the ability to make independent decisions [3]. Figure 1 represents a summary of the circuit involved during comprehension of either written word or spoken word. Written word and spoken word are perceived by the primary visual cortex and primary auditory cortex, respectively. Both signals are relayed to the wernicke's area which is responsible for comprehension. After sense has been made from the spoken or written word, a message is sent to the Broca's area which is responsible for motor processing. In the Broca, the message creates a response and executes by using the motor cortex.

Aphasia assessment and rehabilitation are resource-intensive processes which require the presence of a speech-language pathologist (SLP). This requirement is difficult to fulfil due to the vast number of patients with aphasia (PWAs) and limited resources. Financial difficulties and patients' health conditions also pose difficulty in aphasia assessment and rehabilitation [3]. In the United States of America (USA), about a million people have aphasia, according to research conducted by the National Institute on Deafness and other Communication Disorders (NIDCD) [2]. Another study estimates Indian cases at 3 million [2], while Great Britain has at least 250,000 PWAs [7]. There is a worldwide shortage of SLPs, including advanced countries such as USA, Australia, and China [3, 8, 9], due to the aging population and the need to diagnose children early [10]. SLPs actively manage and care for PWAs by performing functions such as assessment, diagnosis, and treatment [11]. In addition, SLPs classify aphasic syndromes and determine the severity of impairment. These services are resource-intensive [12] where patients are required to meet the SLP physically. Sometimes it is a time consuming and inconvenient for the patient [13, 14]. Furthermore, these services are costly, especially if the duration is long, yet the patients require care [5].

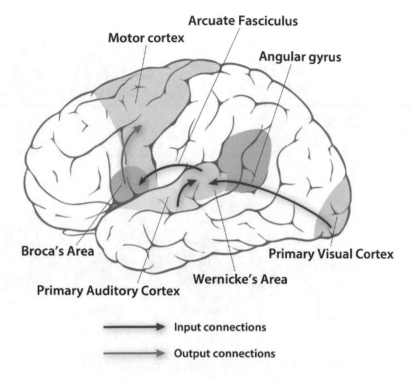

Fig. 1 A summary of the circuitry involved in language processing [6]

Automated speech assessment and therapies are potential solution to all PWAs since enormous number of patients can receive care conveniently. These solutions will also reduce the SLPs' working load, which will improve their quality of care. These forms of therapy also enable patients to self-monitor their speech in-home [3]. Artificial intelligence (AI) algorithms are increasingly being used in automatic aphasia assessment [3]. Machine learning (ML) improves over time as large dataset is available. Therefore, the system can train to assess patients with aphasia in a way comparable to SLPs assessment [15]. Other advantages of ML include a better diagnostic accuracy and early identification of disorders for treatment that may lead to increased success rate in recovery [16]. Machine learning (ML) will therefore increase the efficacy of in-home speech therapy [17].

In this chapter, the application of classical machine learning (CML) and Deep neural networks (DNN) to aphasia assessment and speech therapy will be discussed and presented. Genrally, CML utilizes feature engineering techniques, i.e., feature extraction and feature selection, to process the input, while DNN employ automatic feature detection through layers of algorithms to process input data in the form of 3D image. Figure 2 illustrates the key features of the two machine learning types. This figure demonstare nautral language processing (NLP) as another application to machine learning in the domain of speech and language processing. Figure 2

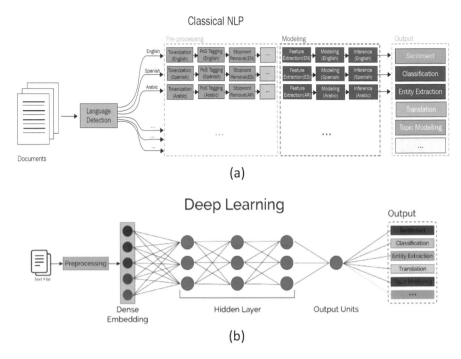

Fig. 2 CML and DNN frameworks. Natural language processing using **a** CML and **b** DNN [18]

(a) shows a langauage identification example of CML that uses feature creation and feature extraction to process the input [18]. On the other hand, Fig. 2b demonstarte the use of DNN in the NLP application [18]. Similar framework can be used in aphasia assessment application with some adjustments to the data type and preprocessing stage. In this chapter, disordered speech will be refered to aphasic speech.

This chapter is organized as follows. Section 2 presents the application of CML and DNN to aphasia assessment. It presents examples of feature extraction methods, classifiers algorithms that can be used within the CML framework, convolutional neural networks (CNN) and recursive neural networks. Section 3 presents practical examples and applications for ML algorithms over aphasia datasets. It also discusses challenges and limitation face automatic aphasia assessment. In Sect. 4, conclusion and future scope is presented.

2   Application of Machine Learning in Aphasia Assessment

Artificial intelligence (AI) algorithms are increasingly being used in automatic aphasia assessment. Machine learning is a unit of AI that automatically improves and learns from experiences [19]. It includes loading data into a computer program, choosing a classification model which fits the data, and allowing the computer to predict the correct class without help from an expert [20]. Explicit computer programming is not a prerequisite for machine learning [21]. The development of algorithm that can be used to learn by themselves and access dataset is a key in machine learning [14, 22]. Machine learning can assess aphasia by defining the severity levels of the speech impairment [12, 23]. In the literature, the two groups of machine learning that applied to assess speech impairment severity levels are deep neural networks (DNN) and classical machine learning (CML) [24]. Machine learning techniques in aphasia are important in decision-making and have a positive impact on patients' lives and their quality of life (QOL) [25]. Commonly, there are three aphasia assessment tasks whereby an SLP performs a comprehensive examination of the patient's communication abilities, including speaking, expressing ideas, understanding language, reading, and writing. These tasks are the discrimination between healthy and aphasic speech, assessing the degree of severity of impairment for aphasic patients, and the classification of aphasia syndromes (such as Global aphasia, Broca's aphasia, Wernicke's aphasia, and amnesic aphasia) [12]. A general framework based on the machine learning, specifically CNN, for the three aphasia assessment tasks is shown in Fig. 3.

The efficacy of the aphasia assessment framework shown in Fig. 3 depends on the quality and size of the datasets, the selection of a ML framework (whether it is a classical machine learning (CML)-based or a deep neural network (DNN)-based

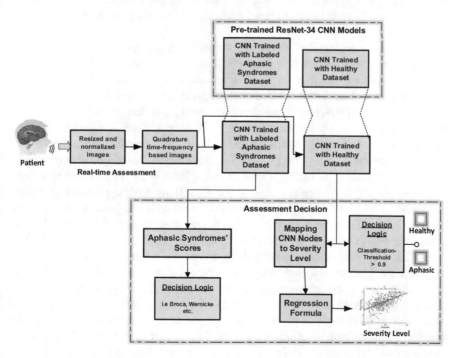

Fig. 3 A general framework for aphasia assessment tasks using CNN based classification method [12]

framework), the selection of an appropriate classifier, and the choice of a suitable training dataset. The training dataset for aphasia assessment tasks includes healthy and aphasic datasets as shown in Fig. 3. The training dataset type (healthy, aphasic, or joint aphasic and healthy datasets) of an ML-based aphasia assessment model determines whether the assessment task is to discriminate between healthy and aphasic speech, assess the severity degree of speech impairment for aphasic patients, or classify the syndrome of aphasia impairment (see Fig. 3). Evaluation, investigation, and description of the effectiveness of the machine learning techniques for aphasia assessment are important in laying a roadmap for their implementation and to provide a guide in laying a foundation for suitable dataset collection [26]. In the next subsections, detail description to the component blocks of CML (feature extraction methods and classifier algorithms) and DNN algorithms will be presented.

2.1   Classical Machine Learning in Aphasia

Rehabilitation and assessment of patients with aphasia (PWA) require applying classical machine learning to assess speech lucidity/ intelligibility features such as fluency, tone, articulation. Speech lucidity features have a major influence on the overall clarity of speech and the understanding of spoken words. Assessment of the speech features is an important prerequisite for the effective assessment of aphasia [12].

Researchers have studied effective automatic speech assessment methods for patients with speech disorders and aphasia [27]. Classical machine learning (CML) is one of these methods that was used to automatically assess speech impairment. Commonly used classifiers in CML include support vector machines (SVMs), Bayes classifier, decision trees, nearest neighbor, k-means clustering, linear regression, and logistic regression [28]. Machine learning, a theory applied in classical machine learning, combines aspects of statistics and computer science with the principal objective of understanding learning [29, 30]. The main objectives of this area of science are to comprehend at an accurate mathematical level what information and capabilities are required to learn and understand algorithms, their application in getting computers to improve their performance from feedback and getting computers to learn from data. This theory aims to help in the understanding of fundamental issues in learning and to help in the creation of better-automated learning methods [30].

In [12, 31], researchers proposed an automatic speech lucidity assessment approach for Mandarin-speaking aphasic patients using an ML-based technique. The method in [31] established a relationship between the severity level of aphasic patients' speech and the three speech lucidity features (articulation, fluency, and tone). The research focused on one of the aphasia assessments tasks that assess the severity of impairment for an aphasic patient. The investigation in [31] also presented a performance comparison between CML learning and DNN algorithms over healthy dataset. In CML simulation, a total of 51 features based on Mel Frequency Cepstral Coefficients (MFCCs), the formant frequencies, and signal energy were calculated for the classification [31]. The feature vectors of all data samples were standardized by subtracting mean and dividing by the standard deviation of each of the data sample's feature vector, to avoid some features from potentially dominating others due to a large magnitude. Several classical machine learning algorithms were evaluated for the classification of the speech signals; specifically, quadratic support vector machine (QSVM), a radial basis function (RBF) kernel SVM, linear discriminant analysis (LDA), random forest and k-nearest neighbours (kNN) were evaluated. Comparative result using the CML and DNN will be presented in the practical application of ML to aphasia assessment section (Sect. 2).

Automatic speech recognition (ASR) in CML consists of the following component blocks; feature extraction stage, feature selection, classifiers, and decision stage that can be used to recognize speech. Disordered speech automatic assessment has a similar blocks component as the ASR which includes (1) defining the essential pathologic aspects (2) feature extraction and feature selection and (3) regression or classification of specific disordered speech components [26]. In the following Subsections, the component blocks of CML framework will be presented.

2.1.1 Feature Extraction Methods

There are number of feature extraction methods, ranging from simple feature as signal energy to a complex time–frequency domain-based feature, that can be applied to speech recognition and speech assessment using speech signal. Some of these methods are discussed in the subsequent sections.

Mel Frequency Cepstral Coefficients (MFCCs)

Initially, this feature was utilized to identify monosyllabic words in continuously and spontaneous speech [28, 32]. It is a replica of the human ear which aims to recreate the working principles of the ears without assuming that the ears are reliable recognizers of a speaker [28]. MFCC's important aspect is that it can recognizes differences of the human ear's important bandwidths using frequency filters logarithmically spaced at high frequencies and linearly at low frequencies, thus retaining vital aspects of human speech signals [28]. Human speech signals contain varying tone frequencies, each tone with a personal pitch and actual frequency (f) Hz, which can be deduced from the Mel scale [28]. The MFCC can be given by

$$Mel(f) = 2595x\log_{10}\left(1 + \frac{f}{700}\right) \tag{1}$$

where $Mel(f)$ is the frequency in Mel scale, f is the signal frequency in Hz and x is the time domain signal. Most human-generated energy sounds have a maximum frequency of 5 kHz which can be accurately captured by MFCC. MFCC is not suitable for the generalizations or when the background sound is present.

Linear Prediction Coefficients (LPC)

Linear prediction coefficients mimic the humans' vocal tract, and it is important feature for robust speech recognition [28]. It assesses speech signals through approximation of formants, mitigating the effect of speech signal and estimating the left residual frequency and concentration [28]. Each of the sampled signals is stated as a direct incorporation of the previous samples. Coefficients of the difference equation characterize formants, LPC approximates the coefficients. This method is widely used as speech analysis technique and formants estimation method [28].

The other variant of LPC is the linear prediction cepstral coefficients (LPCC). The LPCC are the coefficients of the Fourier transformation illustration of LPC. It is commonly used in speech processing as it can symbolize speech characteristics and waveforms [28].

Perceptual Linear Prediction (PLP)

The PLP feature combines linear prediction analysis and spectral analysis [28]. It applies linear predictions for smoothening of the spectrum [28]. It works by combining intensity to loudness, the equal emphasis of information extraction from speech, and critical bands. Initially, it was used in speech recognition by eliminating user-dependent features, but it was mostly utilized in the nonlinear bark scale [28].

Discrete Wavelet Transforms (DWT)

Discrete wavelet transform is based on signal analysis using varying scales in frequency and time domains [28]. Jean Morlet and Alex Grossman introduced a wavelet transform which allows high-frequency identification with increased resolution in the temporal region. It has a reduced duration with an average value of zero. Wavelet transform process can be applied with high efficacy to represent real-life non-stationary signals. It can obtain information both in frequency and time domains from transient signals. Continuous wavelet transforms application can lead to information redundancy which requires large human effort to calculate the translation and scales, limiting its use. The advantage of this wavelet transform is that it offers high temporal resolution analysis that can be used as a robust feature extraction method.

2.1.2 Classical Machine Learning Classifiers

There are number of classifiers that can be used within the CML framework for the application of speech recognition and speech impairment assessment. Selected classifiers are discussed below.

Support Vector Machines (SVMs)

In classical machine learning, SVMs models are used with algorithms that analyze data for regression or classification applications [28]. These models are supervised and reliable prediction methods. They can be used to categorize both labeled and unlabeled classes, enabling the SVMs to perform nonlinear and linear classifications. Moreover, SVMs can be applied in various aphasia assessment tasks such as in the research presented in [12, 31].

SVMs can also be used in the classification of brain cancer patients undergoing transcranial magnet stimulation. In this application, SVMs was used to highlight a significant portion of aphasia [28].

Quadratic Support Vector Machine (QSVM)

The quadratic kernel lacking a nonlinear support vector network is relatively new and can reliably be used in suitable data type classification [28]. It focuses on locating an optimum hyperplane that discriminate and separate between classes among the training dataset [28]. A quadratic support vector machine was used to assess aphasia in patients with brain tumors [33]. It was also used in [31] for the performance comparison between CML and DNN over aphasia dataset.

Compared with the traditional methods, it has a good predictive ability and focuses on finding patterns in the unwieldy and rich data [34].

Random Forest

This algorithm is focused on a collection of subsets of training data that are modeled as tree-based collections [28]. In comparison with single tree techniques, it has reduced variance. Each tree in the subset is trained on a random subject of the full training, and it is less likely that an overfit of the whole ensemble can occur. Random forest classifiers are one of the most accurate tools used in classification challenges. It involves problem setting and solving, prediction of random forest, classification, and regression of data. It can be used for both regressions and classification in aphasia application. The random forest can improve untreated and treated verbs, thus aiding in the data analysis process [35]. The efficacy of random forest can be compared with other assessment algorithms, thus helping clinicians make decisions on their choice of machine learning algorithms in assessing and treating aphasia [12, 31, 36].

2.1.3   Decision on Classifiers' Output

Assigning a classifier's output in an ML framework to correct classes/categories requires a decision logic as a final stage (see Fig. 3). Logics in the decision stage can range from a simple scoring method, as in binary classification, to complex mapping and regression algorithms. This stage in the ML framework is of utmost significance in the aphasia assessment tasks. An example of a decision logic for the binary classification, such as the discrimination between healthy and aphasic speech, is given by the pseudocode in Algorithm 1.

Algorithm 1. Decision Logic for Binary Classification

//This process below usually uses the classifier's output
//Binary Classifier has Single Output Node
1: Start
2: κ ←Classification Threshold (cut-off)
3: C1←Healthy Class
4: C2← Aphasia Class
5: Q←Classifier output//*normalized between 0 and 1*
6: if Q > κ
then
C1   *//the tested speech is healthy*
else
C2   *//the tested speech is aphasic*
7: End

2.2   Deep Neural Network (DNN) in Aphasia

Deep neural networks (DNNs) are a subset of artificial intelligence (AI). DNNs were derived from artificial neural networks (ANNs). Both neural networks, ANN and DNN, consist of hidden layers of algorithms that are used to process input data. ANNs consist of one hidden layer while DNNs contain multiple hidden layers. Therefore, DNNs were more efficient in processing input data due to their large number of layers [37]. In DNN, the number of layers and their size are directly proportional.

Consequently, a higher number of layers make a network complex, larger, and more difficult to train [38]. Like the neurons in our brain, the DNN order of neurons resembles the human neurons order. Neurons pass information to the next level of neurons to form a network that learns with some feedback mechanism. The first level neurons provide output to the second-order neurons, which then offer subsequent output to the next order of neurons. This process continues to pass information to the neurons at the output layer. A typical DNN model is represented in Fig. 4.

The DNN model has input layer (blue), a set of hidden layers with neurons (green), and output layer (purple). The nodes in the three layers are interconnected to aids communication between each neuron. The network operates by discerning a set of patterns from the input data to enable the output classification. The input data must activate several hidden nodes collection. The system is then trained with numerous

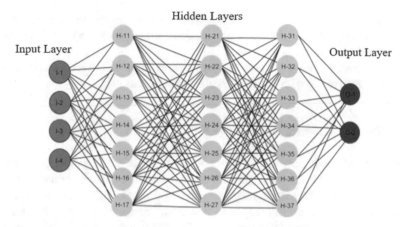

Fig. 4 Representation of the DNN model [39]

inputs to discriminate different patterns [39]. Unlike classical machine learning (CML), DNNs conduct feature extraction from input data/image automatically [39]. In the following subsections, two type of DNN models are presented.

2.2.1 Convolutional Neural Networks (CNN)

Feedforward neural networks include this class of deep neural networks. A CNN consists of input, output, and multiple hidden layers. Feedforward networks (FFNs) do not have bidirectional connections between nodes allowing information to travel in one direction only, i.e., input-hidden layer-output layers, as shown in Fig. 5.

The multiple layers of CNNs enable the computation of non-linearly separable functions [41]. CNNs are majorly applied in visual imagery analysis and speech/audio classification. However, they can be used in speech impairment assessment [12, 15, 31]. It is instrumental in scoring spoken fluency and discriminating between aphasic and healthy speech [12, 15, 31]. A study conducted on Cantonese PWAs revealed that the CNN model performs just as well as the ordinary two-step model in speech assessment. The CNN model offers other advantages such as efficiency

Fig. 5 Representation of the CNN model [40]

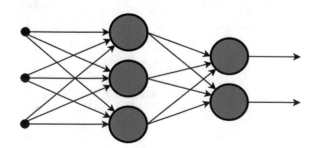

and reducing manual efforts [42]. In [31], the CNN model outperformed the classical machine learning based technique in automatic Chinese-Mandarin words and vowels recognition for aphasia assessment.

CNN is also effective in primary progressive aphasia (PPA) associated with neurodegenerative disorders such as Parkinson's disease [42].

2.2.2 Recursive Neural Networks (RNN)

RNNs deal with arbitrary output/input lengths; while CNN produces fixed-size outputs based on the fixed-size inputs. The internal memory in RNNs enables them to process random sequences of input information [43]. The Output is produced through mathematical execution of the input data and then compares the output value with the correct value to identify an error [43]. The RNNs are used to learning tree-like structures; however, one disadvantage is that the tree structure of the input value must be known beforehand [44]. The input value should be structured as shown at entry points $\times$ 1, $\times$2 [45] (see Fig. 6).

RNNs observe input data at $\times$ 1, $\times$ 2... $\times$ 6 and generate output data [45]. The RNNs are efficient in the analysis of speech and text [46]. They conduct sentimental analysis of audio to identify if it is positive, negative, or neutral and tone of speech [47]. This type of neural network can be used to diagnose aphasia associated with stroke, e.g., Broca's aphasia [48].

Table 1 highlights the differences between CML and DNN networks [49–60].

A limitation in the width of graphical processing unit (GPU) pipelines is one factor that makes deep neural networks slow. Improved GPUs will significantly reduce the processing in real-time [61]. High-performance CPUs also ensure that output is produced in real-time [18].

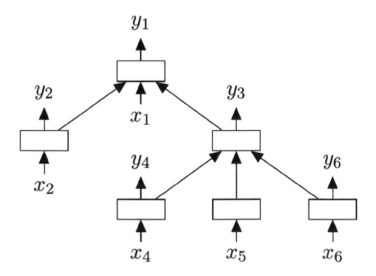

Fig. 6 Representation of the RNN Model [45]

Table 1 Comparison between CML and DNN

Features	CML	DNN
Accuracy	Less accurate	More accurate
Precision	Less precise because there are limitations on the range of input data it can process	More precise, it can handle complex pieces of data owing to the numerous algorithms in the hidden layer
Recall	Recall capacity is slightly lower	It has as a higher recall capacity
Speed	Are faster	They are slower because of a higher load time
complexity	Less complex	More complex. The level of complexity is dependent on several hidden layers
Suitability to real-time processing	Yes. It can produce output in real-time because processing speed is faster	It may not produce output in real-time due to slow processing speed

3 Practical Application of Machine Learning in Aphasia Assessment

In this section, practical application examples of machine learning techniques over healthy and aphasia datasets will be presented. These examples include the utilization of machine learning to assess the degree of severity of impairment for aphasic patients.

3.1 Performance Comparison Between CML and CNN Algorithm Over Healthy Speech Dataset

Performance evaluation of classical machine learning (CML) framework and CNN framework over various aphasia datasets was investigated in [12]. Two separate datasets were constructed to investigate the performance: one containing all Mandarin vowels and words speech data (26 classes) of healthy participants (named *vowels + words* dataset onwards) and the other containing only Mandarin words speech data (20 classes) of healthy participants (named *only-words* dataset onwards). Five-fold cross-validation was used to estimate the performance of the classifiers to classify the two datasets [12, 31].

The performance results for the two frameworks are compared in Fig. 7. The results show that the ResNet-34 CNN framework outperformed all CML algorithms employed to classify the two datasets in terms of the three performance evaluation metrics (Accuracy, Precision, and Recall). Among the CML classification algorithms, the LDA algorithm outperformed the other classifiers [12]. The CNN model has a

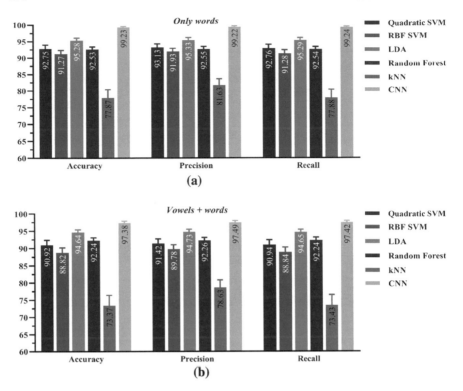

Fig. 7 Performance evaluation for the CNN and classical machine learning (CML) classification frameworks, on (**a**) the Mandarin *Only words* (20 classes) of a healthy dataset and (**b**) the Mandarin *Vowels + Words* (26 classes) of the healthy dataset [12]

higher accuracy of 99.23 ± 0.003% for the *only-words* dataset compared to 95.28 ± 0.79% for the LDA with the same dataset. Also, with a larger class size of 26, the CNN model has 97.39 ± 0.004% accuracy, which is higher than that of the other CML algorithms. Moreover, all classifiers except the kNN performed well on this dataset, and their accuracies exceeded 90% [12].

The purpose of the performance comparative exercise is to assist the selection of a suitable machine learning technique for the assessment of aphasic speech.

3.2 Aphasic Speech Lucidity Assessment Method

In [31], researchers proposed an automatic speech lucidity assessment approach for Mandarin-speaking aphasic patients using a deep learning-based technique. Based on performance comparative results in previous subsection, researcher in [31] chose the ResNet-34 CNN model with hyperbolic T-distribution based time–frequency (TF) images as input to model the Chinese-Mandarin aphasic speech lucidity. The method in [31] established a relationship between the severity level of aphasic patients' speech and three speech lucidity features (articulation, fluency, and tone). The research in [31] focused on one of the previously mentioned aphasia assessments tasks, that assesses the severity of impairment for an aphasic patient.

Figures 8a–c show the scatter representation between the CNN-based aphasic speech model's output probabilities and the 5-point Likert scale scored by five Mandarin linguistic scholars for the articulation, fluency, and tone features, respectively [31]. Results in Fig. 8 show that recruited patients have a wide range of speech severity. This representation aims to investigate the efficiency of the proposed automatic Mandarin aphasic speech model in predicting human scoring. The results show that the relationship between the proposed model and the manual human score by the five Mandarin scholars follow a linear model of the form [31]

$$L_s = \mu M_p + \beta \qquad (2)$$

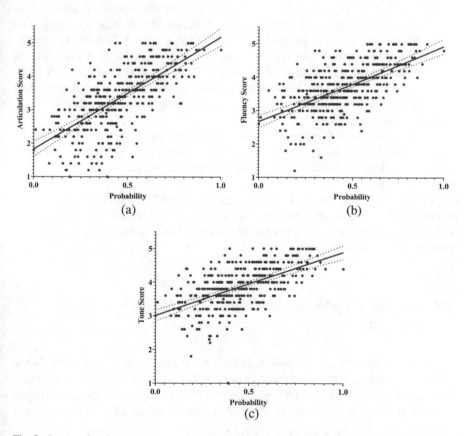

(a)

(b)

(c)

Fig. 8 Scatter plots between the true class probabilities and **a** articulation score, **b** fluency score and **c** tone score. The solid red line in the three figures represents the best fit line using simple linear regression and the dotted red line represents the 99% confidence interval [31]

Table 2 Spearman rank correlation coefficient (CC), 99% confidence interval and the corresponding p-value between the true-class probabilities and the scores of the three fundamental features of Mandarin language: articulation, fluency and tone [31]

Feature	Spearman (CC)	99% confidence interval	p-value (two-tailed)
Articulation	0.7121	0.6402 to 0.7715	<0.0001
Fluency	0.6034	0.5122 to 0.6811	<0.0001
Tone	0.5840	0.4897 to 0.6647	<0.0001

where L_s is the Likert scale from 1 to 5, M_p is the speech lucidity assessment model's output probability, μ is the slope, and β is the y-intercept with the Likert scale axis. Table 2 shows the correlation coefficient (CC) and the 99% confidence interval (CI) for the articulation, fluency, and tone features linear graphs [31].

The articulation results in Fig. 8a show the impact of the aphasia on the recruited patients' articulation system. The model predicted a minimum articulation of 1.826, which is represented in the recruited patients. The reason for that the recruited patients included five patients with dysarthria. Dysarthria is a speech disorder that results in unclear articulation of speech. Such patients were recruited in the research in [31] to examine the performance of the proposed speech model over a wider range of speech disorder patients. Moreover, the model shows its efficacy in responding to dysarthria in the articulation feature while the tone and fluency features were impacted significantly [31].

The tone and fluency results show that these two features have better retention compared to the articulation feature. Tone production is reflected in the fundamental frequency change of pronunciation, which depends on the degree of tension of the vocal cords [31].

The presented two practical examples in this chapter demonstrated the efficacy of machine learning algorithms in the aphasia assessment tasks. However, there are challenges and limitation face the assessment of aphasia. These challenges will be presented in the next section.

3.3 Automatic Aphasia Assessment: Challenges and Limitations

Aphasia assessment using machine learning is a nontrivial task. There are number of challenges and limitation being faced in the development of efficient aphasic speech assessment applications. Careful design of machine learning based assessment application is essential to overcome such issues. Examples of such limitation are listed below.

- Due to the limited availability of speech-language pathologists (SLPs), there is a real problem in constructing the ground-truth labels of the speech dataset of PWA in the absence of standard assessment scores. It is difficult to carry out this task at a considerably large scale. To overcome this problem, a previous study has shown that an untrained listener of a language can evaluate the lucidity of speech close to a specialist level [62]. Therefore, scholars who can speak the language can be recruited to construct the ground-truth.

- An automatic aphasia assessment task that required to discriminate between aphasia syndromes such as Global aphasia, Broca's aphasia, Wernicke's aphasia, and Amnesic aphasia depends on the size of the training dataset from each aphasia's syndrome, which is a real challenge due to domain data scarcity [63].
- Abnormal speech patterns [64], and speaker variability [65] are challenging to any classification problem. Patients with different speech impairment severity levels can resulted in complex and unresolvable common features. To overcome this challenge, a large dataset for each aphasia syndrome with a robust machine learning technique can create distinct boundaries in the features space which will result in high classification accuracy.

4 Conclusion and Future Scope

Considering the amount of time and resources spent on speech assessment using traditional conventional assessment standards such as the CRRCAE, AAT, and BDAE, less time-consuming and cost-effective machine learning speech assessment techniques must be embraced to reduce the assessment time using the same standards. Two machine learning models, the deep neural networks (DNN) and classical machine learning (CML) have been used to design models for speech evaluation in aphasia patients. These models enable a quick diagnosis of aphasia and other speech disorders, thus allowing prompt treatment and rehabilitation of patients with aphasia and other speech disorders. Algorithms selection for speech evaluation should be appropriate for the dataset selected to ensure effective speech assessment and rehabilitation. With the improvement of technology in computer science, statistics and medicine, there is still a need for further research on speech assessment and rehabilitation using machine learning and the best ways this area of study can be improved. A general framework based on machine learning to be used in aphasia assessment tasks was presented in this chapter. Furthermore, challenges and limitations facing an automatic aphasia assessment were presented and discussed.

The recommendation is that more research should be carried in this field so that more reliable and accurate speech assessment methods in patients with aphasia (PWA) and other speech disorders are invented and implemented. Standardized tools for the assessment of aphasic patients should also be formulated. Methods and tools of speech assessment and rehabilitation in patients with speech disorders should also focus on specific speech lucidity features such as fluency, tone, and language and speech clarity. Furthermore, due to the scarcity of the aphasia syndrome dataset, a data collection from this domain is required to improve the accuracy of the CNN based aphasia syndromes assessment and discrimination.

References

1. APHASIA DEFINITIONS [Internet]. Parkinson's Resource Organization. Accessed 11 June 2021. https://www.parkinsonsresource.org/news/articles/aphasia-definitions/ (2021)
2. What Is Aphasia?—Types, causes and treatment [Internet]. NIDCD. Accessed 11 June 2021. https://www.nidcd.nih.gov/health/aphasia (2021)
3. Teshaboyeva, G. Speech defects in young children and ways to overcome them. ACADEMICIA Int. Multidiscip. Res. J. **10**(6), 1761–1767 (2020)
4. Ruksenaite, J., Volkmer, A., Jiang, J., Johnson, J.C., Marshall, C.R., Warren, J.D., Hardy, C.J.: Primary progressive Aphasia: toward a pathophysiological synthesis. Curr. Neurol. Neurosci. Rep. **21**(3), 1–2 (2021)
5. de Aguiar, V., Zhao, Y., Ficek, B. et al. Cognitive and language performance predicts effects of spelling intervention and tDCS in Primary Progressive Aphasia (2019)
6. Tong, Y., Gandour, J., Talavage, T., Wong, D., Dzemidzic, M., Xu, Y., Li, X., Lowe, M.: Neural circuitry underlying sentence-level linguistic prosody. Neuroimage **28**(2), 417–428 (2005)
7. Aphasia Statistics. National Aphasia Association. National Aphasia Association (2016)
8. Chahda, L., Carey, L.B., Mathisen, B.A., Threats, T.: Speech-language pathologists and adult palliative care in Australia. Int. J. Speech Lang. Pathol. **23**(1), 57–69 (2021)
9. Zhang, Z., Xu, Q., Joshi, R.M. A meta-analysis on the effectiveness of intervention in children with primary speech and language delays/disorders: focusing on China and the United States. Clin. Psychol. Psychother. (2020)
10. Horton, R. Systems-based approaches to speech-language pathology service delivery for school age children. In: Cases on Communication Disorders in Culturally Diverse Populations, pp. 113–136. IGI Global (2020)
11. Rumbach, A.F., Clayton, N.A., Muller, M.J., Maitz, P.K.: The speech-language pathologist's role in multidisciplinary burn care: An international perspective. Burns **42**(4), 863–871 (2016)
12. Mahmoud, S., Kumar, A., Li, Y., Tang, Y., Fang, Q.: Performance evaluation of machine learning frameworks for Aphasia assessment. Sensors **21**(8), 2582 (2021)
13. Briffa, C., Porter, J.: A systematic review of the collaborative clinical education model to inform speech-language pathology practice. Int. J. Speech Lang. Pathol. **15**(6), 564–574 (2013)
14. Hickok, G.: The functional neuroanatomy of language. Phys. Life Rev. **6**(3), 121–143 (2009)
15. Jothi, K., Sivaraju, S., Yawalkar, P. AI-based speech-language therapy using speech quality parameters for aphasia person: a comprehensive review. In: 4th International Conference on Electronics, Communication and Aerospace Technology (ICECA), pp. 5382 -5392 (2021)
16. Myers, E. The role of artificial intelligence and machine learning in speech recognition. Rev. https://www.rev.com/blog/artificial-intelligence-machine-learning-speech-recognition (2019)
17. Panch, T., Szolovits, P., Atun, R. Artificial intelligence, machine learning, and health systems. J. Glob. Health **8**(2) (2018)
18. McGonagle, J., Alonso García, J., Mollick, S. Feedforward Neural Networks|Brilliant Math & Science Wiki. Brilliant.org (2021)
19. Kohlschein, C., Schmitt, M., Schuller, B., Jeschke, S., Werner, C. A machine learning-based system for the automatic evaluation of aphasia speech. In: IEEE 19th International Conference on e-Health Networking, Applications and Services (Healthcom) (2017)
20. Fernando, T., Denman, S., Sridharan, S., Fookes, C. Soft+Hardwired attention: an LSTM framework for human trajectory prediction and abnormal event detection. Neural Netw. (2018)
21. Gasparetti, F., De Medio, C., Limongelli, C., Sciarrone, F., Temperini, M.: Prerequisites between learning objects: automatic extraction based on a machine learning approach. Telematics Inform. **35**(3), 595–610 (2018)
22. Qin, Y.: Machine learning based taxonomy and analysis of english learners' translation errors. Int. J. Comput. Assist. Lang. Learn. Teach. **9**(3), 68–83 (2019)
23. Kohlschein, C., Schmitt, M., Schüller, B., Jeschke, S., Werner, C.J. A machine learning based system for the automatic evaluation of aphasia speech. In: 19th IEEE International Conference on e-Health Networking, Applications and Services (Healthcom), pp. 1–6 (2017)

24. Subasi, A. Machine learning techniques. In: Practical Machine Learning for Data Analysis Using Python. Academic Press, Chapter 2-data preprocessing, pp. 27–89 (2020). ISBN 978-0-12-821379-7
25. Nayak, A., Dutta, K. Impacts of machine learning and artificial intelligence on mankind. In: IEEE International Conference on Intelligent Computing and Control (I2C2), pp. 1–3 (2017)
26. Le, D., Licata, K., Persad, C., Provost, E.: Automatic Assessment of Speech Intelligibility for Individuals With Aphasia. IEEE/ACM Trans. Audio Speech Lang. Process. **24**(11), 2187–2199 (2016)
27. Aishwarya, J., Kundapur, P., Kumar, S., Hareesha, K.S. Kannada speech recognition system for Aphasic people. In: International Conference on Advances in Computing, Communications, and Informatics (ICACCI), pp. 1753–1756 (2018)
28. Alim, S.A., Rashid, N.K.A. Some Commonly used Speech Feature Extraction Algorithms, pp. 2–19. IntechOpen (2018)
29. Bzdok, D., Altman, N., Krzywinski, M. Statistics versus machine learning. Nat Methods (2018)
30. Wang, C., Chen, M., Schifano, E., Wu, J., Yan, J.: Statistical methods and computing for big data. Stat Interface **9**(4), 399–414 (2016)
31. Mahmoud, S., Kumar, A., Tang, Y., et al.: An efficient deep learning-based method for speech assessment of mandarin-speaking aphasic patients. IEEE J. Biomed. Health Inf. **24**(11), 3191–3202 (2020)
32. Le, D., Licata, K., Mower Provost, E.: Automatic quantitative analysis of spontaneous aphasic speech. Speech Commun **100**, 1–12 (2018)
33. Latif, G., Iskandar, D.A., Alghazo, J., Butt, M., Khan, A.H. Deep CNN based MR image denoising for tumor segmentation using watershed transform. Int. J. Eng. Technol. **7**(2.3), 37 (2018)
34. Overview of artificial intelligence and role of natural language processing in big data (2021)
35. Lopez-Ruiz, R. From Natural to Artificial Intelligence-Algorithms and Applications, pp. 1–236. IntechOpen (2018)
36. Mousavirad, S.J., Schaefer, G., Jalali, S.M.J., Korovin, I. A benchmark of recent population-based metaheuristic algorithms for multi-layer neural network training. In: Proceedings of the 2020 Genetic and Evolutionary Computation Conference Companion, pp. 1402–1408 (2020)
37. Johnson, J. What is a Deep Neural Network? Deep Nets Explained. BMC blogs (2020)
38. Eckle, K., Schmidt-Hieber, J. A comparison of deep networks with ReLU activation function and linear spline-type methods. Neural Netw. (2019)
39. Bouwmans, T., Javed, S., Sultana, M., Jung, S.: Deep neural network concepts for background subtraction: A systematic review and comparative evaluation. Neural Netw. **117**, 8–66 (2019)
40. Espejo, S., Carmona, R., Domínguez-Castro, R., Rodríguez-Vázquez, A.: A VLSI-oriented continuous-time CNN model. Int. J. Circuit Theory Appl. **24**(3), 341–356 (1996)
41. Akilan, T., Wu, Q.J., Safaei, A., Jiang, W. A late fusion approach for harnessing multi-CNN model high-level features. In: IEEE International Conference on Systems, Man, and Cybernetics (SMC), pp. 566–571 (2017)
42. Qin, Y., Wu, Y., Lee, T., Kong, A.: An end-to-end approach to automatic speech assessment for cantonese-speaking people with Aphasia. J Signal Process Syst **92**(8), 819–830 (2020)
43. Seker, E. Recursive Neural Networks (RvNNs) and Recurrent Neural Networks (RNNs) (2021)
44. Buda, M., Maki, A., Mazurowski, M. A systematic study of the class imbalance problem in convolutional neural networks. Neural Netw. (2018)
45. Nejati, A. Recursive (not recurrent!) neural networks in tensorflow-KDnuggets. KDnuggets (2016)
46. Van, V.D., Thai, T., Nghiem, M.Q. Combining convolution and recursive neural networks for sentiment analysis. In: Proceedings of the 8th International Symposium on Information and Communication Technology, pp. 151–158 (2017)
47. Parisi, G., Kemker, R., Part, J., Kanan, C., Wermter, S.: Continual lifelong learning with neural networks: a review. Neural Netw. **113**, 54–71 (2019)
48. Xiao, L., Liao, B., Li, S., Chen, K.: Nonlinear recurrent neural networks for finite-time solution of general time-varying linear matrix equations. Neural Netw. **98**, 102–113 (2018)

49. Kuo, C.C., Zhang, M., Li, S., Duan, J., Chen, Y.: Interpretable convolutional neural networks via feedforward design. J. Vis. Commun. Image Represent. **60**, 346–359 (2019)
50. TensorFlow. CNN And RNN Difference-Tutorialspoint. Tutorialspoint.com (2021)
51. Sewak, M., Sahay, S.K., Rathore, H. Comparison of deep learning and the classical machine learning algorithm for the malware detection. In: 19th IEEE/ACIS International Conference on Software Engineering, Artificial Intelligence, Networking and Parallel/Distributed Computing (SNPD), pp. 293–296 (2018)
52. Iba, H. Evolutionary approach to machine learning and deep neural networks. Neuro-Evol. Gene Regul. Netw. (2018)
53. Ruder, S. An overview of multi-task learning in deep neural networks. arXiv preprint arXiv: 1706.05098 (2017)
54. Vodrahalli, K., Bhowmik, A.K.: 3D computer vision based on machine learning with deep neural networks: a review. J. Soc. Inform. Display **25**(11), 676–694 (2017)
55. Samek, W., Montavon, G., Lapuschkin, S., Anders, C.J., Müller, K.R. Toward interpretable machine learning: Transparent deep neural networks and beyond. arXiv preprint arXiv:2003. 07631 (2020)
56. Vishnukumar, H.J., Butting, B., Müller, C., Sax, E. Machine learning and deep neural network—Artificial intelligence core for lab and real-world test and validation for ADAS and autonomous vehicles: AI for efficient and quality test and validation. In: IEEE Intelligent Systems Conference (IntelliSys), pp. 714–721 (2017)
57. Rathore, H., Sahay, S.K., Thukral, S., Sewak, M. Detection of malicious android applications: classical machine learning vs. deep neural network integrated with clustering. In: International Conference on Broadband Communications, Networks and Systems, pp. 109–128. Springer, Cham (2020)
58. Molchanov, D., Ashukha, A., Vetrov, D. Variational dropout sparsifies deep neural networks. In: International Conference on Machine Learning, pp. 2498–2507 (2017)
59. Desai, V.S., Crook, J.N., Overstreet, G.A., Jr.: A comparison of neural networks and linear scoring models in the credit union environment. Eur. J. Oper. Res. **95**(1), 24–37 (1996)
60. Guresen, E., Kayakutlu, G.: Definition of artificial neural networks with comparison to other networks. Proc. Comput. Sci. **3**, 426–433 (2011)
61. Fan, J., Ma, C., Zhong, Y. A selective overview of deep learning. Statist. Sci. **36**(2) (2021)
62. Byun, T.M., Halpin, P.F., Szeredi, D.: Online crowdsourcing for efficient rating of speech: a validation study. J. Commun. Disord. **53**, 70–83 (2015)
63. Christensen, H., Cunningham, S., Fox, C., Green, P., Hain, T.A. Comparative study of adaptive, automatic recognition of disordered speech. In: Thirteenth Annual Conference of the International Speech Communication Association, Portland, OR, USA (2012)
64. Mengistu, K.T., Rudzicz, F. Comparing humans and automatic speech recognition systems in recognizing dysarthric speech. Presented at the Advances in Artificial Intelligence, Berlin, Heidelberg (2011)
65. Mustafa, M.B., Rosdi, F., Salim, S.S., Mughal, M.U.: Exploring the influence of general and specific factors on the recognition accuracy of an ASR system for dysarthric speaker. Expert Syst. Appl. **42**(8), 3924–3932 (2015)

A Deep Analysis on the Role of Deep Learning Models Using Generative Adversarial Networks

Alankrita Aggarwal⊙, Shivani Gaba⊙, Shally Nagpal⊙, and Anoopa Arya

Abstract A comparatively novel advance field of deep learning is the Generative Adversarial Network called GAN. These different types of networks if they start working in line with each other and work not to get the better of each other but start working keeping arm to arm connected to each other's world will be different. GAN is a category of machine learning frameworks intended for generation of images in which neural networks challenge each other in format of playing game. One network generates metaphors also called the generator and an additional network attempt to differentiate between the fake and real from the data set called as the discriminator. If suppose a training set is presented this procedure guides to manufacture a innovative information comparable to training set. Pictures created from GAN are same metaphors giving the notion of genuine to any observer having real features. GAN networks work on supervised, unsupervised and for reinforcement learning but earlier it applies or used only unsupervised learning only. This generative network produces candidate and on the other hand, the discriminative network evaluate them. Here producer is a complex neural network and differentiator is a convolution neural network. The GAN network divided into three categories to produce generative model learns and generation of data by probabilistic ideas. Next training of model can be completed in contradictory state. Lastly for training using neural networks, deep learning, and artificial intelligence methods. If generative networks used deep leaning methods then deep learning models can employ a very large amount of dataset, heavily dependent on high-end Machines, tries to solves problem from end to end machines, takes longer time to train means the results are better after getting trained on the other hand takes lesser time to test the data. Applications of GAN networks are increasing day by day as it is touching every sphere of our day today life. Some of the benefits of the deep learning are to creating artificial Intelligence function which mimics the mechanism of human brain in handing out data for decision making. Deep learning if combined with artificial intelligence can be capable of learning data which is considered unlabeled and unstructured. The chapter will relate different models of deep learning and their efficiency can be measured by studying different methods, models and simulation techniques.

A. Aggarwal (✉) · S. Gaba · S. Nagpal · A. Arya
Panipat Institute of Engineering and Technology, Samalkha, Haryana 132101, India

© The Author(s), under exclusive license to Springer Nature Switzerland AG 2022
K. R. Ahmed and H. Hexmoor (eds.), *Blockchain and Deep Learning*,
Studies in Big Data 105, https://doi.org/10.1007/978-3-030-95419-2_9

Keywords Generative adversarial network · GAN · Generative adversarial network in 3D images · Data analytics · Deep learning models · Deep learning · Artificial intelligence

1 Introduction and Scope

A generative adversarial network dependent on sound contrastive judgment and loss function used in GAN and an adversarial network inspired when a blog was written by Olli Niemitalo known as Conditional GAN. An analogous idea was created to reproduce animal behavior by scientists in the early twenties and is a relatively new invention in the deep learning field in which two different networks are used to generate images. Example: Think of the image of shoes (called the generator) and a second one that tries to differentiate between the fake shoes and real shoes from the data set (Called the discriminator). After the model is converged it is hard to discriminate between the produced and the real images as shown in Fig. 1. The association of brain and computer was given by Jon Von Neumann in around 1940s by introducing terms like memory, organ, and neuron and played an important role in the development of the architecture of modern computing and the creation of EDVAC. By the efforts of researchers like Warren McCulloch and Walter Pitts helped in emerging areas of stylized neurons networks with boolean logic. But in later stages, the researcher and Turing Award winner Marvin Minsky along with Seymour Papert showed a class of neural networks where inputs connected directly to outputs but had limitations on its capabilities. The work is extended by David Rumelhart and Ronald J. Williams along with Geoffrey Everest Hinton and suggested an approach called supervised learning which exposed a random network configuration to a training set of input data and suggested initial response have no relationship to the features of input data, an algorithm has to reconfigure the network because the guess scored was against the labels inputted and requires a learning algorithm to reconfigure

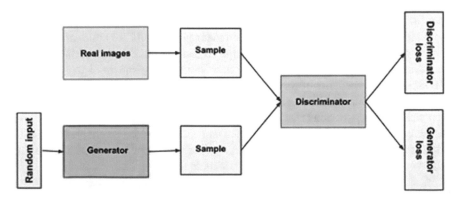

Fig. 1 Block diagram of the generative adversarial network (GAN). *Source* https://developers.goo gle.com/machine-learning/gan/gan_structure

the network to identify features in input data correlates the correct outputs. In the year 1986, the back-propagation algorithm measures the propagated errors backward by neurons connected to the outputs and intermediate hidden neurons between input and output layers to learn efficiently and removing the militations given by Minsky and Papert. Hinton's research coined an area called deep learning inspired by neural networks with multiple layers of hidden neurons to extract higher-level features from input data. Hinton with David Ackley and Terry Sejnowski earlier introduced a class of networks as the Boltzmann machine, which was suited to the layered approach. The area of artificial intelligence started officially around 1956 at a summer conference, the occasion being the celebration of 50th years of AI, sponsored by DARPA at Hanover, New Hampshire held the year 2006 attended by ten think tanks including John McCarthy, Claude Shannon, Marvin Minsky, Arthur Samuel, Trenchard Moore, Ray Solomonoff, Oliver Selfridge, Allen Newell, and Herbert Simon, Moore, Newell & Simon, and term artificial intelligence was devised there along with an introduction to logic theory was given which helps in proving elementary theorems in propositional calculus. The great researcher and mathematician Alan Turing gave the idea on the thought on Computing Machinery and Intelligence and created the thought for the people to think that "Can machines think" which is being replaced from "Can machine do what we (as thinking entities) can do".

2 Working Principal

In this network, the working principle is that neural networks are trying to win each other in any application. The training set is given and the former learn to produce newer data with the same as the training set. If photographs trained with GAN can generate new photographs that are similar to real images having the same characteristics if observed by human observers. Initially, the model based on generative was invented for unsupervised learning but they are also working powerfully for semi-supervised, fully supervised as well as reinforcement learning. The adversarial principle is deployed in association with machine learning in which generative modeling and created models can be simulated from the other theory of networks. For example in mechanical engineering model generation is based on the theory of control. Networks that are based on adversarial learning are used in previous years for training controllers in game theory by blinking in between the iterations. Actual working using GAN started in around 2017 initial faces were created was adopted for image enhancement which put the spotlight on real text rather than accuracy of pixel and produces superior illustration at quality with a higher at high intensification. An example of any random variable is given in a generative network is transformed into an output random variable and can be again reshaped as shown below in Fig. 2.

Figure 3 how an adversarial network works when a random variable is presented to a generator to produce an image and discriminator differentiates between a real image and a comparative score is generated.

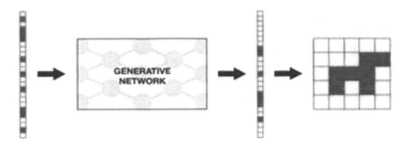

Fig. 2 A random variable transformed into output random variable and again reshaped. *Source* https://towardsdatascience.com/understanding-generative-adversarial-networks-gans-cd6 e4651a29

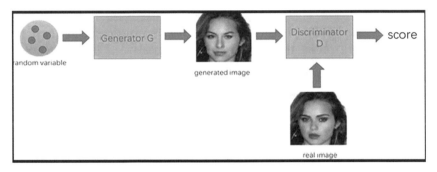

Fig. 3 An example of how an adversarial network works [1]

3 Exclusive Applications of GAN

There is a large vicinity of GAN applications that are escalating quickly with the passage of time.

- Speech to image construction using a Speech2Face model
- Molecules generation protein concerned in fibrosis, cancer & inflammation
- Representation and prediction of the age of the person with the appearance of the face
- Transferring of map styling in cartography or street view the images to augmentation
- Early detection of glaucomatous images avoiding loss of vision
- Photorealistic Images used by Facebook
- Generation of Images of design, clothing, or computer games scenes
- Introduction of motion in videos, 3D models of objects are recreated from images
- To visualize climate change of particular regions, places, or houses
- To approximate a nonlinear optimal control for network training To generate realistic photographs
- Generation of new human poses

- View of front face view generation
- Generation of realistic and human photographs
- Text to Image or Image to Image Translation
- Generation of Image Datasets
- Pictures of human faces generation
- Semantic Image to photo translation
- Generate cartoon characters
- Super-resolution
- Photo imprinting
- Photos to emojis
- Photograph editing
- Image to Photo Translation
- Clothing translation
- Face aging
- Photo blending
- Video Prediction
- Object generation
- Science
- Advertising and Fashion so that photographer hiring is not required, makeup artists and above all preventing the expenses on studios
- Video games: 2D images developed to retain the original level of details and colors for video games
- Classification: images can be classified to forecast the class of the images
- Concerning nasty applications: Images can e developed to mislead people on social media and it must be looked into [2].

4   Literature Review of Existing Work

Yu et al. [3] suggested a network that processes the obscure data with no label and supposition 3D point encoder cloud GAN point encoder by inputting the value and hybrid rebuilding of loss function gauge the dissimilarity between two sets of unordered data.

Go et al. [4] suggested that the efficiency of the network which is trained is achieved by comparison of the generation of holograms in 3D.

Chen et al. [5] proposed architecture of 3D CNN lightweight multi-level structural design is created densely sharp images with wealthy quality details.

Baek et al. [6] suggested GAN produce MR in occluded pixels and trained by images of labels and images without posing labels. The experiments done using object datasets show that our method for hand-only images and compared by HOI data.

Jain et al. [7] proposed a technique known as Poser GAN for a guess of human motion with 3D input human skeleton. A bidirectional GAN framework iterative scheme to predict training than using traditional Euclidean loss.

Cirillo et al. [8] suggested brain tumors images using 3D volume with Generative Adversarial Network segmentation using Vox2Vox model multi-channel 3D MR images and when generator loss is weighted high in comparison to discriminator loss output can be achieved.

Singh et al. [9] developed a framework in which clinical application using GAN is used and helps in the unsupervised picture to picture translation and also the invention of medical images and cross-modality fusion of medical images. Frameworks developed and compared with deep convolutional GAN, laplacian GAN, pix2pix, cycleGAN, and all are combined and a hybrid architecture are created for improved performance.

Islam and Zhang [10] proposed that artificial medical metaphor using generative adversarial networks and created a brain PET image for different stages of Alzheimer's disease namely: normal control, mild cognitive impairment, and Alzheimer's disease.

Zhang et al. [11] worked on the alliance of sequences and modalities whenever a deformable registration of 3D medical images is done using an unsupervised learning and gradient descent and after experimentation, images can cope up with the noise and blurriness and produce with accuracy a deform GAN using a deep learning and adversarial approach to transfer multi-modal similarity to monomodal similarity and thus improving the precision-based registration.

He et al. [12] proposed a solution for ultrasound image resolution by assimilating deep supervised modeling with GAN dependent framework for enabling the end to end encoding and decoding learning. The resultant network is learned the difference between the high and low-resolution images and attention framework using a GAN for the improvisation from ultrasound images. Deep attention GAN is engaged in encoding and decoding end to end for high-resolution images to incarcerate prostate ultrasound images. Attention model used to recover the pertinent in sequence. Differences between low and high-resolution images is validated with more than 20 patients.

Amyar et al. [13] proposed a model that can generate different classes of lesion capable of the small trial size of every lesion applicability of deep convolution generative adversarial network to produce a fast analysis of 3D image from the 2D image [11].

Zhou et al. [14] proposed a method that needed a detailed image registration and produces the HR volume image matching to LR volume image by using optical clearing of images.

Lan et al. [15] assured 3D conditional GAN system using normalization of spectral stabilizes criterion matching of features so that convergence optimization can be achieved. Along with that, the vowels can be separated to establish a relationship a self-attention module is added.

Mokhayeri et al. [16] suggested that in video surveillance performance of faces to be recognized can be improved if the design is created by emulating facial generation using cross-area face emulation with GAN and called CGAN. After several simulations, 3D face model simulations to the images and noise are given to CGAN for refinement of real images.

Taesik et al. [17] suggested a system to record 3D holograms neural network and GAN. A picture conversion from picture not determined on simple white light resource to holographic images by measuring network by comparison of 3D images generated and true holograms of microspheres.

Ye et al. [18] suggested a generative adversarial network focusing enhancement method effective method which confirms the focusing enrichment ability of 2D monochromatic imaging system for creating 3D to reduce the system intricacy for aiming realistic 3D imaging

Jyoti and Zhang [19] proposed a model named enhanced deep Super Resolution GAN to create imagery for the varied phases of brains namely: normal control, mild cognitive impairment, and disease images of Alzheimer.

Wang et al. [20] suggested a regeneration of text with the proposed a comparison between EDSRGAN between SRCNN and methods shows similarity in the ocular. It recovers edge features while another network regenerates using a difference map to differentiate high-frequency texts. Segmented images accuracy is measured by topological phase and permeability depicting that EDSRGAN results in the close match of images. Noise and blurriness can be adapted with augmentation. An improved Generative Adversarial Network called deep super-resolution EDSRGAN trained with SR pictures in 2D and 3D [18].

Lee and Town [21] introduced a Mimicry which is a lightweight PyTorch library that implementation of most famous GANs and The authors provided a comprehension of the different GANs on mostly used datasets by training these GANs having some conditions is done to mitigate the issues of the implementation of model using different frameworks and evaluating when metrics are used.

Jung [22] introduced a GAN Based Texture rendering that needed computations to produce high-resolution texture sampling and shading. In latest years deep learning has been used by scientists on extraction and fusion of content and representation of style from the type of image domains called texture transferring. What is done by the author is inputting to our generative model a 3D rendered image with no texturing results in rendered with a single color. The target image is 3D rendered image with texturing. For example, if we extract from a high-resolution texture the style representation and fuse it with a polygon only a 3D model acting as the content representation.

Oulbacha and Kadoury [23] suggested unsupervised Pseudo 3D Cycle GAN fully convolutional networks with cyclic loss function fusion of CT images with guided surgical MRI images investigative and trainable pre-processing pipeline to normalize MRI data for segmentation of vertebral bodies.

Zhang et al. [24] suggested a 3D block matching algorithm to clear images using threshold wiener filtering solves stained images moulds to get coefficients for getting sparkling images by training latent images through GAN. For assessment successfully remove image noise with improvement in visual effects work on the peak signal-to-noise ratio, structural resemblance, and edge preserve index.

Liu et al. [25] proposed a method using cascade conditional generative adversarial nets for hyperspectral image spatial-spectral sample generation based on the deep learning method. The C^2GAN consists of two stages wherein the first stage class labels and spatial information generation with window size entail feeding random noise is required and the second stage is the spatial-spectral information generation of spectral information bands in the spatial area by giving labels of regions. The datasets samples are derived from universities and are verified and validated to demonstrate better performance.

Zhaoa et al. [26] proposed a Bayesian conditional generative adversarial network with important dropout to obtain the exactness of the image. Uncertain calibration is used by making uncertainty interpretable generated by Bayesian network on samples of datasets of brain tumor compared to traditional Bayesian neural network with Monte Carlo dropout.

Yin et al. [27] presented a DA-GAN frontalization realistic face capturing contextual dependency and local consistency on GAN training prominence pose and illumination discrepancy in images. Here Face frontalization provides an effective and efficient way for face data augmentation and further improves the face recognition performance in extreme pose scenarios.

Yang et al. [28] suggested a deep image network is created as a base method for image segmentation on medical images with a generative adversarial network, machine learning, computer vision for enhanced segmentation for image synthesis, and well semantic segmentation.

Ma et al. [29] proposed a 3D generative model named mesh of wearing clothes doing 3D scans with poses and garments trained with conditional Mesh-VAE-GAN deformation of clothing learned from SMPL body mode.

Shi et al. [30] developed a deep learning-based model to estimate attenuation maps directly from SPECT emission data. The process is also competent of producing reliable attenuation maps to make possible attenuation correction for SPECT only scanners for cardinal perfusion imaging.

Kowalski et al. [31] suggested a ConfigNet model for attribute detection, neural face model authorize domineering figure trained on the real face, synthetic rendering by data to factorize latent space into elements correspond to the traditional rendering pipeline. It unravels aspects, the styling of hair, enlightenment in real facts.

Spick et al. [32] 3D proposed voxel-based 3D generative adversarial network model including shade to produce produced samples by becoming accustomed channels of voxel inputs using unsupervised learning to create a high-class surface to improve the turnaround time tested on the compilation of inputs set of textured models and on data set of 24 variant models of fish and outputs received by generative model shows promising results.

Wang et al. [33] suggested a Signed-distance map (SDM) in which health images and machine learning, and deep learning create a cochlea signed distance map on input parameters leading to improvement in time of computation compared to classical SDM methods.

Tang et al. [34] showed radar-based map deals with the difficulty of signal and map is produced from climatic changes and lighting compatible with nature of sensor localization is prepared when images are placed on ground from FMCW radar.

Minaee et al. [35] proposed assessment of segmentation image with deep learning doing a rigorous review of literature ranging on segmentation of semantic and illustration level, networks covering recurrent networks, multi-scale pyramid based approaches, visual attention generative with adversarial settings.

Zhou et al. [36] implemented defect detection deep learning dependent on imbalance data with universal optimization of GAN can lead to disarrangements so fresh generator and discriminator enquired to generate more discriminate faulty samples from faulty samples using an autoencoding accomplishing to global Optimization to purify disqualified produced samples as of qualified samples for error analysis.

Zihao et al. [37] suggested a proSigned distance map Neural Network and machine learning approaches generates marked distance map depends on input stricture and confirmed deep learning outcome in development compared to conventional production ways.

Tang et al. [38] focused generation of images of small objects images based on the guided scene is tricky so solution by generating a scene with a local context and designing a generative network with semantic maps.

Ye et al. [39] proposed deep learning-dependent e2e wireless communication with conditional GANs using DNNs to channel effect and transmit DNN can be back-propagated from the receiver as Unknown Channels engaged to encoding, decoding, modulation, and demodulation. Judgment of instant channel transfer is necessary to transfer DNN to optimize the receiver gain in decoding during learning the network. Waheed et al. [40] suggested a model that generates synthetic chest X-ray images developing supplementary classifier called CovidGAN and also displays the synthetic images produced from CovidGAN where COVID-19 by acute respiratory syndrome coronavirus that uses augmentation using auxiliary classifier GAN. Model is classified using CNN gives approximate 85% accurate results but accuracy increased by 95% by totaling synthetic images shaped.

Seddik et al. [41] represents data created by generative adversarial nets are informal vectors of strenuous informal vectors and similar to the mixing of Gaussian using deep learning.

Xu et al. [42] suggested a road estimation framework called GAN in which two lanes of two cities for study. Examination pointed approximate traffic by detectors accurately predicted by models by asking data from adjacent units to estimate highway traffic by using the diagram for illustration of the street network and generative adversarial network information to create the road traffic of state in real-time implementation.

Loey et al. [43] showed that a GAN by using three deep learning models for coronavirus exposure in X-ray images and generation of datasets so that all images for COVID-19 can be collected to make as many as high accuracy images for detection of this virus. The dataset was collected from dissimilar sites and can be downloaded. A set of 307 images dataset categorized for four classes. Deep learning models are Alexnet, Google net, and Restnet18 as they are less layered architectures and help in reducing complexity.

Beery et al. [44] suggested an experimental method so that the images can detect and classify the various type of occurrence to solve the type of applications like traffic control by self-driving cars by comparing the with the Adhoc available data and their effect at various axes. Examples of axes are lighting, pose, etc.

Fu et al. [45] proposed a 3D CNN trained network for automatic generation of tumor contours from radio and MR images showing the capability of getting better patient stratification and the continued survival of forecasting the GBM in patients. By applying the deep learning and capture automatic workflow of glioblastoma patients survived by collecting samples of more than three hundred and two hundred samples of the patients survived.

Li et al. [46] proposed an experimental method on 3D CAD object and RGB-D object recovery by learning together with the 3D representation and alignment of the domain by applying unsupervised learning thereby lessening the statistical convergence from domains.

Wang et al. [47] proposed a narrative generative model so that it is possible to create facade videos from images of neutral with or without a label. For example slight smile. The framework image and sequence generator implemented from a deep neural model and GAN along with variable autoencoders on the other hand whereas the sequence generator is conditional labeled recurrent neural network.

Zhu et al. [48] suggested that the images from radiation during CTP risked health problems so, in order to finish CT radiation in CT perfusion imaging is done with the variation with time utilization and elevated price of MR images, CTP imaging has opted.

Alom et al. [49] showed theoretically by using deep learning and supervised, semi-supervised and unsupervised learning also machine learning. DL approaches have been enquired and evaluated in the various applications by using developed frameworks, standard datasets used for implementation and evaluation.

Jin et al. [50] throw light on the automatic creation of facial images using generative adversarial learning for solving the problem of creating facial imagery of anime characters' facial image datasets. The application of Dragon empirical evaluation is done by collecting clean datasets and a website was made for the pre-trained model.

Wu et al. [51] created a fresh framework called 3D Generative Adversarial Network creating 3D objects convolutional networks, and GANs. Three-tier method using adversarial principle instead of the heuristic generator to detain structure of producing high excellence 3D object with producer maps by a minute dimension to 3D objects to sample objects without any orientation. The adversarial discriminator shapes and recognizes 3D images with unsupervised learning over supervised learning provisions Mittal et al. [52] suggested that deep learning methods can be used to segment the brain medical images.

Kaur et al. [53] created an efficient algorithm for the detection of objects associating analyzing the back and foreground brain images.

Verma et al. and Dash et al. [54, 55] showed that the identification and analysis of biomedical and health diseases get easier by using machine learning intelligence techniques and deep learning techniques.

Mittal [56] suggested that the detection of pneumonia on chest X rays images can be done using convolutions dynamic capsule routing.

Aggarwal et al. [57–60] suggested the various methods to control risks in various projects and the use of random forest in minimizing the risk also in GAN networks Gaba et al. [61] and Sharma et al. [62] explained the role of genetic cryptography and machine learning in various types of ad-hoc networks.

5 Some of the Famous Datasets Available of GAN

- https://github.com/timzhang642/3D-Machine-Learning
- http://homepages.inf.ed.ac.uk/rbf/CVonline/Imagedbase.htm
- www.kaggle.com/tags/gan.

6 Introduction of Deep Learning

This type of learning can be thought of as neural networks of a large number of parameters and different coating in network architectures. Also, it is like the next field of machine learning algorithms which are inspired by the configuration and function of the brain known as neural networks. Initially, the deep learning concept was just a concept of neural networks and nothing else around the last twenty years but the perspective was something more than neural networks. The idea of deep learning generated by Andrew Ng founder of Coursera and the chief scientist at Baidu Research both formed a company Google's brain was actually a large neural network and the creation of deep learning technologies is available on many platforms. Inventors of deep learning have already written much about the concept and he started with describing deep learning in the context of traditional ANN later years it has been unfolded that deep learning is self-taught based on supervised (or learning from labeled data) as well as unsupervised feature learning. Moreover, the field of unsupervised is also maturing for dealing with the availability of non-labeled data. One more researcher Jeff Dean a Google Senior Fellow at Google has been concerned for scaling of adoption of deep learning and development of major deep learning software i.e DistBelief, now known as TensorFlow. The highlight is on the scalability of neural networks representing results will be optimized larger data and greater models so in return more computation power is required for training. If deep learning methods are compared with traditional methods that are reaching a level of the plateau.

6.1 Objective of Deep Learning

To implement the applications like images, waveforms having speech, symbols in the natural language of artificial intelligence applications by reception a large variety of data so wealthy and hierarchical models representing probability distributions can be built. An important characteristic of deep learning is in differential models communicated to highly dimensional inputs sent to a class of labels. Such features are dependent on backpropagation and dropout algorithms using piecewise linear units and gradient behavior is good. Such a generative model whose foundation is deep learning impact is smaller as estimate inflexible probabilistic computation is hard to find.

7   Deep Learning Models

Famous Deep learning methods AI Practitioners need to apply in various applications. The motivation of deep learning is from the human brain how it refines the information. Basically, the deep learning area of machine learning teaches the computer to refine inputs through the layers in order to predict and classify information. The output is in the form of images, sound, or text. The deep learning techniques are classified as supervised, unsupervised, and semi-supervised and also a reinforcement deep learning i.e an amalgamation of supervised and unsupervised. Types of deep learning methods are as shown in Fig. 4 as follows.

Various types of deep learning methods can be classified on the basis of learning they are using i.e. supervised, unsupervised, semi-supervised, and reinforcement which is depicted in Fig. 5 and Table 1 respectively.

Fig. 4 Venn diagram of types of learnings. *Source* https://www.slideshare.net/ssuser77ee21/convolutional-neural-network-in-practice

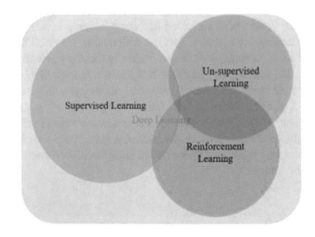

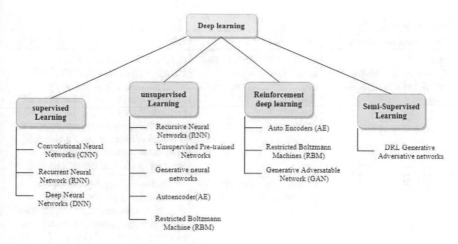

Fig. 5 Deep learning models

7.1 Deep Learning Dependent on Supervised Learning

- Convolutional Neural Networks (CNN)
- Recurrent Neural Networks (RNN)
- Deep Neural Networks (DNN).

Table 1 Comparisons of deep learning models with GAN in image transformation

Deep learning models	Improvements	Advantages/Disadvantages/Scenarios
CovidGAN [40]	Used for improved Covid-19 detection to data augmentation using Auxiliary classifier GAN	More accurate, easy convergence generates diverse samples/ Adjustments of parameters for training different data Model collapse gradients disappear/Explode/Suitable for the syndrome of respiration coronavirus 2 i.e. SARS-CoV-2
SC-GAN [15]	For spectral normalization and multi-model neuroimaging 3D, self-attention is done with conditional GAN	Image synthesis and enabling less scan period to improve image/Problems occur with 2D images/For neuroimaging synthesis based on deep learning/ For neuroimaging synthesis based on deep learning
Deform-GAN[1]	Unsupervised Learning Model for Deformable Registration	Can cope with nonfunctional data, intensity relations, noise, and blur/Sometimes background labeling is required during training/Other GAN models do not converge and models collapse

(continued)

Table 1 (continued)

Deep learning models	Improvements	Advantages/Disadvantages/Scenarios
Deep Learning–dependents E2E Wireless Communication [39]	Using DNN provides end to end wireless communication system is done	Successful on additive white Gaussian noise channels, Rayleigh fading channels, and frequency-selective channels/For an end to end communication a data-driven DNNs is built/When transmitters are designed it is very difficult to get the channel State information with time and location in wireless communications
3D-GAN[8]	Object shaping using a 3D Generative-Adversarial	Generates high-quality 3D objects and object recognition based on features/Sometimes slow learning/To combine high-quality 3D objects use of an adversarial rather than traditional heuristic criteria so that object structure implicitly can be captured

7.2 Deep Learning Models Dependent on Unsupervised Learning

- Recursive Neural Networks (RNN)
- Unsupervised Pre-trained Networks
- Generative neural networks
- Autoencoder(AE)
- Restricted Boltzmann Machine (RBM).

7.3 Deep Learning Models Based on Semi-Supervised Learning

- DRL Generative Adversative networks.

7.4 Deep Learning Models based on Reinforcement Deep Learning (RDL)

- Auto Encoders (AE),
- Restricted Boltzmann Machines (RBM),
- Generative Adversatable Network (GAN).

8 Common Datasets for Deep Learning

- MNIST is the most popular deep learning dataset
- MS-COCO is a dataset for object detection, segmentation, and captioning datasets
- ImageNet
- Open Images Dataset
- VisualQA
- House Numbers (SVHN)
- CIFAR-10
- Fashion-MNIST
- IMDB Reviews
- VoxCeleb
- Twitter Sentiment Analysis
- Age Detection of Indian Actors
- Urban Sound Classification.

9 Common Datasets for 3D Object Generation Using GAN

- 2D-to-3D Deformable Sketch
- ANN_SIFT1M
- CIFAR-10
- CLEF-IP 2011
- Contour Drawing Dataset
- DeepFashion.

10 Algorithmic Developments

The proposed algorithm for the operation of 3D patterns of image patterns with GAN is given below as steps:

- To initiate the random Generator (G) and Discriminator (G) to produce (fake X (label Y = 0) -→ (Y,Y) and real pair (label Y = 1) alternatively
- Create the type of images using the generator
- Now actual labeled 2D images are trained using the discriminator labeled as Y = 1 and 3D Images produced labeled as Y = 0
- Freeze weights by discriminator and pile for Generator
- Train the piled arrangement using images produced by labeled forcibly as Y = 1.
- Iterate and go to step 2.

11   Research Methodology

1.   To analyze and examine the thorough blow of 2D to 3D images with application
     of GAN to predict the numerous instances.

     a.   Related dataset to be fetched lively.
     b.   Build the Benchmark with important points and features.

2.   To analyses the threshold and suitability score.For that to do the threshold
     analyses and suitability of attributes.
3.   An amalgamation of the forming of data images and the simulation results.
4.   To generate the Dataset and apply the process of Cleaning and Pre-processing
     of data from the dataset.

     a.   To apply data analytics and forming 3D images.
     b.   Images Creation with GAN.

5.   To establish training with GAN

     a.   To train and test for Image Segmentation on data.

6.   To examine and combine of exactness of image generation.
7.   To form the images and patterns of analysis.

   The presented steps show flow in the pattern including analysis of harsh brunt
of deep learning using GAN for prediction of multiple illustrations, analysis of
threshold, acceptability of score, amalgamation into model & data formation, dataset
generation, data cleaning, data pre-processing with activation of predictive analytics.

12   Conclusion and Future Scope

The chapter inspects numerous applications of GAN after revising the deepness
of GAN and deep learning, their applicability in previous years with the usage of
numerous significant learning models: be it supervised, unsupervised, and reinforcement learning. In addition, various datasets and frameworks on deep learning can be
used to depict the performance of problems related to deep learning. They offered an
effort that can be implemented for real-world datasets in the direction of expansion
of immunization and alleviate prevailing respiratory diseases like COVID-19. The
Deep learning and GAN models can be used in the areas of healthcare, pharmaceutics,
biotechnology, bioinformatics, computational biology, Computer-aided health care,
drug design, structural biology, and immunology. In the coming future, the models of
deep learning can be used to mine large volumes of data to identify patterns and extract
features from complex unsupervised data without involving humans and creating an
important tool for Big Data analysis for estimating unknown future results. Last but
not least improvements can be made in Deep learning and GAN Models by balancing
the loss between the generator and the discriminator as it provides an elusive solution.

It can be done by maintaining a static ratio between a number of gradient descent iterations on the discriminator and the generator respectively.

References

1. https://en.wikipedia.org/wiki/Generative_adversarial_network
2. Aggarwal, A., Mittal, M., Battineni, G. Generative adversarial network: An overview of theory and applications. Int. J. Inf. Manage. Data Insights 100004 (2021)
3. Yu, Y., Huang, Z., Li, F., Zhang, H., Le, X. Point Encoder GAN: a deep learning model for 3D point cloud inpainting. Neurocomputing **384**,192–199 (2020)
4. Go, T., Lee, S., You, D., Lee, S.J. Deep learning-based hologram generation using a white light source. Sci. Rep. **10**(1), 1–12 (2020)
5. Chen, Y., Christodoulou, A.G., Zhou, Z., Shi, F., Xie, Y., Li, D. MRI super-resolution with GAN and 3D multi-level DenseNet: smaller, faster, and better. arXiv preprint arXiv:2003.01217 (2020)
6. Baek, S., Kim, K.I., Kim, T.K. Weakly-supervised domain adaptation via GAN and mesh model for estimating 3D hand poses interacting objects. In: Proceedings of the IEEE/CVF Conference on Computer Vision and Pattern Recognition, pp. 6121–6131 (2020)
7. Jain, D.K., Zareapoor, M., Jain, R., Kathuria, A., Bachhety, S. GAN-Poser: an improvised bidirectional GAN model for human motion prediction. Neural Comput. Appli. 1–13 (2020)
8. Cirillo, M. D., Abramian, D., Eklund, A. Vox2Vox: 3D-GAN for brain tumour segmentation. arXiv preprint arXiv:2003.13653 (2020)
9. Singh, N.K., Raza, K. Medical image generation using generative adversarial networks. arXiv preprint arXiv:2005.10687 (2020).
10. Islam, J., Zhang, Y.: GAN-based synthetic brain PET image generation. Brain Inf. **7**, 1–12 (2020)
11. Zhang, X., Jian, W., Chen, Y., Yang, S. Deform-GAN: an unsupervised learning model for deformable registration. arXiv preprint arXiv:2002.11430 (2020)
12. He, X., Lei, Y., Liu, Y., Tian, Z., Wang, T., Curran, W.J., Yang, X. Deep attentional GAN-based high-resolution ultrasound imaging. In: Medical Imaging 2020: Ultrasonic Imaging and Tomography (Vol. 11319, p. 113190B). International Society for Optics and Photonics (2020)
13. Amyar, A. et al. RADIOGAN: deep convolutional conditional generative adversarial network to generate PET images. arXiv preprint arXiv:2003.08663 (2020)
14. Zhou, H., Cai, R., Quan, T., Liu, S., Li, S., Huang, Q., Zeng, S.: 3D high resolution generative deep-learning network for fluorescence microscopy imaging. Opt. Lett. **45**(7), 1695–1698 (2020)
15. Lan, H., Toga, A.W., Sepehrband, F., Alzheimer Disease Neuroimaging Initiative. SC-GAN: 3D self-attention conditional GAN with spectral normalization for multi- modal neuroimaging synthesis. bioRxiv (2020)
16. Mokhayeri, F., Kamali, K., Granger, E. (2020). Cross-domain face synthesis using a controllable GAN. In: The IEEE Winter Conference on Applications of Computer Vision (pp. 252–260)
17. Deep learning-based hologram generation using a white light source. Sci. Rep. (Nature Publisher Group) **10**(1) (2020)
18. Ye, G., Zhang, Z., Ding, L., Li, Y., Zhu, Y. GAN-based focusing-enhancement method for monochromatic synthetic aperture imaging. IEEE Sensors J. (2020)
19. Jyoti, I., Zhang, Y. GAN-based synthetic brain PET image generation. Brain Inf. **7**(1) (2020)
20. Wang, Y.D., Armstrong, R.T., Mostaghimi, P. Boosting resolution and recovering texture of 2D and 3D micro-CT images with deep learning. Water Resour. Res. **56**(1), e2019WR026052 (2020)
21. Lee, K.S., Town, C. Mimicry: towards the reproducibility of GAN research. arXiv preprint arXiv:2005.02494 (2020)

22. Jung, J. RenderGAN: GAN based texture rendering
23. Oulbacha, R., Kadoury, S. MRI to CT synthesis of the lumbar spine from a Pseudo-3D cycle GAN. In: 2020 IEEE 17th International Symposium on Biomedical Imaging (ISBI), pp. 1784–1787. IEEE (2020)
24. Zhang, S., Wang, L., Chang, C., Liu, C., Zhang, L., Cui, H. An image denoising method based on BM4D and GAN in 3D shearlet domain. Math. Probl. Eng. (2020)
25. Liu, X., Qiao, Y., Xiong, Y., Cai, Z., Liu, P.: Cascade conditional generative adversarial nets for spatial-spectral hyperspectral sample generation. Inf. Sci. **63**(140306), 1–140306 (2020)
26. Zhaoa, G., Meyerand, M.E., Birn, R.M. Bayesian conditional GAN for MRI brain image synthesis. arXiv preprint arXiv:2005.11875 (2020)
27. Yin, Y., Jiang, S., Robinson, J.P., Fu, Y. Dual-attention GAN for large-pose face frontalization. arXiv preprint arXiv:2002.07227 (2020)
28. Yang, D., Xiong, T., Xu, D., Zhou, S.K. Segmentation using adversarial image- to-image networks. In: Handbook of Medical Image Computing and Computer Assisted Intervention, pp. 165–182. Academic Press (2020)
29. Ma, Q., Yang, J., Ranjan, A., Pujades, S., Pons-Moll, G., Tang, S., Black, M.J. Learning to dress 3d people in generative clothing. In: Proceedings of the IEEE/CVF Conference on Computer Vision and Pattern Recognition, pp. 6469–6478 (2020)
30. Shi, L., Onofrey, J.A., Liu, H., Liu, Y.H., Liu, C. Deep learning-based attenuation map generation for myocardial perfusion SPECT. Eur. J. Nuclear Med. Mol. Imaging 1–13 (2020)
31. Kowalski, M., Garbin, S.J., Estellers, V., Baltrušaitis, T., Johnson, M., Shotton, J. CONFIG: controllable neural face image generation. arXiv preprint arXiv:2005.02671 (2020)
32. Spick, R., Demediuk, S., Alfred Walker, J. Naive Mesh-to-Mesh Coloured Model Generation using 3D GANs. In: Proceedings of the Australasian Computer Science Week Multiconference, pp. 1–6 (2020)
33. Wang, Z., Vandersteen, C., Demarcy, T., Gnansia, D., Raffaelli, C., Guevara, N., Delingette, H. A deep learning based fast signed distance map generation. arXiv preprint arXiv:2005.12662 (2020)
34. Tang, T.Y., De Martini, D., Barnes, D., Newman, P.: RSL-Net: localising in satellite images from radar on the ground. IEEE Robot. Autom. Lett. **5**(2), 1087–1094 (2020)
35. Minaee, S., Boykov, Y., Porikli, F., Plaza, A., Kehtarnavaz, N., Terzopoulos, D. Image segmentation using deep learning: a survey. arXiv preprint arXiv:2001.05566 (2020)
36. Zhou, F., Yang, S., Fujita, H., Chen, D., Wen, C. Deep learning fault diagnosis method based on global optimization GAN for unbalanced data. Knowl. Based Syst. **187**, 104837 (2020)
37. Zihao, W.A.N.G., Vandersteen, C., Demarcy, T., Gnansia, D., Raffaelli, C., Guevara, N., Delingette, H. A deep learning based fast signed distance map generation
38. Tang, H., Xu, D., Yan, Y., Torr, P.H., Sebe, N. Local class-specific and global image-level generative adversarial networks for semantic-guided scene generation. In: Proceedings of the IEEE/CVF Conference on Computer Vision and Pattern Recognition, pp. 7870–7879 (2020)
39. Ye, H., Liang, L., Li, G.Y., Juang, B.H.: Deep learning-based end-to-end wireless communication systems with conditional GANs as unknown channels. IEEE Trans. Wireless Commun. **19**(5), 3133–3143 (2020)
40. Waheed, A., Goyal, M., Gupta, D., Khanna, A., Al-Turjman, F., Pinheiro, P.R. Covidgan: Data augmentation using auxiliary classifier gan for improved covid-19 detection. IEEE Access **8**, 91916–91923 (2020)
41. Seddik, M.E.A., Louart, C., Tamaazousti, M., Couillet, R. Random matrix theory proves that deep learning representations of gan-data behave as gaussian mixtures. arXiv preprint arXiv: 2001.08370 (2020)
42. Xu, D., Wei, C., Peng, P., Xuan, Q., Guo, H. GE-GAN: a novel deep learning framework for road traffic state estimation. Transp. Res. Part C Emerg. Technol. **117**, 102635 (2020)
43. Loey, M., Smarandache, F., Khalifa, N.E.M. Within the lack of chest COVID-19 X-ray dataset: a novel detection model based on GAN and deep transfer learning. Symmetry **12**(4), 651 (2020)
44. Beery, S., Liu, Y., Morris, D., Piavis, J., Kapoor, A., Joshi, N., Perona, P. Synthetic examples improve generalization for rare classes. In: The IEEE Winter Conference on Applications of Computer Vision, pp. 863–873 (2020)

45. Fu, J., Singhrao, K., Zhong, X., Gao, Y., Qi, S., Yang, Y., Lewis, J.H. An automatic deep learning-based workflow for glioblastoma survival prediction using pre- operative multimodal MR images. arXiv preprint arXiv:2001.11155 (2020)

46. Li, W.H., Xiang, S., Nie, W.Z., Song, D., Liu, A.A., Li, X.Y., Hao, T. Joint deep feature learning and unsupervised visual domain adaptation for cross-domain 3D object retrieval. Inf. Process. Manage. **57**(5), 102275 (2020)

47. Wang, W., Alameda-Pineda, X., Xu, D., Ricci, E., Sebe, N. Learning how to smile: expression video generation with conditional adversarial recurrent nets. IEEE Trans. Multimedia (2020)

48. Zhu, H., Tong, D., Zhang, L., Wang, S., Wu, W., Tang, H., Li, B.: Temporally downsampled cerebral CT perfusion image restoration using deep residual learning. Int. J. Comput. Assist. Radiol. Surg. **15**(2), 193–201 (2020)

49. Alom, M.Z., Taha, T.M., Yakopcic, C., Westberg, S., Sidike, P., Nasrin, M.S., Asari, V.K. The history began from alexnet: A comprehensive survey on deep learning approaches. arXiv preprint arXiv:1803.01164 (2018)

50. Jin, Y. et al. Towards the automatic anime characters creation with generative adversarial networks" demonstrates the training and use of a GAN for generating faces of anime characters (i.e. Japanese comic book characters) (2017)

51. Wu, J., Zhang, C., Xue, T., Freeman, B., Tenenbaum, J. Learning a probabilistic latent space of object shapes via 3d generative-adversarial modeling. In: Advances in Neural Information Processing Systems, pp. 82–90 (2016)

52. Mittal, M., Arora, M., Pandey, T., Goyal, L.M. Image segmentation using deep learning techniques in medical images. In: Advancement of Machine Intelligence in Interactive Medical Image Analysis, pp. 41–63. Springer, Singapore (2020)

53. Kaur, B., Sharma, M., Mittal, M., Verma, A., Goyal, L.M., Hemanth, D.J.: An improved salient object detection algorithm combining background and foreground connectivity for brain image analysis. Comput. Electr. Eng. **71**, 692–703 (2018)

54. Verma, O.P., Roy, S., Pandey, S.C., Mittal, M. (eds.) Advancement of Machine Intelligence in Interactive Medical Image Analysis. Springer Nature (2019)

55. Dash, S., Acharya, B.R., Mittal, M., Abraham, A. Deep Learning Techniques for Biomedical and Health Informatics. In: . Kelemen, A. (ed.). Springer Nature (2020)

56. Mittal, A., Kumar, D., Mittal, M., Saba, T., Abunadi, I., Rehman, A., Roy, S.: Detecting pneumonia using convolutions and dynamic capsule routing for chest X-ray images. Sensors **20**(4), 1068 (2020)

57. Aggarwal, A., Dhindsa, K.S., Suri, P.K.: A pragmatic assessment of approaches and paradigms in software risk management frameworks. Int. J. Nat. Comput. Res. (IJNCR) **9**(1), 13–26 (2020)

58. Aggarwal, A., Dhindsa, K.S., Suri, P.K.: Performance-aware approach for software risk management using random forest algorithm. Int. J. Softw. Innov. (IJSI) **9**(1), 12–19 (2021)

59. Aggarwal, A., Gaba, S., & Mittal, M. A comparative investigation of consensus algorithms in collaboration with IoT and blockchain. In: Transforming Cybersecurity Solutions Using Blockchain, 115p (2021)

60. Aggarwal, A., Gaba, S., Nagpal, S., Vig, B. Bio-Inspired Routing in VANET. In: Cloud and IoT Based Vehicular Ad-Hoc Networks, 199p (2021)

61. Gaba, S., Aggarwal, A., Nagpal, S. Role of machine learning for Ad Hoc networks. In: Cloud and IoT Based Vehicular Ad-Hoc Networks, 269p (2021)

62. Sharma, A., Kumar, S., Gaba, S., Singla, S., et al. A genetic improved quantum cryptography model to optimize network communication, Int. J. Innov. Technol. Exploring Eng. (IJITEE) **8**(9S), 256–259 (2019)

A Mediator Approach for a Semantic Integration of Heterogeneous Proteomics Data Sources

Chaimaa Messaoudi, Rachida Fissoune, and Hassan Badir

Abstract With the advance of high-throughput technologies, biological data sources are growing at an exponential rate. Data integration systems that combine data from heterogeneous sources help biologists to investigate the outcomes of their experiments. However, the heterogeneity of the different data sources, at the syntactic, schema, and semantic level, still holds considerable challenges for achieving interoperability among biological data sources. In this chapter, a new semantic data integration system, which uses a mediator approach, is proposed. This system offers a unified interface for query processing and data exploration on four well-known proteomic data sources: UniProt (protein annotation), String (protein-protein interaction), PDB (protein structure), and PubMed (biomedical citation). We use a domain ontology that allows the user to formulate its queries in terms defined in the ontology. We present a query rewriting algorithm that, using the annotated ontology, converts queries posed over the ontology to queries over the sources. This architecture takes advantage of the Apache Spark framework to perform the query rewriting and execution needed to question the integrated data sources.

1 Introduction

Biological databases have grown exponentially during the last decades, especially with the advance in high-throughput omics technologies. The volume and types of data generated by research laboratories have increased significantly in recent years. In 2010, the European Bioinformatics Institute (EBI) data source had 400 million

C. Messaoudi (✉)
System and Data Engineering Team, Abdelmalek Essaadi University,
BP 1818, 90000 Tangier, Morocco
e-mail: chaimaa.messaoudi@etu.uae.ac.ma

R. Fissoune · H. Badir
Engineering Team, Abdelmalek Essaadi University, BP 1818, 90000 Tangier, Morocco
e-mail: rfissoune@uae.ac.ma

H. Badir
e-mail: badir.hassan@uae.ac.ma

© The Author(s), under exclusive license to Springer Nature Switzerland AG 2022   199
K. R. Ahmed and H. Hexmoor (eds.), *Blockchain and Deep Learning*,
Studies in Big Data 105, https://doi.org/10.1007/978-3-030-95419-2_10

entries. In 2014, it exceeded one billion entries [34]. Biological databases are now considered big data [31, 39], emphasizing their large volume, their wide variety of data types stored, and their velocity at which the data is generated, collected and processed. Moreover, in the field of proteomics, the volume of data has grown widely. Nowadays, biologists are faced with questioning different data sources and deal with their growing heterogeneity to gather data and respond to a specific research question. In past decades, significant research work was developed to integrating multiple data sources to offer the biologist unified access to these sources.

Data integration refers to the problem of combining data from multiple sources in a unified way by providing to the user a single integrated source over a single global schema. While research in data integration has been going on for decades, several challenges still exist and particularly the accommodation of the heterogeneity of the data. These heterogeneities range from having various syntax (e.g. file formats, access protocols) where data formats can vary from compressed to flat and other formats (e.g. JSON, BAM, MITAB) to heterogeneity in the schema where data can be structured (e.g. relational tables), semi-structured (e.g. JSON, XML), or unstructured (e.g. texts), and finally semantics heterogeneity where the meanings and interpretations of attributes are required to connect different sources.

There are few approaches developed to solve problems of database integration. Two common strategies for the integration of biological databases are the following.

Data warehouse approach is defined in [21] as a subject-oriented, integrated, time-variant, and non-volatile collection of data in support of management's decision-making process. Building a data warehouse consists of locally materializing the data retrieved from the sources, transforming them to make them compatible with the previously defined global schema, making allowance for redundancies and complementarities, then executing queries on the consolidated data. Some of the most well-established systems given in the literature are: **Genomics Unified Schema (GUS)** [15] project at the University of Pennsylvania is one of the first data warehouse designed for the integration of sequence databases (GenBank/EMBL/DDBJ, dbEST, and SWISS-Prot). GUS uses a relational data model where tables hold the nucleotide and amino acid sequences along with associated annotation and are implemented over the Oracle and PostgreSQL DBMSs. The **YeastHub** [12] is a project for establishing a Web application based on an RDF data warehouse to support the integration of different types of yeast genome data. The application permits the user to register a dataset and convert it into RDF format if it is in tabular format. Once the datasets are loaded into the repository, they can be queried. Similar work is **YeastMine** [2] an integrated data warehouse for the budding yeast Saccharomyces cerevisiae. Further, **InterMine** [22, 33] is a data warehouse system for the integration and analysis of different types of biological databases, which models biological entities (such as genes and proteins) as objects that are described by a set of attributes; and their relationships with other objects are modeled as references. The InterMine system allows for the integration of different types of biological databases, and it comes pre-equipped with data integration features that can directly parse the data from commonly used data formats and sources (such as UniProt, OBO, FASTA, and

BioPAX). InterMine also allows users to design their data parsers. **TargetMine** [10, 11] is a customised version of the core InterMine data model. It was first developed for target discovery and prioritization of candidate genes, especially in early-stage drug discovery. The authors aim to transform TargetMine into an integrative data analysis platform that can more effectively interpret information-rich omics datasets for biological knowledge discovery. Data warehousing approach is limited by the problem of maintenance [1]. Researchers face challenges in keeping the data up to date and consistent. Additionally, updating the warehouse can be costly.

The Mediator approach or virtual integration consists in defining a mediation system between the agent (human or software) who asks a request and all the potentially relevant sources accessible via the Web to answer the request [41]. In doing so, it hides from the various heterogeneities by allowing the user to express queries over multiple data sources as if they were a single data source. The main components of a mediation system are the Mediator and a Wrapper for each data source. The mediator deals with data source distribution while the wrappers deal with data source heterogeneity and autonomy. Given a user query against the global schema, the mediator transparently decomposes it into constituent local subqueries against the appropriate sources, collects partial query results from the sources, and after due postprocessing, reports the combined results to the user. There are two predominant techniques to map the source schemas to the global schema: Local-as-view (LAV) and global-as-view (GAV). Several biomedical integration systems have been adopting the mediator-wrapper architecture. The first mediators developed in the computer community exploited the very fashionable object-oriented technology and did not consider semantic knowledge. For example, the **TSIMMIS** [9] system which is based on an Object Exchange Model (OEM) object-oriented language for the global schema and views, and a query language called OEM-QL which is an SQL-like language for dealing with labels and object nesting. This system follows a GAV approach. **K2/Kleisli** [15] is one of the first data integration projects developed at the University of Pennsylvania. K2 is the new version of an old project dating from 1997: Kleisli. The data model chosen for K2 is the object model. The K2/Kleisli supports several high-level query languages like the Collection Programming Language (CPL), a nested relational version of SQL and OQL. However, K2/Kleisli does not use any ontology over which a user can formulate queries and does not provide information about data provenance. Other examples of traditional systems are Garlic [8] and Interbase [5].

TAMBIS [36] follows the same direction as the K2 project on the management heterogeneity problems. However, it is considered to be the first project to perform an ontology-based integration in the era of bioinformatics and molecular biology. It offers an integration based on the TaO ontology which is complex and capable of describing all the concepts encountered in biological data sources. TAMBIS gives access to five biomedical sources (Swiss-prot, Enzyme, Cath, blast and Prosite). It provides users with a graphical query interface for query formulation where they have to browse over the multiple concepts defined in the ontology in order to construct their query. While this method is powerful, creating queries is not minor and may be complicated for biologists to understand the data structure of an ontology. Similar work is **SEMEDA** [23] an ontology-based semantic integration of biological sources such as Swiss-prot, OMIM, RZPD, EMP, MDDB, and KEGG.

Biomediator [6] a federated biological data integration system based on XML. The central element of the BioMediator system is its knowledge base, which consists of descriptions of the various data sources, the source mappings, and the mediation schema. The system also includes wrappers that perform syntactic translations by translating the returned results into an XML document, a meta-adapter that conducts semantic translations by mapping that XML document to the mediation schema, and a query processor that queries (using XQuery language) the mediation schema. More recently, similar work is **YeastMed** [4] a mediator-based system for the integration of yeast data sources such as SGD, YEASTRACT, MIPS-CYGD, BioGRID, and PhosphoGRID. YeastMed presents a domain ontology that plays the role of the global schema and supports the user queries. **Zhang et al**. [43] develop an ontology-based semantic data integration system to help integrative data analysis of cancer survival. They adjust existing semantic resources, for instance, the National Cancer Institute (NCI) Thesaurus in order to create an ontology named Ontology for Cancer Research Variables (OCRV). The framework adopt a GAV approach and mapping axioms to connect the data elements in different sources to the ontology.

In this chapter, we will mainly cover semantic mediator-based data integration. Integration at the semantic level indicates presence of an explicit formal description of the data semantic properties in terms of the subject domain ontology [20]. In our previous paper [27], we modeled a semantic mediator-based data integration approach to integrate four different data sources to support proteomics data integration. In this extended book chapter, we significantly expanded our query execution mechanism. The semantic data integration techniques are defined. Additionally, we present a case study of a semantic data integration system in proteomics. It deploy the mediator-based approach aiming to provide transparent access to different proteomics data sources. The system provides a unified interface between the user who submits a query and a set of four data sources accessible via network. It relies on the query rewriting engine to perform the query transformation needed to question the data sources. These sources are UniProt [38], PubMed [30], PDB [29] and StringDB [37]. They provide complementary data on biological entities that are protein annotation, biomedical citation, protein structure, and protein-protein interaction, respectively.

This chapter is structured as follows. In Sect. 2, we give the background knowledge needed to understand the terminology used in this chapter. Section 3 presents the major components and feature of a semantic data integration system for proteomics data. Section 4 presents a performance evaluation of the system. Finally, we conclude and present future works in Sect. 5.

2 Modeling a Semantic Biological Data Integration System

2.1 Characteristic of Biological Data Sources

Biological data sources are a collections of life science information from scientific surveys, high throughput experimental technologies, available literature and computer analyzes [3]. Most biological data sources are publicly available on websites that categorize the data that biologists can browse online. Among the characteristics of biological data sources we find:

Volume

In recent years, technological developments in mass spectrometers, liquid chromatography and next-generation sequencing technologies have led to a quantum leap in the amounts of data and the speed of their generation. These advancements have led to an exponential increase in the volume of biomedical data as illustrated in Fig. 1. In fact, there has been a great increase in the sequences present in UniProtKB [38] during the last decade. After removing the redundant data in 2015, we are seeing exponential growth in the number and volume of data. For example from 2016 to 2020 the size as well as the number of data has almost tripled. Moreover, biomedical data is expected to grow very rapidly over the next few years, reaching over one Zettabytes per year by 2025 [35].

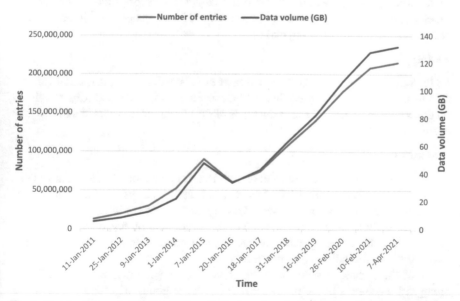

Fig. 1 Size development of UniprotKB knowledge base from January 2011 to April 2021. Blue line: Volume in gigabytes. Orange line: Number of entries (*Source* https://www.uniprot.org/statistics/)

Velocity

Speed describes how often data is generated, captured, and shared. Recent technological developments such as new high throughput sequencing technologies have resulted in the production of billions of DNA sequence data every day at relatively low cost.

Heterogeneity

The heterogeneity of biological data sources has several aspects. First, it can come from the heterogeneity in schema. Each data source adopts its own model or data schema that differs from one source to another. We find the relational model, for example in the database ENSEMBL [14], the object model, like the data warehouses Gedaw [19], and the semi-structured XML models in UniProt.

Second, biological data are of several types. They consist of sequences, graphs, geometric information, images and many more. For example, scientific literature, images and other free text documents are typically stored in unstructured or semi-structured formats (raw txt files, HTML or XML files, binary files). Another example is BLAST (the basic local alignment search tool), which is the most frequently used program in the molecular biology community. It requires a specific format (FASTA) for the input sequence and outputs a list of sequence alignments per pair.

Finally, the semantic heterogeneity where the meanings and interpretations of attributes are the principal of the effort required to connect different sources. For instance, there are very often several names (synonyms) for the same gene or for the same protein, not only within the same source but also across sources and species. Likewise, it is possible that two proteins or two different genes have the same name or a name in common (homonyms).

Variety

As biological research improves and greater understanding rises, it is usual for new data to be obtained that contradicts old data. Frequently, new data organizational schemes start to be crucial, even new types of data or completely new databases may become crucial [13].

Autonomy

Biological data sources are considered autonomous since, most of them have been developed and managed by individual research groups or research institutions. At first, they were developed for individual use by these groups or institutions, and even when they demonstrated to bring value to the larger community, data management practices specific to these groups have remained. Hence, biological databases always have their own formalism and infrastructure.

All of these characteristics poses challenges to integrating data across different biological databases. In the next subsection, we are going to present the mediator architecture as a solution for biological data integration.

2.2   Mediator-Based Integration

In general, data integration involves combining data residing in different sources to provide users with a unified view of it [24]. It is defined using two schemas: the global schema, the source schema, and correspondences between these two schemas, called mapping.

Definition A data integration system I is a triple $< G, S, M >$ where [7]

- G is the global schema, expressed in the relational model, possibly with constraints.
- S is the source schema, also expressed in the relational model.
- M is the mapping between G and S, constituted by a set of assertions of the form $q_S \subseteq q_G$, where q_S and q_G are two queries of the same arity, respectively over the source schema S et sur le schéma global S.

Querying autonomous and heterogeneous data sources is challenging. In biological data integration, there are two main approaches to resolve this problem: DataWare-housing [21], and Mediators and Wrappers [41]. In this section, we focus on the Mediator approach, as this is what we use in this work.

According to [41], the Mediator approach consists in defining a mediation system between the agent (human or software) who asks a request and all the potentially relevant sources accessible via the Web to answer the request. The goal is to give the user the illusion of interrogating a homogeneous and centralized system by avoiding him having to find the relevant data sources for his query, to interrogate them one by one in their specific language and according to their local schema, and combine the information obtained. The main components of a mediation system are the Mediator and the Wrapper for each data source. The Mediator includes an intermediate layer called global schema G (see Definition 1). It creates a connection between the different accessible data sources. In addition, It can be a model of the application domain of the system. It provides an explicit formal description of the semantic data properties in terms of a subject domain ontology, serving as a support for the expression of queries. In this chapter, by ontology, we use the widely accepted definition in data and knowledge engineering giving by Gruber [18]: "a formal, explicit specification of a shared conceptualization". According to this definition, an ontology specifies a common vocabulary for researchers wishing to share the basic terms and relationships of a given domain. It models knowledge by incorporating definitions, concepts of a domain, and their relationships. Thus, ontologies allow the modeling, sharing and reuse of knowledge. Ontologies have been extensively used to deal with semantic data integration [43]. Their importance in biology is significant, repositories like OBO Foundry [32] include over 200 ontologies, Bioportal [40] more than 800 biomedical ontologies, and Ontobee [42] past 200 ontologies.

The schema of a data source, known as a source schema S, represents the structures serving as a container for storing data in this source. The Wrapper contains technical and data model details of the source. The Mediator cannot directly assess the requests made to him because the data is stored in a distributed manner in independent sources.

It only has abstract views of the data source. These views are pre-defined queries expressed as a function of the global schema in a particular language, called a view language.

To deal with the heterogeneous nature of data sources, Wrappers translate the common query language into the data source query language. Each data source has an associated wrapper that exports information about the source schema, data, and query processing capabilities. The Mediator consolidates the information provided by the wrappers in a unified view using mapping formalisms. There are three mapping formalism proposed in the literature: Global-As-View (GAV) [25], Local-As-View (LAV) [25], and Global-Local-As-View (GLAV) [17].

In the GAV approach, the global schema is defined as a set of views on the data sources. Meaning, the mapping M associates to each element g in G a query q_S on S. Thus, a GAV mapping is a set of assertions, one for each g element of G, of the form $q_S \subseteq gs$.

The LAV approach adopts the opposite approach to GAV which consists in defining the data sources schemas to be integrated, according to the overall schema. The mapping M associates with each element s of the source schema S a query q_G on G. Thus, an LAV mapping is a set of assertions, one for each s element of S, of the form $s \subseteq q_G$.

The GLAV approach is a combination of the GAV and LAV formalisms. It offers the expressive power of both techniques, aiming to overcome their limits. GLAV is based on the idea that a view on the schemas of the integrated data sources should be characterized in terms of the view on the overall schema of the Mediator [16].

In the following, we first provide a high-level description of our data integration approach, then we evaluate the performance of the system.

3   Case Study: A Semantic Integration of Proteomics Data Sources

To discern clearly the problem of biological data integration from different sources, consider the following simple biological question to obtain information on one or more proteins: "What are the sequence and the protein-protein interaction network where the protein structure identifier '1FIK' is involved?" Answering this question often requires accessing at least three different sources:

1. A protein structure database like PDB [29], where we will retrieve the UniProt identifier of the protein.
2. A protein annotation database such as UniProt [38] where we will extract the sequence and the identifier of the interaction network.
3. A protein-protein interaction database like String [37] where we will unfold the protein-protein interaction network.

Giving these sources, the according query system and the data model may vary. The biologist needs to question every single data source. Meaning, they should be familiar with at least one query language. Moreover, to recover relevant entries, he/she is led to use the specific identifiers (IDs) and names (synonyms) of the protein. This is where dealing with issues ranging from heterogeneous database schemas, data redundancy, numerous independent identifiers, and differing curation standards and practices is a time-consuming task and can decrease the advantage that can be drawn from the available sources. Instead, an integrated view would offer to the biologist unified access to these sources, without him/her knowing any details regarding the data sources. Our challenge is to capture the growing heterogeneity and breadth of information and make it easily accessible for the biologist. Moreover, we seek to optimize query performance by using Apache Spark to perform the query transformation and execution needed to question the integrated data sources.

3.1 System Overview

Integrated Proteomics Data System (IPDS) [27] is a semantic mediator-based system using Spark for the integration of heterogeneous proteomics data sources such as:

- UniProt (Universal Protein Resource) [38] provides the scientific community with a comprehensive, high-quality, and freely accessible resource of protein sequence and functional information. UniProt Knowledgebase (UniProtKB) counts over 120 million proteins and supports several application programming interfaces (APIs) to query its data. In the UniProt REST API, all resources are accessible using simple URLs (REST). The data is available in text, XML, RDF, FASTA, GFF, and tab-separated formats.
- String (Search Tool for the Retrieval of Interacting Genes/Proteins) [37] is a biological database and Web resource of known and predicted protein-protein interactions. Its latest version 11 contains information about 24.6 million proteins from 5090 organisms. String offers API calls by making URI format and returns data in JSON, tsv, tsv-no-header, psi-mi, and psi-mi-tab formats.
- PubMed [30] is a free search engine accessing primarily the MEDLINE database of references and abstracts on life sciences and biomedical topics. PubMed consists of over 30 million citations for biomedical literature from MEDLINE, life science journals, and online books. PubMed citations and abstracts include the fields of biomedicine and health, covering portions of the life sciences, behavioral sciences, chemical sciences, and bioengineering. It also provides access via RESTful APIs using simple URLs and returns data in a tab-delimited text file, XML, or JSON formats.
- Protein Data Bank (PDB) [29] is a database for the three-dimensional structural data of large biological molecules such as proteins and nucleic acids. To date, PDB counts 156101 biological macromolecular structures. PDB supports RESTful (Representational state transfer) Web services to access its data. The RESTful Web services accept queries in PDB-specific XML format and return a list of IDs.

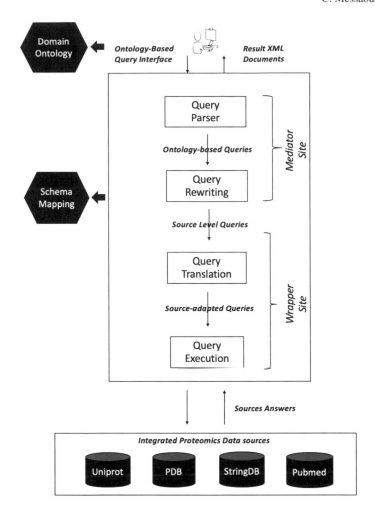

Fig. 2 Overview of the mediator-based data integration architecture applied to UniProt, PDB, StringDB, and PubMed

In order to achieve interoperability between UniProt, String, PDB, and PubMed we have chosen a mediator based integration approach. The advantage of a mediator approach is to avoid any management of data updates since the data remains in the sources, unlike maintaining a centralized data warehouse which is a fastidious and time-consuming task. This advantage is significant in the context of biological data that are progressing rapidly. Figure 2 presents the architecture of the system.

In order to overcome the semantic challenge of biomedical data sources, we develop a domain-specific ontology (further described in Sect. 3.2) as the global schema. It allows the user to formulate its queries in terms defined in the ontology. Moreover, to solve the syntactic heterogeneity, we rely on an ontology-based query

Fig. 3 User interface

interface to query all data sources. Figure 3 presents the search form that allows biologists to express their request in natural language. For instance, by giving specific information about protein annotation, protein-protein interaction (e.g. a required score, network limit), structure annotation and features (e.g. structure title, experimental method), protein sequence and bibliographic reference. This search form make use of the ontology terms, which are presented in natural language to facilitate the query formulation process for biologists most of whom are not used to work with knowledge representation and query languages. Users formulate their queries by selecting items from the form field. These items have their equivalents in the ontology (concepts and properties) and are written in natural language.

3.2   Domain Ontology

The aim of our system is to provide a single access point to heterogeneous proteomics data sources. Accordingly, we developed a domain ontology to support user queries. Each element in the search form has its equivalents in the ontology. To provide semantic encapsulation of the integrated proteomics data sources, the ontology has been established from scratch by harmonizing the various data source schemas into a single coherent ontology. It characterizes a concept hierarchy, which is a classification of all the biological entities utilized by the system. It represents a knowledge model that captures biological and bioinformatics knowledge in a simple hierarchical conceptual framework constrained by parent-child relationships, where a child is defined as a subset of a parent's elements. Each child inherits all of its parent's properties, although, has more specific properties of its own. We classified ontology concepts into two types:

- The biological-based concepts class which is a combination of all the classes modeling biological entities in the integrated sources. For example, *SubCellular-Location* concepts are a superclass of 12 classes representing different types of SubCellular Localization (e.g. Cytoplasm and Golgi).
- The source-based concepts class represents concepts related to each source. For example, the data property *hasID* represents the four integrated data sources IDs, which are hasPDBID, hasPubmedID, hasStringID, and hasUniprotID

Moreover, ontology defines two types of properties to bring further semantic information about the concepts. The first is characterized by a set of object properties that models the relationships between two individuals belonging to one or two different classes of the ontology. The second type defines data properties, which are relationships joining an individual to a literal data.

To further illustrate the role of properties in conveying semantics to the developed ontology, we illustrate a real-world example in Fig. 4. The *PRO1* gene of Arabidopsis thaliana corresponds to the *Q42449* UniProt ID and can also be found in PDB and StringDB because UniProt cross-references PDB and StringDB. This protein is having the name *Profilin-1* and an interaction score of 0.999. Figure 4 also shows concepts such as *Gene*, *Protein*, and *Interaction*, and some object properties such as *codeFor* and *codedBy* that link *Gene* to *Protein*. Next, the property *hasInteraction* and its inverse *hasInteractor* link *Protein* to *Interaction*. We can also observe the datatype properties: *hasScore*, *hasProteinName*, *hasID*, and *hasGeneName* linking *Interaction*, *Protein*, and *Gene*, respectively, to literal values of type String.

Additionally, using this ontology example, we can answer for instance this query: What is the protein-protein interaction involved in the "PRO1" gene product of the species "Arabidopsis thaliana"? This question refers to pairs of gene and protein (those in Arabidopsis) and their interaction network. Apart from the *Organism* class, which is not explicitly presented, all of the information is already captured by Fig. 4.

The system ontology is defined in the W3C Web Ontology Language (OWL) [26] and it is constructed using the Protégé tool [28]. Figure 5 shows a screenshot of the ontology.

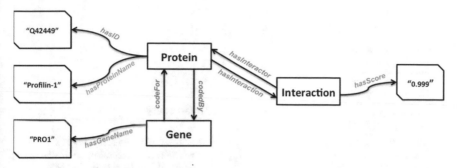

Fig. 4 A schematic representation of three biological concepts linked by four object properties (blue), three datatype properties (orange) linking concepts to values of type String and two parent-child relationships

Fig. 5 IPDS ontology using the Protégé tool

3.3 Mediator Site

Once the mediator receives the submitted query, the query is reformulated by the query parser to a set of queries in terms defined by the ontology. Each element of the form fields has its equivalents in the ontology. For instance, the datatype property *hasStrucID* in the ontology is interpreted in the form fields as *Structure ID* and the object property *hasProteinSequence* is translated as *Protein Sequence* in the search interface. To clarify the role of the Query Parser, Fig. 6 shows the formulation of a query to extract the sequences and the interaction network of a protein having as identifier of its structure "1FIK".

Once the Query Parser has reformulated the user query, the mediator uses the schema mappings defined for the application domain to translate the query from the

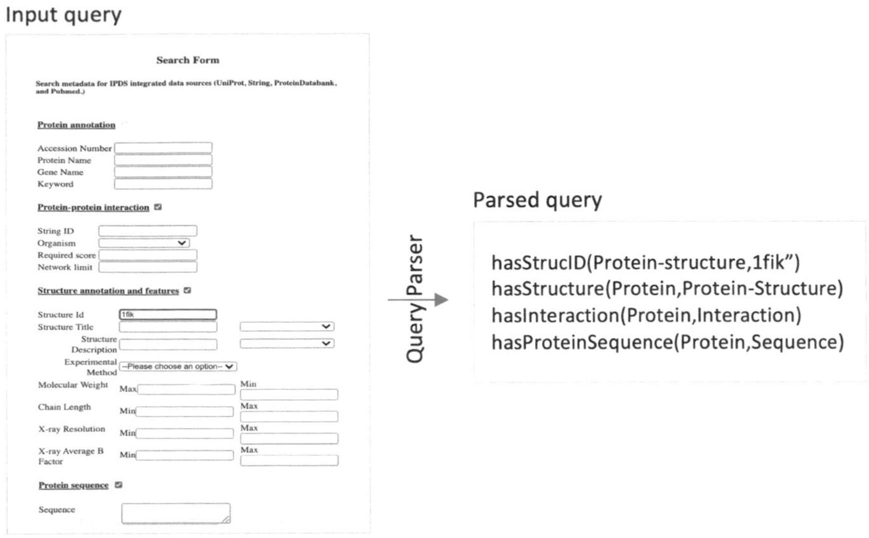

Input query

Fig. 6 Example of a query based on ontology terms

ontology schema into an executable query over the source schemas. This procedure is called query rewriting.

However, the schema mappings is needed to describe the relationships between the sources schemas and the domain schema. These mappings are described through a set of declarative logical rules. Indeed, the mediator uses a set of declarative rules in order to define how the predicates of the domain schema relate to predicates of the source's schema. Our system is designed to decompose queries on GAV approach-based mappings. This means each concept and property in the ontology is a view defined in terms of the source schemas' elements. This view specifies how to obtain instances of the mediated schema elements from sources. In this context, the mapping rules we have used are defined as pairs (P, Q). P is the value expressed in the ontology terms and Q is the corresponding expression defined in terms of the source schema. Two types of mappings have been defined:

1. Datatype property mapping: it maps ontology datatype properties to source schemas. The mapping form is (Ontology-Class-Name.Ontology-Property-Name, Value-Element-Source). The following example concerns with the datatype property *hasProteinName* of the ontology class *Protein* in the source schema UniProt: (*Protein.hasProteinName, name*).
2. Object property mapping: it maps ontology object properties to source schemas. It has the following form: (Ontology-Class-Name.Ontology-Property-Name, Spark-XML-Attribut, Data-Source-Name, Value-Element-Source). The following example shows how the object property *hasInteraction* is mapped to the source schema UniProt: (*Protein.hasInteraction, _id,STRING,hasID*).

In addition to the mapping rules, the algorithm takes as input a conjunctive query expressed in the ontology terms and returns as output a set of sub-queries structured in a Spark graph according to their execution orders. The steps of the algorithm are described as follows:

Algorithm 1: Query Rewriting

Input: A query in the ontology terms and the mapping rules.
Output: A Spark graph of query execution.
1. Classify the obtained query predicates into two groups based on predicates with or without values. The group G1 will contain the datatype property predicates and the second group G2 will have the object property predicates.
2. Add the class type predicates in each group based on the argument of each property type predicate. Once added remove these arguments.
3. Add in G1 the list of missing class predicates from G2. G1 will construct the vertices Dataframe of the graph.
4. Sort the class type predicates of G2 starting with those having values in G1. G2 will form the edges Dataframe.
5. Create the Spark graph (G1,G2).
6. Query each vertex with values and replace it with its equivalent in the mapping list.
7. Query each edge and replace it with its equivalent in the mapping list.

The algorithm will return as output a Spark graph of query execution. The edges and vertices of the Spark graph are of type Dataframe. The edges of the Spark graph are formed from the object predicates, while the vertices of the Spark graph are instances of the class predicates and have datatype property predicates as an attribute. Figure 7 presents an example of the Spark query execution graph formulated in Fig. 6.

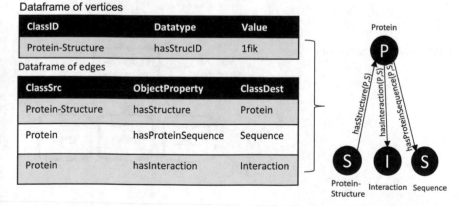

Dataframe of vertices

ClassID	Datatype	Value
Protein-Structure	hasStrucID	1fik

Dataframe of edges

ClassSrc	ObjectProperty	ClassDest
Protein-Structure	hasStructure	Protein
Protein	hasProteinSequence	Sequence
Protein	hasInteraction	Interaction

Fig. 7 Example of a query execution graph

3.4   Wrapper Site

After the query rewriting, the wrappers translate the Spark graph of the queries execution into the native query language of the data sources. In other words, the wrappers translate the vertices Dataframe into a native request for each data source (XML document for PDB and Parameter/Value document for UniProt, String and Pubmed).

This process is performed using two Spark APIs:

- Spark-XML which provides the data source "com.databricks.spark.xml" allowing to write Spark DataFrame in an XML file. In this case, we will translate the vertices Dataframe into XML file when it comes to the PDB data source (see Fig. 8).
- Spark-CSV which provides the data source "com.databricks.spark.csv" allowing to write Spark DataFrame in a CSV file (see Fig. 9).

Thus, these queries will be executed by calling the data services in the order determined by the execution graph. As execution progresses, the query executor generates instances for the variables of the subqueries to be executed from the results of the already executed subqueries.

The results returned by the data services are XML files whose structures are pre-defined by an XML schema. We translate these files into a Dataframe with SPARK-XML to be able to easily extract and obtain the desired results as shown in Fig. 10.

Fig. 8 Spark XML translator

Fig. 9 Spark CSV translator

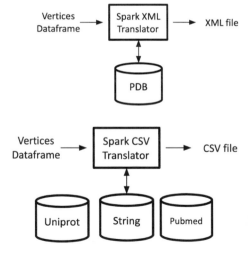

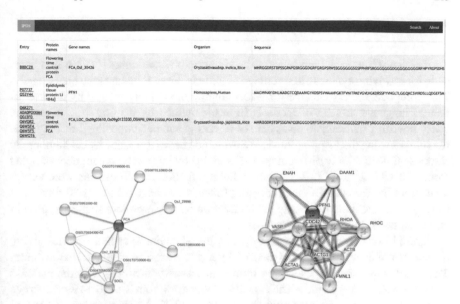

Fig. 10 Result display in IPDS

4 System Evaluation

We evaluate the performance of the system over four public proteomics data sources (UniProt, PDB, String and PubMed). We achieve a performance study of the system to reveal how run times behave towards the increase of query size (number of fields). This performance study is performed on the run times of two major aspects of the system architecture: (i) Mediator site and (ii) Wrapper site. The user formulates a query, which is first rewritten by the query parser to a set of queries with respect to the ontology and then reformulated and decomposed over the source schema using the mapping assertions. After the query rewriting, the Wrappers parse the source-level queries and generates a source-adapted queries, which are forwarded to the data service of each data source for execution. All run times shown for each scenario corresponds to the average time of thirty consecutive executions. The experiments are performed on an Intel i5 3 GHz machine with 8 GB RAM. We implemented our framework in Java and used Apache Spark 2.3.0 for the query execution in standalone deploy mode provided by Spark. Besides, we used HDFS 3.1.1 as a storage layer for the system.

4.1 Mediator Site Performance

The Mediator is a crucial module in a data integration system like IPDS. It is responsible for resolving syntactic and semantic conflicts. First, it features the query parser. This component is responsible for reformulating the request entered by the user in natural language to a set of requests in terms defined by the ontology. Then, the query rewriting mechanism takes care of rewriting and decomposing the query over the sources using the mapping rules. We evaluated its performance by calculating the response time of twenty different queries and this by increasing the number of fields from 3 to 13 (e.g. 3, 5, 7, 9, 11 and 13 fields). A total of 120 queries were tested. To capture the variation in the processing times, we repeated the tests 30 times. All input queries are fixed to display the interaction network, protein sequence, protein structure, and bibliographic references.

Figure 11 reveals run times performed by the Mediator to execute queries giving the number of their fields (from 3 to 13). Mean values are also displayed in the figure. The result shows that the execution time increases exponentially with the increase in the number of query fields. The execution time goes from 2.12 (0.14)s on average (standard deviation) for a query with three fields to 22.22 (2.13)s on average (standard deviation) for a query with thirteen fields. The standard deviations also increase with the mean values. This is expected because the rewriting get more complex when having more fields. In fact, the query parser and the query rewriting mechanism take longer to process a query with a large number of fields.

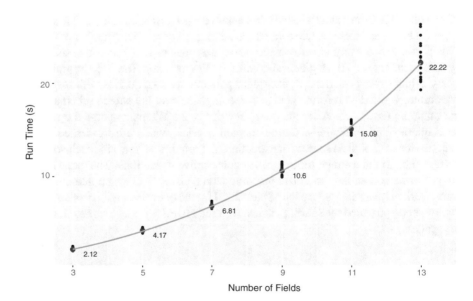

Fig. 11 Execution time for the Mediator with regard to the increase of the query size

4.2   Wrapper Site Performance

The second aspect of IPDS system performance is at the Wrapper site. These include
the translation of queries at the source level and the generation of native queries
for each data source to execute them. The results returned by the data services are
formulated and returned to the user. We evaluated this performance by calculating
the execution time of twenty different queries and this by increasing the number of
fields from 3 to 13 (e.g. 3, 5, 7, 9, 11 and 13 fields). A total of 120 queries were
tested. To capture the variation in processing times, we repeated the tests 30 times.

Figure 12 displays the execution times performed by the wrappers for twenty
queries by varying the number of their fields (from 3 to 13). The result shows that
the execution time increases slightly with the increase in the number of fields. The
execution time goes from 6.83 (7.75)s on average (standard deviation) for a query with
three fields to 6.44 (0.58)s on average (standard deviation) for a query with thirteen
fields. We observe that the average execution time does not change significantly
depending on the number of fields. When a small number of fields are used the
number of responses may become large. As a result, we can observe longer time for
some queries. For example, in Fig. 12, we observe 3 queries having 3 fields, but their
execution time is significantly higher than the other 3 fields queries times.

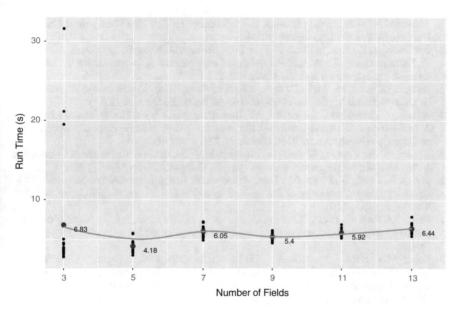

Fig. 12 Execution time for the wrappers with regard to the increase of the query size

5 Conclusions and Future Works

Data integration over heterogeneous biological databases is an important factor for gaining new biological insights in the proteomics era. Here, we introduced, a semantic mediator-based system to proteomics data integration. This approach offers a new query processing mechanism having as components a query parser, a query rewriting algorithm, a query translator and execution mechanism. This approach leverages the Apache Spark framework to perform the necessary transformation and query execution. It is established on a mediator system integrating four sources of heterogeneous proteomic data: UniProt (protein annotation), String (protein-protein interaction), PDB (protein structure) and PubMed (biomedical citation). In the future, we plan to include more data sources in the mediated system, focusing primarily on publicly available pathway databases such as Kyoto Encyclopedia of Genes and Genomes (KEGG). In a second step, it would be interesting to extend the coverage of different query languages such as SPARQL since most biological data sources start to offer SPARQL access points.

References

1. Abiteboul, S., Manolescu, I., Rousset, M.-C., Senellart, P.: Web Data Management. Cambridge University Press, Philippe Rigaux (2011)
2. Balakrishnan, R., Park, J., Karra, K., Hitz, B.C., Binkley, G., Hong, E.L., Sullivan, J., Micklem, G., Michael Cherry, J.: Yeastminean integrated data warehouse for saccharomyces cerevisiae data as a multipurpose tool-kit. Database (2012)
3. Bhatt, V.D., Patel, M., Joshi, C.G.: An insight of biological databases used in bioinformatics. In: Current Trends in Bioinformatics: An Insight, pp. 3–25. Springer (2018)
4. Briache, A., Marrakchi, K., Kerzazi, A., Navas-Delgado, I., Rossi Hassani, B.D., Lairini, K., Aldana-Montes, J.F.: Transparent mediation-based access to multiple yeast data sources using an ontology driven interface. BMC Bioinform. 13(1), S7 (2012)
5. Bukhres, O.A., Chen, J., Weimin, D., Elmagarmid, A.K., Pezzoli, R.: Interbase: an execution environment for heterogeneous software systems. Computer 26(8), 57–69 (1993)
6. Cadag, E., Louie, B., Myler, P.J., Tarczy-Hornoch, P.: Biomediator data integration and inference for functional annotation of anonymous sequences. In: Biocomputing 2007, pp. 343–354. World Scientific (2007)
7. Calì, A., Calvanese, D., De Giacomo, G., Lenzerini, M.: On the expressive power of data integration systems. In: International Conference on Conceptual Modeling, pp. 338–350. Springer (2002)
8. Carey, M.J., Haas, L.M., Schwarz, P.M., Arya, M., Cody, W.F., Fagin, R., Flickner, M., Luniewski, A.W., Niblack, W., Petkovic, D., et al.: Towards heterogeneous multimedia information systems: the garlic approach. In: Proceedings RIDE-DOM'95. Fifth International Workshop on Research Issues in Data Engineering-Distributed Object Management, pp. 124–131. IEEE (1995)
9. Chawathe, S., Garcia-Molina, H., Hammer, J., Ireland, K., Papakonstantinou, Y., Ullman, J., Widom, J.: The Tsimmis project: integration of heterogenous information sources (1994)
10. Chen, Y.-A., Tripathi, L.P., Fujiwara, T., Kameyama, T., Itoh, M.N., Mizuguchi, K.: The targetmine data warehouse: enhancement and updates. Front. Genet. 10, 934 (2019)
11. Chen, Y.-A., Tripathi, L.P., Mizuguchi, K.: Targetmine, an integrated data warehouse for candidate gene prioritisation and target discovery. PloS one 6(3), e17844 (2011)

12. Cheung, K.-H., Yip, K.Y., Smith, A., Deknikker, R., Masiar, A., Gerstein, M.: YeastHub: a semantic web use case for integrating data in the life sciences domain. Bioinformatics **21**(Suppl 1), i85–i96 (2005)
13. Chung, S.Y., Wooley, J.C.: Challenges faced in the integration of biological information. In: Bioinformatics, pp. 11–34. Elsevier (2003)
14. Cunningham, F., Achuthan, P., Akanni, W., Allen, J., Ridwan Amode, M., Armean, I.M., Bennett, R., Bhai, J., Billis, K., Boddu, S., et al.: Ensembl 2019. Nucleic Acids Res. **47**(D1), D745–D751 (2019)
15. Davidson, S.B., Crabtree, J., Brunk, B.P., Schug, J., Tannen, V., Christian Overton, G., Stoeckert, C.J.: K2/Kleisli and GUS: experiments in integrated access to genomic data sources. IBM Syst. J. **40**(2), 512–531 (2001)
16. Doan, A.H., Halevy, A., Ives, Z.: Principles of Data Integration. Elsevier (2012)
17. Friedman, M., Levy, A.Y., Millstein, T.D., et al.: Navigational plans for data integration. AAAI/IAAI 67–73 (1999)
18. Gruber, T.R.: A translation approach to portable ontology specifications. Knowl. Acquis. **5**(2), 199–220 (1993)
19. Guerin, É., Marquet, G., Burgun, A., Loréal, O., Berti-Équille, L., Leser, U., Moussouni, F.: Integrating and warehousing liver gene expression data and related biomedical resources in GEDAW. In: International Workshop on Data Integration in the Life Sciences, pp. 158–174. Springer (2005)
20. Gusenkov, A., Bukharaev, N., Birialtsev, E.: On ontology based data integration: problems and solutions. J. Phys. Conf. Ser. (IOP Publishing) **1203**, 012059 (2019)
21. Inmon, W.H.: Building the Data Warehouse. Wiley (2005)
22. Kalderimis, A., Lyne, R., Butano, D., Contrino, S., Lyne, M., Heimbach, J., Fengyuan, H., Smith, R., Štěpán, R., Sullivan, J., et al.: Intermine: extensive web services for modern biology. Nucleic Acids Res. **42**(W1), W468–W472 (2014)
23. Köhler, J., Philippi, S., Lange, M.: SEMEDA: ontology based semantic integration of biological databases. Bioinformatics **19**(18), 2420–2427 (2003)
24. Lenzerini, M.: Data integration: a theoretical perspective. In: Proceedings of the Twenty-First ACM SIGMOD-SIGACT-SIGART Symposium on Principles of Database Systems, pp. 233–246. ACM (2002)
25. Levy, A.Y.: Logic-based techniques in data integration. In: Logic-Based Artificial Intelligence, pp. 575–595. Springer (2000)
26. McGuinness, D.L., Harmelen, F.V., et al.: OWL web ontology language overview. W3C Recommendation **10**(10) (2004)
27. Messaoudi, C., Fissoune, R., Badir, H.: IPDS: a semantic mediator-based system using spark for the integration of heterogeneous proteomics data sources. In: Concurrency and Computation: Practice and Experience, p. e5814 (2020)
28. Musen, M.A., et al.: The protégé project: a look back and a look forward. AI Matters **1**(4), 4 (2015)
29. Protein data bank: the single global archive for 3D macromolecular structure data. Nucleic Acids Res. **47**(D1), D520–D528 (2018)
30. Pubmed, a free search engine accessing primarily the medline database of references and abstracts on life sciences and biomedical topics. https://www.ncbi.nlm.nih.gov/pubmed/
31. Sima, A.C., Stockinger, K., de Farias, T.M., Gil, M.: Semantic integration and enrichment of heterogeneous biological databases. In: Evolutionary Genomics, pp. 655–690. Springer (2019)
32. Smith, B., Ashburner, M., Rosse, C., Bard, J., Bug, W., Ceusters, W., Goldberg, L.J., Eilbeck, K., Ireland, A., Mungall, C.J., et al.: The OBO foundry: coordinated evolution of ontologies to support biomedical data integration. Nat. Biotechnol. **25**(11), 1251–1255 (2007)
33. Smith, R.N., Aleksic, J., Butano, D., Carr, A., Contrino, S., Hu, F., Lyne, M., Lyne, R., Kalderimis, A., Rutherford, K., et al.: Intermine: a flexible data warehouse system for the integration and analysis of heterogeneous biological data. Bioinformatics **28**(23), 3163–3165 (2012)
34. Squizzato, S., Mi Park, Y., Buso, N., Gur, T., Cowley, A., Li, W., Uludag, M., Pundir, S., Cham, J.A., McWilliam, H., et al. The EBI search engine: providing search and retrieval functionality for biological data from EMBL-EBI. Nucleic Acids Res. **43**(W1), W585–W588 (2015)

35. Stephens, Z.D., Lee, S.Y., Faghri, F., Campbell, R.H., Zhai, C., Efron, M.J., Iyer, R., Schatz, M.C., Sinha, S., Robinson, G.E.: Big data: astronomical or genomical? PLoS Biol. **13**(7), e1002195 (2015)
36. Stevens, R., Baker, P., Bechhofer, S., Ng, G., Jacoby, A., Paton, N.W., Goble, C.A., Brass, A.: TAMBIS: transparent access to multiple bioinformatics information sources. Bioinformatics **16**(2), 184–186 (2000)
37. Szklarczyk, D., Franceschini, A., Wyder, S., Forslund, K., Heller, D., Huerta-Cepas, J., Simonovic, M., Roth, A., Santos, A., Tsafou, K.P., et al.: String v10: protein–protein interaction networks, integrated over the tree of life. Nucleic Acids Res. **43**(D1), D447–D452 (2014)
38. UniProt Consortium: UniProt: a worldwide hub of protein knowledge. Nucleic Acids Res. **47**(D1), D506–D515 (2018)
39. Vidal, M.-E., Endris, K.M., Jozashoori, S., Karim, F., Palma, G.: Semantic data integration of big biomedical data for supporting personalised medicine. In: Current Trends in Semantic Web Technologies: Theory and Practice, pp. 25–56. Springer (2019)
40. Whetzel, P.L., Noy, N.F., Shah, N.H., Alexander, P.R., Nyulas, C., Tudorache, T., Musen, M.A.: Bioportal: enhanced functionality via new web services from the national center for biomedical ontology to access and use ontologies in software applications. Nucleic Acids Res. **39**(Suppl 2), W541–W545 (2011)
41. Wiederhold, G.: Mediators in the architecture of future information systems. IEEE Comput. **25**(3), 38–49 (1992)
42. Xiang, Z., Mungall, C., Ruttenberg, A., He, Y.: Ontobee: a linked data server and browser for ontology terms. In: ICBO (2011)
43. Zhang, H., Guo, Y., Li, Q., George, T.J., Shenkman, E., Modave, F., Bian, J.: An ontology-guided semantic data integration framework to support integrative data analysis of cancer survival. BMC Med. Inf. Decis. Making **18**(2), 41 (2018)

Blockchain for Technological and Mindset Revolution: A Qualitative Study in the Sultanate of Oman

Salam Saif Said Al-Riyami and Smitha Sunil Kumaran Nair

Abstract Blockchain defined as a decentralized and distributed public ledger that records and replicates information stored in blocks spread across several computers in a network represents the utmost innovative technology in the Fourth Industrial Revolution era. While the technology was developed to improve online transactions, it has witnessed a variety of uses based on its core attributes of security, transparency, and decentralization. However, this technology has not been widely adopted in the Sultanate of Oman. The present study aims to explore the adoption of blockchain technology in the Sultanate by investigating barriers and enablers and finally propose a framework for the adoption of blockchain technology. The qualitative approach undertaken involved interviews conducted with key stakeholders from both private and public sectors of Oman with keen insight or experience in blockchain technology. The content analysis was employed with themes and categories identified. The study found out that the critical barriers were lack of awareness, lack of potential application, lack of experience and expertise, and lack of sufficient infrastructure. In addition, the lack of government regulation and policy barriers the adoption. The enablers include the government's commitment to creating more awareness about the technology and provide necessary resources and infrastructure in addition to the presence of a defined accountable and responsible authority for the policy and regulations. Three recommendations are put forward to serve as a framework for the adoption of blockchain in the future which can also help the nation transform into a technological hub in the region. Designing and implementing a strategy that mitigates the barriers and setting up incentives and enablers for the technology is an approach that is suggested, to improve the adoption of blockchain technology.

Keywords Barriers · Blockchain · Enablers · Framework · Sultanate of Oman

S. S. S. Al-Riyami · S. S. K. Nair (✉)
Centre for Postgraduate Studies, Middle East College, Rusayl, Muscat, Sultanate of Oman
e-mail: smitha@mec.edu.om

© The Author(s), under exclusive license to Springer Nature Switzerland AG 2022
K. R. Ahmed and H. Hexmoor (eds.), *Blockchain and Deep Learning*,
Studies in Big Data 105, https://doi.org/10.1007/978-3-030-95419-2_11

1 Introduction

Blockchain Technology is a decentralized and distributed public ledger that records, and replicates information stored in blocks spread across several computers in a network. Blockchain technology improves security and transparency in the recording and storing of information [19]. This technology has become a common trend in the present 4th Industrial Revolution era [5]. Businesses and individuals have realized the benefits they can derive from this technology that no other innovation can offer. The said benefits include faster transactions, transparency, and improved security [40]. However, the technology has not been adopted nationwide which prompted to undertake the present study to explore the barriers and enablers of its adoption.

1.1 Blockchain in the Sultanate of Oman

Sultanate of Oman has made significant steps in expediting the adoption of blockchain technology in the country. The government has realized the benefits that a decentralized and distributed ledger systems could bring into businesses in various sectors. One of the efforts made towards expediting the adoption of the technology attribute to conducting a symposium that features stakeholders from both public and private sectors. Oman conducted its first blockchain symposium in 2017 [8]. The symposium was organized by various government bodies such as the Transfer of Science, Knowledge, and Technology Office at the Ministry of Foreign Affairs (MOFA), Capital Markets Authority (CMA), and Information Technology Authority (ITA). The event attracted both local and international experts on blockchain technology, where issues regarding adoption were discussed. The symposium achieved high ends in terms of kickstarting the adoption of blockchain technology in Oman. To begin with, it proposed the raising of awareness and recognition of the technology intending to help people appreciate its unprecedented opportunities. Secondly, it proposed the use of technology in the government to set a foundation and attract other industry stakeholders in the future.

The second effort made by Oman towards expediting the adoption of blockchain technology is to create an organization that should deal with blockchain adoption issues. During the symposium in 2017, the government resorted to create a body called Blockchain Solutions and Services (BSS) to handle all issues regarding blockchain technology [8]. The organization's core task will be to develop the critical infrastructure required to support the adoption of blockchain technology. Additionally, it will help to spread awareness of the technology and improve nationwide adoption by consolidating and sharing information to the public and other relevant stakeholders. Apart from forming the BSS, the government also created the Blockchain Club designed to bring together individuals and organizations that have vast knowledge and experience with blockchain technology [8]. The club's membership is free to all individuals and organizations who consider themselves savvy enough to share something about blockchain technology. The ultimate objective of the club is to increase awareness of blockchain technology to the masses and provide the needed momentum for an expedited adoption.

The third effort made by the government to support the adoption of blockchain technology in Oman is to hold a conference where the progress made after the symposium held in 2017 is evaluated. In 2018, the Business Process Outsourcing Service (BPOS) together with ITA organized a two-day conference to discuss the progress made in the adoption efforts after the symposium [8]. The goal of the conference was to build on the recommendations made during the symposium by developing strategies for achieving the desired outcomes. ITA found benefit of the strategies that would help the government to digitally transform so that it can set the foundation necessary for the adoption of blockchain technology by other stakeholders.

Lastly, key industries have already embraced the adoption of blockchain technology to improve operations. A key example is Bank Dhofar, which joined a global enterprise blockchain network (Ripple) to improve its payment services [8]. The network provides solutions such as instant global payments, frictionless payments, and secure online transactions. Muscat Clearing and Depository Company (MCD) is another early adopter of blockchain technology. The company has chosen to implement an electronic subscription system that will offer a superb investment experience to its customers [8]. The system will have benefits such as being accessible worldwide, being error-free, and being secure. Investors will be able to carry out all their desired transactions through the system, such as receiving dividends and being able to access all the information and services required to make informed investment decisions.

It is evident that the Sultanate is making efforts to improve the adoption of blockchain technology in the country. It is also evident that stakeholders in the private sector have taken the initiative to lead the pack in exploiting the benefits and opportunities brought forth by the technology. Nevertheless, further efforts are required to expedite the process and ensure nationwide adoption.

1.2 Problem Statement

Developments in the twenty-first century have been mainly underpinned by advancements in technology. Since the last decade of the twentieth century, it became evident that technology had an important role to play in advancing civilization to the next level [27]. As the Internet became a commonplace, it is now an inescapable fact that technology will be the foundation for all future advancements. One technology that has proved beneficial with its number of use cases is the blockchain technology. The technology has become popular in the last two decades and is now used considerably in the developed countries [1]. For instance, big banks and businesses accept Bitcoin, a currency born out of blockchain development. Developing countries have used blockchain technology to improve speed and transparency in the transactions. There are also plans to use the technology for other use cases, such as keeping patient records [6]. However, there needs to be use cases to be developed. For instance, financial sector requires regulations and policy to deal with cryptocurrency in the Sultanate.

Additionally, both the private and public sectors in the Sultanate have not been proactive in adopting blockchain technology. In this regard, there is a need to investigate the barriers that prevent public and private enterprises in Oman from adopting the technology. Moreover, there is a need to investigate the factors that enable the adoption of technology in order to promote its adoption. Consequently, this research targets to investigate the barriers and enablers of blockchain technology in Oman and also aims to determine how the Sultanate of Oman can become a regional hub for innovation and business development pertaining to blockchain technology by proposing a blockchain adoption framework for the Sultanate of Oman.

2 Literature Review

2.1 Technology Adoption Theories

The impact of technology on the social, cultural, and economic welfare of people cannot be overlooked. In this regard, researchers have endeavored to establish theories that predict the spread or use of technology [18]. While many theories have been proposed so far, they can generally be categorized into five distinct groups. The five groups are diffusion theories, user acceptance theories, decision-making theories, personality theories, and organizational structure theories.

2.1.1 Diffusion Theories

Diffusion theories focus on the fact that technologies are designed to improve how tasks and functions are carried out in the contemporary setting. The history and evolution of technology show that the need to simplify the handling of every-day tasks has been a driving motive [26]. For instance, some technologies are designed to improve efficiency. Others design for productivity. As a result, it becomes inevitable that technologies will always be adopted since they enhance activities that promote a higher standard of living. Diffusion theories consider the adoption of technologies to be inevitable in any given environment provided the technology serves a useful purpose. The technology is said to diffuse as it moves from being a new development to a regular item that cannot be ignored.

Diffusion theories have two significant and distinctive characteristics when it comes to how they explain the adoption of technology. Fundamentally, researchers of the theories assume that all technologies play a crucial role in society, a role that cannot be ignored. However, while the technology is new, its utility is not reasonably evident to potential users. As a result, it takes time and resources to make the technology known to potential users. Since technologies are assumed to play a vital function, diffusion theories postulate that they will ultimately be adopted even if it takes a longer time. For instance, [20] argues that the time it takes for new technology to move towards adoption can view in terms of a life-cycle that starts with the promotion of the technology and ends with its eventual acceptance. Aspects that will impact the life-cycle of the technology will be crucial to the path it takes towards adoptions. While some technologies are readily accepted (diffuse easily), others have to undergo several iterations and significant changes before they become mainstream.

Secondly, diffusion theories assume that there exists frameworks in an environment to promote the adoption of technology. Technology adoption is not always guaranteed even when the said technology promises to address an important issue or provide a specialized function that would drastically improve people's lives. There are social, political, and economic factors that may impede the adoption of the technology. For instance, expensive technologies may not be adopted quickly in an environment where cheaper, albeit less efficient alternatives are found. Diffusion theories presuppose that there will always be positive factors to promote the adoption of the said technology [25]. Diffusion theories are relevant in the sense that they explain the adoption of technologies that are not impeded by social, political, or economic factors. Consequently, they are ideal in tracking progress, especially when the adoption of a given technology is key to the welfare of the people.

2.1.2 User Acceptance Theories

User acceptance theories are another category of technology adoption theories. The theories built upon the rationale that individuals may show interest in a given technology based on the value they hope to get from adoption. As pointed out by [38], there is a group of consumers that readily seek out new technologies in order for them to be the first ones to benefit from their usefulness. Additionally, some technologies stand out in terms of their utility and usability. As a result, individuals have sometimes shown the initiative to adopt technologies with little to no influence from external sources. User acceptance theories focus on such a premise and postulate technologies that would be adopted once they are accepted by their intended audience.

Several characteristics are unique to user acceptance theories. The most significant characteristic is the fact that they pre-suppose attributes that should exist before the technology is accepted. To begin with, the target individuals must show a significant degree of interest in the said technology. Acceptance occurs when there is little undue influence from the environment for people to accept a particular technology. It has argued in the past that a population may consider adopting a given technology if it meets their interests in a way that it cannot be ignored [12]. Therefore, the presence of knowledgeable and interested parties is crucial to ensuring that technology is accepted.

Secondly, acceptance theories emphasize the need for a framework to support acceptance. While interested parties may show interest in technology before it becomes mainstream, a framework is required to ensure that such interest maintains in the long-term. Acceptance can support a framework that emphasizes the utility of a given technology to a specific population. In the past, governments have used such a strategy to ensure that citizens move towards a new direction with regards to embracing technology. As pointed out by [35], a framework is crucial to ensuring that individuals recognize and appreciate the usefulness of a given technology in their lives. Taking the example of blockchain technology, a desirable framework for user acceptance include the provision of necessary infrastructure to support the technology and regulations that guide how it will be used in its various use cases.

Thirdly, it argues that user acceptance theories also lay focus on the prevailing economic and social conditions of a population, and how new technology can lead to massive improvements. Humanity strives to improve itself daily by using the knowledge gained from the past to make better decisions in the future [29]. In this regard, individuals have always found themselves cognizant of new developments and how they can impact their lives positively. When adopting such a perspective, it is possible to see how users can accept technology since everyone remains with the goal of improving themselves and bettering their conditions. User acceptance theories go as far as recognizing such a reality and its impact on how individuals embrace a given technology.

User acceptance theories are relevant in any discussion that promotes the acceptance of new technology. The theories make it clear that there are individuals who will be interested in the new technology based on their interests. Therefore, it is essential to consider such a population and how they can be helped to influence others towards adopting technology [12].

2.1.3   Decision-Making Theories

Decision-making theories represent the third category of technology adoption theories. Decision-making theories focus on the fact that both individuals and organizations may seek out new technologies based on their interests or a problem they need to address in their environment. Conventionally, new technology is usually desired with the desired function in mind and a target population [25]. The developers of a new technology recognize a need in the market and a population that may benefit when such a need arises. However, it is not always the case that there is a technology for a need that an individual or organization possesses. For instance, organizations have come to rely on automation to improve efficiency and the speed at which they carry out several of their daily tasks. Additionally, automation has also proved critical towards cost reduction and ensuring that companies remain competitive and sustainable in the long-term. While there are several technologies in the market to address such a need, a company may fail to find one that exactly fit to their situation. As a result, they will proactively seek such technology to ensure that they can improve their operations as desired [37].

Despite sharing similar characteristics, decision-making theories are diverse when compared to technology adoption theories in any other category. One theory focuses on rational choice and how both individuals and organizations can make a conscious decision to seek out new technology for their benefits. Proponents of the rational choice theory argue that society has grown increasingly conscious of the benefits of technology and will continue to make deliberate decisions to seek out new developments. For instance, in light of the sustainability challenges currently plaguing the business world, companies are looking for ways of embracing new technologies to address the challenges [31]. Some companies have adopted a green policy to reduce their overall environmental impact and climate change. Others are embracing the use of technology to outsource employees from various parts of the world. Therefore, according to rational choice theory, there remains a need for both individuals and organizations to seek out new technologies that will improve adoption.

Decision-making theories are also diverse in the sense that others seek to address existing challenges that have affected organizations for years. One such challenge is the risk management. Companies exist in a tumultuous environment where risks, both natural and human-made, can reverse the gain made in trying to establish the organization. As a result, risk management is one area that is critical to their operations. Proper risk management ensures that an entity will be able to respond to any eventuality without suffering catastrophic damage. Some decision making theories focus on how risks can be handled better through the use of technology. For instance, in the transport and logistics management, tracking of shipments using the Internet and the Global Positioning System is now a mainstream [29]. The practice reduces the risk of losing cargo while in transit or failing to prepare for late deliveries. Therefore, organizations can make a rational choice to adopt a technology based on an issue they need to address in their operations.

Decision-making theories are still relevant when discussing technology adoption. The theories are ideal in situations where both individuals and organizations make it a mission to seek new technologies that address a pressing need decisions made by companies to use blockchain technology explained by the theory. As pointed out earlier, technology brings out new desirable attributes such as transparency, accountability, and security. Therefore, an organization seeking to bolster its security or promote transparency and accountability in its operations may end up choosing blockchain technology [12].

2.1.4 Personality Theories

Personality theories represent another category of technology adoption theories. Proponents of the theories argue that some technology may pique the interest of individuals because they fit a particular aspect of their personality. As pointed out by [10], some technological innovations have gone beyond providing a specific desirable function to individuals. Such technologies have become a personal statement, while others have gone to becoming cultural icons. Smartphones are a great example of such technologies. While viewed in the past as devices that help one to communicate and organize their lives, smartphones now represent personal objects that are also bought because they fit the personality of a buyer [25]. Companies have gone as far as differentiating themselves with regards to their service and product offerings to the point they attract different sets of consumers. Personality theories recognize and emphasize such a relationship where technology is acceptable naturally because it resonates with the personality of an individual.

Personality theories are not as many when compared to other categories of technology adoption theories. Due to this fact, they address minimal occasions where the function and utility of a given technology is not the only thing that influences adoption. At present, technologies that can be explained by the theories are limited. They primarily consist of items that have significant cultural significance even though they deliver on specific functions. As pointed out by [37], there is a current debate on whether technologies that fit the theory designed by the personality of users in mind or the appeal is just by accident. For instance, a cult item that is bought more for its personality attributes rather than its actual function view in light of the theory. However, it may not be possible to tell whether the item drew the personality of interested individuals or its utility and function made it desirable to a specific group of people.

Personality theories hold some relevance, even in the context of blockchain technology. Since its inception, blockchain technology has maintained three key attributes that are lacking in many systems that perform similar functions. One key attribute is decentralization. The decentralized nature of the technology makes it a desirable technology for users who cherish democracy and are not comfortable with the fact that governments and organizations control much of the resources people need to lead quality lives. Therefore, an individual may take it personally when embracing the use of blockchain technology since it promotes freedom achieved through decentralization. The second key attribute is transparency. Transactions carried out using the blockchain technology are transparent in the sense that they record in verifiable ledgers. Such transparency is lacking in many industries such as the banking industry. Thirdly, blockchain technology is secure in the sense that records are stored in more than one location so that they cannot be altered and affect the integrity of existing records. The level of integrity possible with blockchain technology is unparalleled. In this regard, individuals who consider transparency and integrity to be reliable attributes may adopt the technology directly because it fits their personality and values. Therefore, personality theories are applicable in the context of adopting blockchain technologies.

2.1.5 Organizational Structure Theories

The last category of technology adoption theories is organizational structure theories. The theories focus on the strategic interest that organizations may have, which may promote the adoption of the technology. It is evident that organizations are at the frontline of developing and promoting the adoption of new technologies. As pointed out by [28], such technologies may represent a strategic interest for the organization. Therefore, despite the benefits that the technology will bring to users, it will also be able to sustain the operations of the entity in the future. Organizational structure theories focus on how technology is adopted based on the strategic interests of organizations.

Organizational structure theories have various distinguishing attributes. To begin with, they recognize the fact that organizations have more resources and avenues to promote the adoption of new technology than individuals. Technologies such as the use of credit cards were developed and promoted by organizations that had their strategic interest in play. The same said for mobile money payments and electronic payments, which have become widely accepted in today's world of commerce. The power of organizations to develop and push for the adoption of new technology can be used to explain how some technologies are embraced by society.

In the context of blockchain technology, the theories are highly relevant. It has shown that blockchain technology has a variety of use cases where it will be a disruptive technology use. For instance, in the financial world, the technology can allow faster transactions since a third entity, such as a bank, is not needed to verify transactions. In healthcare, the system uses to securely store and share patient information for improved outcomes. The benefits that the technology brings to any use case are significant enough to be disruptive. In this regard, organizations can make strategic mission to embrace technology and promote its use among the people. In doing so, both individuals and organizations benefit. Therefore, organizational structure theories are still relevant today when discussing the adoption of new technologies.

2.2 Barriers to the Adoption of Blockchain Technology

2.2.1 Knowledge and Information About Blockchain

Blockchain is a relatively new technology when compared to other technologies that have similar use cases. As a result, there is limited awareness among both organizations and the general public. The limited awareness is due to the lack of sufficient knowledge and information about the technology. As pointed out by [23], the adoption and implementation of a new technology heavily depends upon the knowledge that the organizations and potential users have on the subject. Knowledge and information create an atmosphere where there is expertise to implement the technology successfully [23]. Additionally, awareness increases use cases and make it more likely that the general public will embrace the technology. Blockchain technology has suffered limited implementation because there is not sufficient knowledge and information on the subject.

2.2.2 Fear of Herd Behaviour

Herd behavior and the bandwagon effect are notions that have been studied extensively by researchers [3, 4]. Concerning how notions influence decision-making in organizations, herd behavior refers to the phenomenon where group of people choose to follow a popular idea or path even when their instincts tell them otherwise. The phenomenon usually leads to failure as an organization ends up making decisions that are not in line with its goals and current state [21]. The bandwagon effect refers to a situation where an individual or a group of people choose to join in a popular idea due to the fear of losing out. For instance, having an online presence especially among businesses was a popular idea in the late 90's when the internet had taken hold. However, not all businesses with an online presence succeeded [3]. In the case of blockchain technology, herd behavior and the bandwagon effect have been major barriers. Organizations fear embracing technology since they are not certain of it is permanence. They fear to lose out in the long-term after the technology has outlived its utility. As a result, there are fewer companies that have shown genuine interest in adopting blockchain technology.

2.2.3   Regulatory Challenges

Regulations are critical to the successful adoption and implementation of new technologies. Regulations provide a framework that evens out the playing field to allow businesses to compete fairly [32]. Additionally, a regulatory framework provides the necessary guidelines that organizations need to follow to adopt and use a new technology successfully. In the case of blockchain technology, regulation has been the main challenge. Due to its newness, bodies tasked with regulating various industries have not caught up with the benefits and challenges that technology brings [9]. For instance, in the banking sector, there are no regulations to dictate a form of blockchain technology that should be adopted and who will be in control of user data. Lack of sufficient regulation in digital currencies limits the adoption of blockchain technology, especially by large organizations [1]. The regulatory challenges that new technologies bring have been a hindrance to their successful adoption and implementation.

2.2.4   Privacy Protection and Security

The fame and recognition that blockchain technology has received stems from a number of its core features that are radically different from any other technology in existence. However, the core features that give the technology its strengths also serve as a barrier, especially when privacy protection and security issues are considered. Blockchain is a technology designed to be open and transparent [5]. The openness means that individuals in the chain have access to data stored in the blocks. The transparency increases trust since the authenticity of any data stored can be verified by a host of individuals. However, openness possesses a challenge to businesses that have traditionally thrived on limiting the access to data. For instance, banks limit access to information regarding user accounts and transactions to the users themselves and a few individuals [30]. Nevertheless, if blockchain technology used in the same uses case; the banks face the challenge of protecting user data privacy while remaining transparent enough to comply with the core values and virtues of blockchain technology. As a result, organizations that have relied on controlled access to user data have experienced challenges in adopting blockchain technology [13].

Secondly, blockchain technology is decentralized. There is no single individual or entity that has control over its operations. While such a feature is highly desirable among the early use cases of the technology, the decentralization poses a significant challenge to organizations that have traditionally thrived through centralized control. Governments provide an excellent example of entities that may be reluctant to adopt the technology due to its decentralized nature [11]. Governments have thrived by being in control of most of its activities. Blockchain technology threatens such as tradition by spreading control to entities beyond the government. As a result, it has been difficult for governments and other organizations relying on centralization to adopt blockchain technology.

Lastly, the privacy that blockchain technology brings with regards to the real identity of its users has also been a significant cause for concern among organizations. Illegal activities that exploit the anonymity provided by blockchain technology has significantly affected adoption [15].

2.2.5 Implementation Challenges

New technologies are fraught with a host of implementation challenges, and blockchain technology has also been a victim. As earlier pointed out, knowledge and information regarding the technology are not as extensive as desired. As a result, adopters face the challenge of understanding the several use cases of the technology and which approach is best to use during implementation [24]. Additionally, organizations such as banks already have systems in place that carry out most of the activities that blockchain technology seeks to replace. There is a challenge as to how well the new technology can be integrated to the old one and phase out the old one gradually [24]. Implementation challenges are a significant barrier since organizations lack the motivation to learn more about the technology and commit to using it in their day-to-day activities.

2.2.6 Beginning Costs

Cost is yet another challenge that comes with the adoption of blockchain technology. Research shows that the adoption and successful implementation of blockchain technology in contemporary organizations is a complex and novel affair [22]. The complexity is brought about by its newness and lack of sufficient knowledge and expertise to guarantee a successful implementation. Additionally, a significant investment in technology infrastructure is required to adopt and implement blockchain technology [24]. The said challenges increase the initial costs required to adopt the technology. It is unclear whether the implementation costs will reduce in the future, considering that the technology is still evolving and other factors that are not relevant today may become relevant tomorrow. Overcoming the initial costs will go a long way towards promoting adoption.

2.2.7   Anonymity

Blockchain technology provides users the ability to use pseudonyms rather than their real identity to carry out transactions. Since the architecture of the technology is open, the use of pseudonyms provides users with the privacy they require. However, it has been shown that privacy compromise through tricks, such as reporting one for an illegal activity where they will require to reveal their actual identity for investigation [22]. The fact that privacy on the platform may be compromised affects proposed use cases such as in banking sector where user account information cannot be shared with others.

2.2.8   Trust Issues

While blockchain technology has created a reputation for security and privacy, the privacy aspect has yielded a trust issue. Since individuals can transact anonymously on the platform, others have come to distrust the platform since it creates an environment for perpetuating illegal activities [2]. The fears confirm on various occasions where criminal elements such as hackers have used the platform to receive ransom for their illegal activities [34]. Therefore, organizations are hesitant to adopt the technology since they do not trust their privacy features to be exploited for illegal activities.

2.2.9   Scalability of the Technology

The design of blockchain technology limits the number of transactions that can be carried out within a minute for purposes of providing enough time for verification. At present, the two most popular blockchain platforms, Bitcoin and Etherium, allow 7 and 15 transactions per minute, respectively [24]. This limitation creates a scalability challenge, especially in industries where up to 1000 transactions can be handled in a minute. As a result, organizations in various industries have been hesitant to adopt the technology since it may not scale well to their needs in the future.

2.2.10   Energy Efficiency and Consumption

Blockchain technology relies on the computational power of computers in a network to carry out basic functions such as validating transactions and mining cryptocurrencies in cases where such is required. The activity leads to high energy consumption, especially when the network grows and more computational power is required to verify and increased number of transactions and mine cryptocurrencies [24]. The energy use is a significant concern, especially for industries that may handle thousands of transactions in a minute, such as the banking industry. Fears of being energy inefficient in the future have discouraged organizations from adopting the technology.

2.2.11 Usability

Current users of blockchain technology, such as users of Bitcoin, have praised its security features while critiquing its ease of use [1]. Blockchain is a sophisticated technology that may be challenging in normal transactions. For instance, records entered cannot be altered. Therefore, there is no easy way to reverse errors, such as banking errors, once they have made [14]. Secondly, private and public keys consisting of alphanumeric characters generated using sophisticated encryption algorithms, are required to validate transactions [23]. The use of such keys may prove difficult to a majority who have been used to simple steps, such as entering a simple password to validate a transaction. Therefore, the ease of using blockchain has been a hindrance to its adoption.

2.2.12 Storage Concerns

One of the security features of blockchain technology is that information or data replicates in all the nodes of the network. Consequently, it is not easy to lose data or compromise one node to affect the integrity of the whole network. While such a measure increases security and reliability, it also creates storage concerns. The replication of data takes up more space especially when it considers that information has to be stored in all the available nodes in the network [42]. Storage concerns discourage organizations from embracing technology since there is the possibility of poor scalability and inefficiencies in the future.

2.2.13 The Consensus of 51%

While blockchain technology touted for its security, there is the risk of compromising the whole platform through a 51% attack. Blockchain security arises from the fact that there is no one single node that controls the rest of the network. Information is duplicated in the chain, thereby, eliminating incidents where data is stolen or manipulated by people who gain control of a network or a database. When a single piece of data corrupt in the blockchain, it can easily correct through references to other nodes that have stored similar information as a security measure [42]. However, there is still the risk of gaining control of the whole network if one can amass 51% computing power on the platform. While such a fete is difficult, especially with an extensive network consisting of multiple computers, it is still a likely event [42]. The

51% consensus has dissuaded organizations from adopting the technology since they do not want to face the hazard of losing control of their network and data to criminal elements.

2.2.14   Maturity of the Technology

Blockchain technology also suffers from the fact that it is new. Initially, the technology was created to promote a limited number of online transactions by providing a digital currency acceptable to users online. However, current use cases were later proposed, such as those used in the bank and the healthcare industry. Since the technology is not designed with such use cases in mind, it lacks maturity related to how well it can meet the needs of the industry. As a result, organizations have been reluctant to adopt the technology since they fear that it may never meet their needs well, both in the present and in the future. At present, the maturity problem is addressed by creating new forms of blockchain technology that can meet the needs of contemporary nature. Due to its open nature, organizations can take technology and modify it to suit their needs without any repercussions. However, such an approach leads to fragmentation, which may affect collaboration in the future [24]. In this regard, organizations need to convince that blockchain is a mature technology to increase adoption.

2.3   State-of-Art of Blockchain Technology Adoption

Among the Middle East nations, United Arab Emirates (UAE) has been very active in pushing the technology adoption in various sectors of the country. Smart Dubai aims to function the government transactions on blockchain platforms [39]. The research conducted by a researcher affiliated to universities in Sweden and Denmark evidences that there had been attempts made by several Middle East countries including Jordan, Egypt, Qatar, Kingdom of Saudi Arabia, UAE, Kuwait and Oman on implementing blockchain technology in various sectors including banking, insurance, education, government/public sector and real estate. However, successful implementation is sparse. The study concludes figuring out the challenges and opportunities from an organization and nation focussed perspective versus regulatory and technology responses. There are four major factors that attribute to the challenges and opportunities namely the regulation or policy, education or awareness, collaboration and the culture [16]. Every country finds enablers and barriers to the adoption of a new technology like blockchain. Literature review apparently shows that the scope of adopting blockchain technology in potential sectors is huge.

2.4   Current State of Blockchain Technology in Oman

2.4.1   Blockchain Infrastructure

A government-owned entity named BSS was established in the Sultanate of Oman in November 2017. The purpose of establishment was to create awareness among the public about the benefits and uses of blockchain technology, as well as enable the Oman government to establish its blockchain trading currency. The BSS sought to launch initiatives to partner with local companies so that they would be willing to be part of the blockchain infrastructure that would be set up by the government. Introducing Blockchain technology in the Oman Capital Markets would require a shift in trend towards allowing the trade of digital currencies in the Sultanate. A lot of investors and stakeholders would have to be consulted and involved in this decision [36]. Above all, the government was determined to ensure that the citizens are well aware of the perceived benefits, and they get on board with the plan.

The first approach by the BSS was to establish symposiums throughout the country. These symposiums were funded by private sectors but represented by government agencies and authorities. The approach was to have all stakeholders present their memoranda on how to approach the transition to the blockchain [33]. The economic foundation of the Oman government has been on merchant trade. In order to establish a concise trading unit that would interest the country, it would be prudent to assign a digital value to the conventional trading entities [41]. The American investors who participate in blockchain transactions were invited in the blockchain symposiums that took place during 2018 and continue to date [29]. This would be of help to advise the public sectors on how the implementation of the technology has been successful in the United States, especially the adoption of the Bitcoin cryptocurrency.

2.4.2   The Proliferation of E-Governance Services

The approach by Oman government to ensure that the blockchain technology is fully adopted as soon as possible began with a clamor for e-government. The development of e-government services would require a conversion of value and trade in the multi-billion sectors to ensure a secured system of exchange between government and contractors. The need for blockchain is realized as a step to guarantee better and safer transactions that would similarly assign a value to the underlying government entities [41]. Government adoption of e-services and the use of blockchain would, in turn, lead to a national trend in the use of the cryptocurrency in the country. The choice of cryptocurrencies accepted by the government discusses over time, even with the potential of an Oman owned cryptocurrency were suggested [31]. The trend in ensuring a full transition to government e-services would make the country a leader in the proliferation of blockchain technology within the region.

E-government services ensure smooth and reliable access to aspects such as permits and licenses to trade with the government or the public. The digitalization of government services would require a secure process that would protect the interest of the public. The Oman government, under the BSS, is committed to using blockchain technology in the digitalization of its services. With the intention of setting-up a secure channel for communication and portal for a request of services, there was the necessity of advanced cryptic algorithms applied in commonly secure currencies such as Bitcoin and Ethereum. These currencies would form the basis for which the blockchain technology would be used in the country's digitalization efforts [37]. Likewise, the government is committed to allowing the public to voice their opinions and suggestions as to the transitioning of the government services to a modern electronic form that would reach all citizens faster and secured through blockchain technology [41].

2.4.3   A Crypto-Currency for Oman

Even though over ten different cryptocurrencies currently competing for market acceptance with the two main currencies; Bitcoin and Ethereum, the government should place and start its cryptocurrency. Currently, the nation is making the most significant strides in the Middle East concerning the adoption of blockchain technology [31]. In conjunction with India, the Oman government can develop a currency cantered mainly on government financial services. It may be a difficult and expensive venture but based on global trading dynamics; cryptocurrencies will be the future of currency. The opportunity for the government to join in the trend and gain from early exposure cannot be undermined. It is essential to take advantage of the opportunities in global cryptocurrency trading as the market demands continue to increase. Doing so ensures that the government of Oman secures its place among top trade currencies in the world.

Developing a cryptocurrency requires a lot of research, collaboration, policy and regulation, and data mining [31]. These are efforts that the Oman government is already taking; thanks to the establishment of the BSS. Perhaps, there is a need of more expertise on board. Regardless, the involvement of an influential government such as Oman in the mining of cryptocurrencies is a welcome trend among technology experts in the Middle East and Asia [41]. The government of Oman in creating interest among investors and indeed, raising the country's investment profile as the research on digitalization of government services continues. It is, however, essential to ensure that action is taken sooner so that the process of adopting a blockchain technology pioneer in the region. Oman needs to take charge of the process and ensure that they are bold in the decisions as cryptocurrencies have also been known to be risky. It is nonetheless a risk worth venturing in the current era [29].

3 Research Methodology

Researchers can employ several approaches to carry out a study and generate data necessary for analysis. The methodologies include qualitative approach, quantitative approach, correlational design, and quasi-experimental design among others. The appropriateness of a chosen methodology to the goals of a study is crucial towards a successful data collection exercise.

3.1 Research Design

A qualitative approach is chosen as the method of choice for the study. Qualitative research involves the collection of data from observations and human experiences with the goal of gaining insight into a topic of interest. There are several factors that make a qualitative research design appropriate for this study. The goal of the study is to establish barriers and enablers to the adoption of blockchain technology in Oman. The information to be obtained from participants will include their views and opinions based on either their experience or what they have observed over the time. A qualitative research approach is appropriate in such a situation where information to be gathered involves people's opinions and views regarding a subject.

Secondly, a qualitative research approach lends itself well to fluidity when it comes to both the analysis and interpretation of the collected data. There are a variety of tools that can be used to analyze data collected from qualitative research. They range from simple empirical deduction to sophisticated statistical analysis. It is usually the duty of the researcher to determine the appropriate data analysis tool or approach to use when evaluating qualitative data. The selection should base on the type of data that has collected and the goals of the research. The fluid nature of data analysis in qualitative research provides a leeway to focus on data collection without the need to worry about fitting data to a particular analysis tool. As a result, the approach will make it easy to gather as much useful information from the respondents and select an appropriate analysis methodology for evaluation.

Additionally, the fluidity seen in qualitative research makes it easy to capture context, especially in studies where the interpretation of the data is heavily reliant on the context. When gathering the views and opinions of people on a subject, the context is of importance. For instance, people loyal to a given government may have positive views with regards to its operations as opposed to those who have no attachment. As a result, it becomes crucial for researchers to maintain context in particular studies to improve the validity and reliability of their findings.

Lastly, qualitative research is appropriate in situations where it would not be possible to obtain empirical data. Empirical data always favored by researchers since it yields objective results and avoids bias that may result from people sharing their opinions or views. However, it is not possible to gather empirical data in all types of studies. As an example, it is a current study where the goal is to identify barriers and enablers to the adoption of blockchain technology in Oman. The barriers and enablers can be identified easily by seeking the opinions and views of individuals with significant experience or ties to blockchain technology. Therefore, a qualitative research approach is ideal for the proposed study and only qualitative data has been gathered from the respondents.

3.2 Sampling

The participants of the study were the individuals drawn from both the private and the public sectors. The goal was to identify participants with keen insight or experience in blockchain technology in Oman. The said individuals were to offer reliable views and opinions, which had been used as a basis for the findings of the research. Notably, private and public organizations with experience or ties to blockchain technology were identified. In Oman, the government has shown significant interest in the technology for the past five years. It has gone to the level of organizing a conference with the goal of exploring how blockchain technology can benefit both public and private organizations in the country. Therefore, the public organizations that were chosen are tied to the government and its efforts to increase the adoption of the technology among the public.

In the case of private sector organizations, it shows previously that blockchain technology can find use in industries such as healthcare, finance, and logistics management among others. In this regard, private sector organizations were chosen based on whether blockchain technology has a use case in their industry. The questions on whether they have adopted blockchain technology and what factors have either hindered or enabled the adoption were included.

After identifying the organizations, the next step was to identify individuals in the chosen organizations for being the actual participants. Individuals were selected based on their power to make decisions in the organizations. The adoption of blockchain technology is an initiative that will primarily stem from the decisions made by upper management. It is the CEOs and Directors of a company who will decide whether blockchain technology is of any benefit to them and whether they should use it. The said individuals, therefore, will have strong knowledge of why a particular entity had adopted or has failed to adopt blockchain technology. Therefore, the views on the subject will be reliable and insightful.

Lastly, the sampling process will conclude by an enumeration of all the individuals who have shown interest in taking part in the study. Due to time and resource constraint issues, the research targets to sample a maximum of 40 individuals and a minimum of 30. Therefore, 40 individuals will be selected from the pool of potential candidates to form the sample population. The criteria used during the selection involved selecting diverse people so that the collected data can be representative of the whole population.

3.3 Data Collection and Analysis

Data was collected through interviews. The sample population were interviewed to obtain their responses with regards to the barriers and enablers of blockchain technology in Sultanate of Oman. To begin with, a questionnaire to guide the interviewing process was created. While questions were being asked to every participant, the interviews slightly varied based on the follow-up questions raised to the participants and their responses to the base questions. Necessary permission had been obtained from the respective organizations to carry out the interviews. The process of obtaining permission involved creating a schedule for each of the interviews and signing the confidentiality agreement form. Finally, tools to be used during the interviews had been tested to guarantee that they function as desired. Data from every interview had been assembled separately for analysis.

4 Data Results and Analysis

The analysis began with the responses from the interviewees. As mentioned earlier, the goal of the study was to identify both public and private sector companies so that a broader perspective on the implementation of blockchain technology in Oman is realized. Among the seven companies identified, two of them were private sectors while the remaining five were government-affiliated organizations. In the course of selecting the companies, it was noted that fewer private companies were willing to participate in the interview compared to public entities.

The interviews were scheduled and conducted during July 2019. After obtaining the necessary permission and gathering the required tools, the interviews were carried out and lasted between 40 to 90 min. It is important to note that the people who took part in the interview for the selected organizations were not widely varied in terms of position and knowledge regarding the respective companies and blockchain technologies. All of them were high-level management who could be relied upon to provide credible and valuable information. They included CEOs, Deputy CEOs, Directors, Research analysts, and Application support specialists. In some interviews, there were more than one individual in the room. However, the questions were directed to a primary individual while the other members provided supportive information.

While the questionnaire used to guide the interviews contained a variety of questions, the interviews themselves guide by themes that reflected the objectives of the research. It included barriers to the adoption of blockchain technology, enablers, and effort that the Sultanate of Oman can make to increase the adoption and become a regional technological hub. Care was taken to ensure that the three themes were addressed in each interview. A summary of the responses provided by interviewees with regards to the mentioned themes is summarized in subsequent sections.

4.1 Analysis

The results of the study illustrated in Fig. 1 provide several key insights into the adoption of blockchain technology in the Sultanate of Oman in terms of barriers and enablers. The insights draw from the factors that either enable or hinder the adoption of blockchain technology in Oman. The first insight is that a few companies have attempted to adopt the technology. However, there are more barriers to adoption than enablers as observed. It is apparent that companies have faced different challenges while making an attempt in the adoption of the technology. Developing nations that have established reliable procedures for adopting blockchain technology have a significant number of organizations that have gained experience through past attempts [27]. The Sultanate of Oman will reach to such a level when more companies attempt to adopt the technology successfully.

The second insight is that there is little motivation among private companies to adopt blockchain technology compared to public organizations. This is evident from the willingness of participation in the interview. BSS has a key focus on developing blockchain technology solutions in the country and the region at large. Additionally, public organizations exhibited more knowledge and experience regarding the technology compared to private entities. In developing countries where the adoption of blockchain technology has been positive, the private sector has played a significant role in providing and influencing the general public to use the technology [36]. The Sultanate of Oman lacks a private sector that heavily invests in embracing new and disruptive technologies such as blockchain technology.

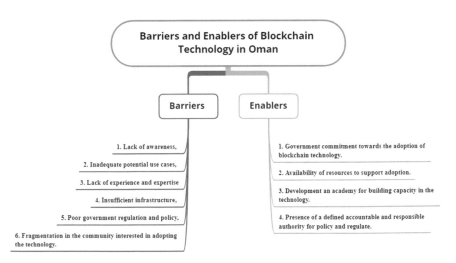

Fig. 1 Barriers and enablers of blockchain technology in Oman

4.1.1 Research Question (RQ) 1: What Are the Enablers to the Adoption and Development of Blockchain Technology in the Sultanate of Oman?

Establishing factors that acted as enablers to the adoption of blockchain technology was a key objective of the research. During the interviews, it was noted that among the selected organizations, majority participated very little in the adoption and implementation of a blockchain technology solution. As a result, knowledge of the enablers limits as compared to the barriers. It is not possible to reliably identify potential enabling factors unless an organization has sufficient experience in the implementation of a blockchain technology solution. Nevertheless, two critical enablers identified are the government commitment and the availability of resources.

Government commitment is the most commonly mentioned enabler to the adoption of blockchain technology. The organizations recognized the fact that the government of Oman has invested enough time and resources in making blockchain technology a reality in the country. An example of the government's commitment towards the endeavor is the staging of conferences geared towards increasing the knowledge of blockchain technology [8]. The government has held two significant conferences so far where guests from developing countries were invited to share their knowledge and experience on the technology. Another conference was held to bring key stakeholders together and chart out a path that will see the increased adoption of blockchain technology [8]. The government has plans to conduct more conferences in the future in order to check the past goals and monitor progress on adoption efforts. The government's commitment is evident in establishing a public organization under the information technology ministry to deal with issues regarding the adoption of blockchain technology. The organization serves as the go-to source for both public and private entities seeking to know more about blockchain technology or implement the desired solution.

The second enabler identified during the interviews is the availability of resources to support adoption. Compared to the Western world, blockchain is a relatively new technology in the Sultanate of Oman. In the past 3–5 years, resources that could aid adoption and implementation were limited [8]. However, with efforts from the government, public sector, and the private sector, there exists resources that can aid adoption, currently. For instance, ITA has developed a variety of frameworks that can aid implementation in various industries. While the frameworks are yet to receive sufficient testing, their availability motivated organizations to adopt blockchain technology.

4.1.2 RQ2: What Are the Barriers to the Adoption and Development of Blockchain Technology in the Sultanate of Oman?

Establishing key barriers that hindered the adoption of blockchain technology in Oman was a vital objective of the research. Several barriers were mentioned by the interviewees. The crucial barrier is lack of awareness. Blockchain is relatively a new technology even to the developed world [3]. Additionally, it's expanded use cases that have currently been its main attraction, such as in the financial and healthcare industries, have not been implemented on a full scale even in developed countries. Moreover, the knowledge people possess regarding technology is limited. As pointed out by [28], the general public has to be sufficiently aware of a new technology before they can be ready to embrace it once been implemented by public or private organizations.

During the interviews, both public and private sector organizations mentioned that there was minimal awareness of the technology in the Sultanate of Oman. The lack of awareness was at two levels. The first level was that of organizations and the other stakeholders who are required to aid the implementation of such a technology. A majority of public and private organizations in Oman do not have sufficient knowledge with regards to blockchain technology. They do not understand its practical use cases in their industries and the benefits they would derive from the technology, if implemented. As a result, they have little motivation to adopt the technology. The second level is the lack of awareness among the general public. Organizations exist to serve the interests of the public. The production of goods and services relies on the ability of the people to use/consume the said goods and services. Additionally, the needs of the population influence organizations to adopt new ways of doing things so that they can remain in-line with people's preferences. In the case of blockchain technology in Oman, the general public have little awareness. As a result, organizations are not motivated to adopt a technology that the public does not understand.

The second key barrier identified during the interviews was a lack of potential use cases. Blockchain technology was initially designed to act as a decentralized exchange of value on the Internet [3]. However, due to the desirable core attributes of the technology, such as security and transparency, it has found other potential use cases, such as exchange of value in the finance industry and storing of patient records in the healthcare industries. However, in the Sultanate of Oman, the interviewees mentioned that there were no potential use cases. The fact that Oman is still a young and growing economy can explain the lack of potential use cases. While developing countries currently face significant challenges that can be addressed by the use of blockchain technology, the Sultanate of Oman still has young and growing industries that are not well poised for disruption.

The third barrier identified was the lack of experience and expertise in the technology. This was especially the case for private companies that had shown an interest in adopting the technology and public companies that had taken part in increasing awareness of the technology. Interviewees mentioned that no company in Oman had implemented the technology on a reliable scale. Additionally, they mentioned that the organizations lack expertise in the area. For instance, the Director of a public organization mentioned that they are dissuaded from adopting blockchain technology because they could not find reliable frameworks for implementation and companies or individuals who could carry out the implementation and provide support services in the future. It was also mentioned that a few companies that were involved with technology were merely testing its practicality in limited areas. As a result, there is little expertise in blockchain technology available in Oman which discourages interested parties from adopting the technology.

The fourth barrier identified was the lack of sufficient infrastructure and public policy to support adoption. The Sultanate of Oman differs significantly from developing nations that have commenced the adoption of blockchain technology in various industries. One of the key differences is that the government in Oman has a significant influence on the establishment of infrastructure and policy to support the adoption of anything new. In other developed nations, the private sector plays a significant role in developing critical infrastructure and influencing policy to support the adoption of new technologies. While the government of Oman has committed to enabling the adoption of technology, there lacks sufficient infrastructure and policy to support adoption.

One area where policy and regulation were keys was the open-source nature of blockchain technology which required that data gathered over time be protected and used appropriately. The lack of regulation raises fear among organizations which deal with sensitive data from the public. Setting up a policy that would provide a standard of how the adoption will lead to data protection and security in the long-term was identified as a critical step towards encouraging adoption.

The final barrier identified was the fragmentation brought about by the lack of collaboration between interested parties and stakeholders in blockchain technology. As mentioned earlier, most companies involved with blockchain technology in Oman are primarily testing its practicality in a limited use case. Since different companies have different needs, there is fragmentation in the potential solutions developed for the technology. As a result, there is the concern that implementations may end up being incompatible in the future especially in the current world where technology has unified several practices and activities in organizations across varying industries. Addressing the fragmentation through a collaborative effort in developing the solutions will help allay the fears and promote adoption.

4.1.3 RQ3: What Are the Recommendations Suggested to Improve the Quality of Use and Wide Acceptance of Blockchain Technology?

Yet another objective of the research was to identify ways through which the Sultanate of Oman could increase the adoption of blockchain technology and become a regional hub. Three essential suggestions were obtained during the interviews. The first suggestion was the government's initiative to develop a policy that would guide the adoption of blockchain technology in the country. The role of the policy will be to provide a baseline for implementation and address fears relating to the adoption of the technology, such as the protection and control of user data. In most industries where blockchain technology has a potential for initiating a disruption, centralization of activities has been the norm. The centralization helps to make it clear who has control over user data and who is responsible for fair use. However, with the decentralized nature of blockchain technology, the protection and control of user data is a vital issue. Setting up a policy will help alleviate such fears and encourage adoption. Additionally, policy changes can be used to address future problems that may result from the use of technology.

The second suggestion was that there should be initiatives to create knowledge and expertise in blockchain technology. Organizations rely on the availability of knowledge and expertise in a given technology so that they can implement it reliably [17]. The availability of knowledge and expertise means that organizations can depend upon individuals to implement required technologies and provide the needed support in the long-term. The Sultanate of Oman lacks companies that can provide blockchain technology solutions for a variety of industries and aid in implementing and providing support in the future. It also lacks developers who can provide niche solutions, such as the development of applications that can be used by the general public to access services build under blockchain technology. Efforts to increase knowledge, expertise, and experience in the country will go a long way towards improving adoption and making the country a regional hub for technology.

The final suggestion provided during the interviews was that the government should provide sufficient infrastructure and resources to support the adoption of blockchain technology. Infrastructure and resources required to adopt blockchain technology includes reliable and robust internet and communication networks and increased awareness on the subject. The government should educate the masses on the technology so that they can provide an incentive for organizations to adopt the technology. Taking such an approach will increase the adoption and transform Oman into a regional hub for technology.

4.2 A Framework to Aid the Adoption of Blockchain Technology

The results of the study, coupled with the identified insights provide a basis that can be used to develop a framework that aids the adoption of blockchain technology in the Sultanate of Oman. Additionally, the framework can help transform the country into a regional hub for technology. The framework rests on three key pillars that have been informed by the findings of the study (Table 1). **The First Pillar** is the setting up of a policy that will guide the adoption of blockchain technology by both private and public entities. The policy will help address fears associated with blockchain technology, such as the protection and control of user data, and provide a standard that can be adopted by all organizations.

The Second Pillar is the provision of infrastructure and resources to support the adoption of blockchain technology in the country. It is evident that, unlike the case in developed countries where the private sector is at the frontline of new technologies, the government of Oman holds much power in the direction that the country should take regarding new technologies. In essence, both the public and private industries

Table 1 Three pillars of adoption framework

Pillars	Importance
Setting up a policy that will guide the adoption of blockchain technology	The policy will address fears associated with blockchain technology, such as the protection and control of user data, and provide a standard that can be used as a basis for adoption by all organizations
Provision of infrastructure and resources to support the adoption of blockchain technology	The government of Oman holds significant influence in the adoption of new technologies. Therefore, it should be at the forefront of providing the necessary infrastructure and resources to encourage adoption
Increase the awareness of blockchain technology in the country	It will increase knowledge and expertise about blockchain technology in the country, thereby encouraging adoption

rely on the government's approval to adopt transformative technologies. The government achieves such a goal by providing the necessary infrastructure and resources to support the new trend. While the government has already shown its commitment so far, more efforts shall be directed towards creating the necessary infrastructure and developing the required resources to aid adoption.

The Third Pillar is the increase of awareness and the development of knowledge and expertise in the country. Lack of awareness, knowledge, and expertise are cited as one of the critical barriers to the adoption of blockchain technology. Both private and public sector organizations lack individuals whom they can rely upon to implement the technology and provide the desired support in the future. While past conferences have gone a long way towards welcoming experts to the region, the transfer of knowledge and expertise however, yet to happen on a reliable scale. In this regard, there is a need for the creation of awareness and the development of knowledge and expertise to aid in adopting the technology. Figure 2 illustrates the framework of prioritization for business processes. The suggested framework informed by the barriers identified in the study, enablers, and the suggestions provided for transforming Oman into a regional hub for technology is depicted in Fig. 3. The framework will help Oman achieve significant strides in adopting blockchain technology.

4.3 Where Can Blockchain Technology Be Adopted?

The findings of the study provide insight into possible sectors where blockchain technology can be adopted. It is a fact that the adoption by every sector would depend upon the issues addressed by government that serve as barriers to adoption. One of the main sectors is the financial industry. Banking and finance rely upon the accurate recording and storing of user information, and the facilitation of faster transactions. At present, blockchain technology provides an unprecedented level of security and

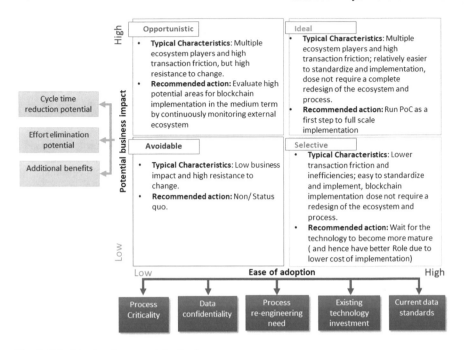

Fig. 2 Prioritization framework for business processes

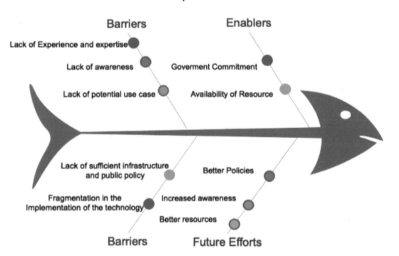

Fig. 3 Fishbone diagram for illustrating the barriers, enablers and future efforts

reliability which can protect information stored by the banks. It also provides faster transaction times which will benefit constomers. Therefore, the finance industry is one sector that can benefit from the adoption of blockchain technology. However, the government needs to put appropriate regulations in place that will provide guidelines for the control of data stored in blockchain.

The second sector is the healthcare. Healthcare sector can make a significant stride by adopting blockchain technology as a way to store patient information securely and promote easy access when needed. However, the government still needs to implement policies that guide how such data can be controlled to encourage healthcare facilities to adopt the technology.

Another promising sector in the country is the logistic and supply chain industry. In the last five years, the government focused and established many subsidiaries of the ASYAD group. In the present business world, information of an organization's exchanges is overwhelmingly secretly put away and, as a rule, no ace record of all exercises is accessible. On the other hand, this data is regularly dispersed crosswise over internal capacities and specialty units. For instance, the frequent stock exchange executed in microseconds, the exchange of the stock can take any longer, basically in light of the fact that the gatherings engaged with the stored network have no entrance to one another's records and cannot naturally check are the advantages, indeed, possessed and can be moved through hyperledger platform which is an open-source implementation of spread ledger framework [7].

5 Conclusion

Blockchain represents one of the most revolutionary technologies to have come out of the twenty-first century. The technology has become a common trend in the developed world. Businesses and individuals have realized the benefits they can derive from technology that no other innovation can offer. The said benefits include faster transactions, transparency, and improved security. However, the adoption of the technology has remained behind despite its usefulness and a variety of use cases.

In this study, the adoption of blockchain technology in the Sultanate of Oman is investigated. The study targets to increase adoption by proposing a framework informed by the current enablers and barriers of technology in the country. It was observed that the main barriers are lack of awareness, lack of potential use cases, lack of experience and expertise, lack of sufficient infrastructure, weak government regulation and policy, and fragmentation in the community interested in adopting the technology. The enablers included government commitment towards creating more awareness and providing the necessary infrastructure. The framework proposed for blockchain adoption based on three pillars include firstly, the government should develop a policy to guide the adoption of blockchain in the country secondly, create a pool of expertise in blockchain technology and provide awareness to the community and finally, the government should provide sufficient infrastructure and resources to support the adoption of blockchain technology.

Future research investigation is necessary to validate a conclusion that can be drawn from this study. However, future research shall examine strategically in policy and regulation of implementing cryptocurrency also in the country. In this context, future research could examine the possible way of decreasing the barriers of adopting blockchain technology and how one can utilise the available enablers to create the awareness among the people. The framework developed as a research outcome of the proposed study shall aid the adoption of blockchain technology, and how the country can become a regional hub for technologies of 4th Industrial Revoution and attract forging investments and professionals to establish business and experiment potential use cases in Oman.

References

1. Abramova, S., Böhme, R.: Perceived benefit and risk as multidimensional determinants of bitcoin use: a quantitative exploratory study. In: 2016 International Conference on Information Systems, ICIS 2016, (Zohar 2015), pp. 1–20 (2016). https://doi.org/10.17705/4icis.00001
2. Avital, M., Beck, R., King, J.L., Rossi, M., Teigland, R., Kursh, S.R., Gold, N.A.: Jumping on the blockchain bandwagon: lessons of the past and outlook to the future. Bus. Educ. Innov. J. **8**(2), 1–12 (2018). https://doi.org/10.1007/s00203-007-0327-5
3. Avital, M., Beck, R., King, J.L., Rossi, M., Teigland, R.: Jumping on the blockchain bandwagon: lessons of the past and outlook to the future. In: 2016 International Conference on Information Systems, ICIS 2016, pp. 1–6 (2016)
4. Banerjee, A.V.: A simple model of herd behavior. Q. J. Econ. **107**(3), 797–817 (1992). https://doi.org/10.2307/2118364
5. Beck, R., Czepluch, J.S., Lollike, N., Malone, S.: Blockchain-the gateway to trust-free cryptographic transactions. In: ECIS 2016 Proceedings, pp. 1–15 (2016)
6. Böhme, R., Christin, N., Edelman, B., Moore, T.: Bitcoin: economics, technology, and governance. J. Econ. Perspect. **29**(2), 213–238 (2015). https://doi.org/10.1257/jep.29.2.213
7. Cachin, C.: Architecture of the hyperledger blockchain fabric. In: Workshop on Distributed Cryptocurrencies and Consensus Ledgers (DCCL 2016) (2016). https://www.zurich.ibm.com/dccl/papers/cachin_dccl
8. CBO: Financial Stability Report. Muscat (2018)
9. Christopher, C.: Whack-A-Mole: Why Prosecuting Digital Currency Exchanges Won't Stop Online Money Laundering, pp. 1–36 (2014)
10. Comin, D., Hobijn, B.: Cross-country technology adoption: making the theories face the facts. J. Monet. Econ. **51**(1), 39–83 (2004). https://doi.org/10.1016/j.jmoneco.2003.07.003
11. Dhillon, G.: Money laundering and technology enabled crime: a cultural analysis. In: AMCIS 2016: Surfing the IT Innovation Wave - 22nd Americas Conference on Information Systems, pp. 1–9 (2016)
12. Disspain, C., Szyndler, P.: Navigating the multi-stakeholder morass: the past, present and future of internet governance. Telecommun. J. Aust. **63**(3), 1–12 (2013). https://doi.org/10.7790/tja.v63i2.431
13. Egelund-Müller, B., Elsman, M., Henglein, F., Ross, O.: Automated execution of financial contracts on blockchains. Bus. Inf. Syst. Eng. **59**(6), 457–467 (2017). https://doi.org/10.1007/s12599-017-0507-z
14. Fabian, B., Ermakova, T., Sander, U.: Anonymity in bitcoin - the users' perspective. In: 2016 International Conference on Information Systems, ICIS 2016, pp. 1–12 (2016)
15. Fabian, B., Ermakova, T., Sander, U.: Anonymity in Bitcoin the users' perspective e-mail and web tracking empirical analysis, privacy impact, countermeasures view project privacy

policies-readability, content, impact view project. In: ICIS 2016 Proceedings, pp. 1–12 (2016). Accessed on https://aisel.aisnet.org/cgi/viewcontent.cgi?article=1040&context=icis2016% 0Ahttps://www.scopus.com/inward/record.uri?eid=2-s2.0-85019460049&partnerID=40& md5=0d6a90ae395672cd63d5efa6f8f35f22

16. Ghazawneh, A.: Blockchain in the middle east: challenges and opportunities. In: MCIS Proceedings, pp. 1–13 (2019). Accessed on http://www.itais.org/ITAIS-MCIS2019_pub/ITA ISandMCIS2019-pages/pdf/85.pdf

17. Goldenfein, J., Hunter, D., Brennan, D., Dyball, A., Hawthorn, R., Kingsley, J., Wan, A., et al.: Blockchains, orphan works, and the public domain. Colum. J.L. & Arts, **1**, 41 (2017). Accessed on https://perma.cc/4B6X-9ZUD

18. Jackson, T., Daly, H., Mckibben, B., Robinson, M., Sukhdev, P.: Prosperity Without Growth Economics For a Ainite Planet, 1st edn. Earthscan, London (2009)

19. Kamble, S., Gunasekaran, A., Arha, H.: Understanding the Blockchain technology adoption in supply chains-Indian context. Int. J. Prod. Res. 1–25 (2018). https://doi.org/10.1080/002 07543.2018.1518610

20. Lai, P.: The literature review of technology adoption models and theories for the novelty technology. J. Inf. Syst. Technol. Manag. **14**(1), 21–38 (2017). https://doi.org/10.4301/S1807-17752017000100002

21. Leibenstein, H.: Theory of consumers' demand. Q. J. Econ. **64**(2), 183–207 (1950)

22. Manski, S.: Building the blockchain world: technological commonwealth or just more of the same? Strateg. Chang. **26**(5), 511–522 (2017). https://doi.org/10.1002/jsc.2151

23. Mori, T.: Financial technology: blockchain and securities settlement. J. Secur. Oper. Custody **8**(3), 208–227 (2016)

24. Mougayar, W.: The Business Blockchain: Promise, Practice, and Application (2016)

25. Norta, A., Vedeshin, A., Rand, H., Tobies, S., Rull, A., Poola, M., Rull, T.: Self-aware agent-supported contract management on blockchains for legal accountability. https://Whitepaper. Agrello.Org/Agrello_Self-Aware_Whitepaper.Pdf, (Cc), pp. 1–35 (2017). Accessed on https:// docs.agrello.org/Agrello-Self-Aware_Whitepaper-English.pdf

26. Oliveira, T., Martins, M.F.: Literature review of information technology adoption models at firm level. Electron. J. Inf. Syst. Eval. **14**(1), 110–121 (2011). https://doi.org/10.1017/CBO978 1107415324.004

27. Ølnes, S.: Beyond bitcoin enabling smart government using blockchain technology. In: Scholl, H.J., Glassey, O., Janssen, M., Klievink, B., Lindgren, I., Parycek, P., Sá Soares, D., et al. (eds.) Electronic Government, pp. 253–264. Springer International Publishing, Cham (2016)

28. Patrick, Y., Chau, K.: Information technology acceptance by individual professionals: a model comparison approach. Decis. Sci. **32**(4), 1–14 (2001)

29. Queiroz, M.M., Fosso Wamba, S.: Blockchain adoption challenges in supply chain: an empirical investigation of the main drivers in India and the USA. Int. J. Inf. Manag. **46**, 70–82 (2019). https://doi.org/10.1016/J.IJINFOMGT.2018.11.021

30. Rainer, Christin, N., Edelman, B., Moore, T.: Bitcoin: economics, technology, and governance. J. Econ. Perspect. **29**(2), 213–238 (2015). Accessed on http://www.jstor.org/stable/24292130

31. Rashid, S.K.: Potential of Waqf in contemporary world. J. King Abdulaziz University, Islamic Econ. **31**(2), 53–69 (2018). https://doi.org/10.4197/Islec.31-2.4

32. Risius, M., Spohrer, K.: A blockchain research framework: what we (don't) know, where we go from here, and how we will get there. Bus. Inf. Syst. Eng. **59**(6), 385–409 (2017). https:// doi.org/10.1007/s12599-017-0506-0

33. Saadeh, Y.: The effects of blockchain implementation on cyber risks mitigating strategies in the financial sector in the United Arab Emirates, pp. 1–121 (2018). Accessed on https://bsp ace.buid.ac.ae/bitstream/1234/1239/1/2014303006.pdf

34. Sas, C., Khairuddin, I.E.: Design for trust: an exploration of the challenges and opportunities of bitcoin users. In: Conference on Human Factors in Computing Systems - Proceedings, 2017-May, pp. 6499–6510 (2017). https://doi.org/10.1145/3025453.3025886

35. Straub, E.T.: Understanding technology adoption: theory and future directions for informal learning. Rev. Educ. Res. **79**(2), 625–649 (2009). https://doi.org/10.3102/0034654308325896

36. Sullivan, C., Burger, E.: Blockchain, digital identity, e-government. In: Business Transformation Through Blockchain, pp. 233–258 (2019). https://doi.org/10.1007/978-3-319-990 58-3_9

37. Tarhini, A., Arachchilage, N.A.G., Masa'deh, R., Abbasi, M.S.: A critical review of theories and models of technology adoption and acceptance in information system research. Int. J. Technol. Diffus. **6**(4), 58–77 (2015). https://doi.org/10.4018/IJTD.2015100104

38. Venkatesh, V., Thong, J.Y.L., Xu, X.: Consumer acceptance and use of information technology: extending the unified theory of acceptance and use of technology. Manag. Inf. Syst. **36**(1), 157–178 (2012). https://doi.org/10.2307/41410412

39. WASS, S.: Dubai launches platform to push blockchain adoption (2018). Accessed on MENA website: https://www.gtreview.com/news/mena/dubai-launches-platform-to-push-blockchain-adoption/

40. Yli-Huumo, J., Ko, D., Choi, S., Park, S., Smolander, K.: Where is current research on blockchain technology?—A systematic review. PLoS ONE **11**(10), e0163477 (2016). https://doi.org/10.1371/journal.pone.0163477

41. Zamani, E., He, Y., Phillips, M.: On the security risks of the blockchain. J. Comput. Inf. Syst. **60**(6), 495–506 (2018). https://doi.org/10.1080/08874417.2018.1538709

42. Zohar, A.: The myths, the hype, and the true worth of bitcoins. Commun. ACM **58**(9) (2015). https://doi.org/10.1145/2701411

Recent Advancement and Challenges in Deep Learning, Big Data in Bioinformatics

Ajay Sharma and Raj Kumar

Abstract More data have been produced in recent years than in the thousands of years of human history. This data represents an important gold mine for policymakers in terms of commercial value and reference material. But much of this value is untapped, worse, wrongly comprehended as long as it is impossible to use the tools needed to process the stunning amount of information. In this book chapter, we will examine how machine learning can give us a glimpse of the patterns in Big Data and obtain key information in all fields of biology, healthcare. An analysis, ineffectiveness storage, and depth of learning algorithms in this field are essential to the electronic equipment which generates an anonymous scale of data referring to diversity and veracity. The architecture of Hadoop-based maps and deep learning algorithms like Convolutional Neural Network, Recurrent Neural Network have transformed the way we analyze massive data. The role, impact, and prospect of deep learning algorithms, reinforcement learning to manage big data in the area of bioinformatics, computer aided drug design, structural biology and computational biology are discussed in this book chapter. In last section author has discussed about the role of deep learning in next generation sequencing, biomedical image processing and drug discovery and molecular modelling and dynamics studies.

Keywords Big Data · Healthcare · Deep learning · Reinforcement learning · Convolutional neural network (CNN) · Map-reduce · Hadoop · Bioinformatics · Structural biology · Drug design

1 Introduction

The data is everywhere as the time pass and become history and to keep history in some organized manner human being store data somewhere. The data is in an unstructured, unprocessed form. A human being applies some technique to extract some information from this data to enhance his understanding of that particular field.

A. Sharma · R. Kumar (✉)
Departmetn of Biotechnology and Bioinformatics, Jaypee University of Information Technology (JUIT), Waknaghat Solan 173234, Himachal Pradesh, India

© The Author(s), under exclusive license to Springer Nature Switzerland AG 2022
K. R. Ahmed and H. Hexmoor (eds.), *Blockchain and Deep Learning*,
Studies in Big Data 105, https://doi.org/10.1007/978-3-030-95419-2_12

As biological problems/disease remains the major culprit of human life and humans try to understand the mechanism behind these natural phenomena. Humans begin, with the help of artificial techniques, to use the computer to understand the biological problem. Advances have been made in every field, including biotechnology, medical, life sciences, and IT. During a biological experiment of microarray, sequencing lot of data has been generated by the NGS sequencer machines. To use the enormous amount of data we need some sort of system to store, analyze it, so we later see in this context how the field of biotechnology and medical science is transformed by machine learning [1, 2]. In the current book chapter, we will learn about big data and deep learning. Let's take a brief introduction to machine learning, Deep Learning, and big data analysis and their use cases in the field of bioinformatics and other medical science fields.

Big Data: The data which could be unstructured or structured and generated on the day to day basis from the various recourses with the help of manual feeding to a system or via some IoT electronic sensor (heat, temp, pressure, etc.) and having the 5V's properties (volume, variability, veracity, velocity, value) are referred to as the big data. In the 1990s first time John Mashey coined the term big data but it gains the momentum in early 2000 when the computational power is going on increasing and more and more data is generated from various resources (e.g.: Facebook, YouTube, NCBI (National Centre for Biotechnology Information), PDB (Protein Data Bank), DDBJ (DNA Data Bank of Japan))etc. Later we will learn more about big data that how it is used in bioinformatics [3]. In the section no 2 you will learn more about how big data is important in bioinformatics, how the different V's (**Volume, Velocity, Variety, Veracity, Value**) of big data is related with the bioinformatics.

Machine Learning: Machine learning (ML) is a subset of artificial intelligence that makes the system learn from the input data (Features) and able to user to build, train models. Based on these models one can make the prediction of the unknown data sets. The typical ML models are generally classified into supervised, unsupervised, semi-supervised learning. Machine learning is an old concept and Arthor Samuel coined this phrase during 1952. Since 1950 and till 1990 we don't have so much computational power and don't have access to data. We see a great emergence in ML algorithms as the internet become popular during the late1990s and early 2000s and people have the access to the data, the great improvement of hardware by the semiconductor industry. In this book chapter, we will learn how ML algorithms will be used to handle biological data.

2 Big Data in Bioinformatics

Bioinformatics is a multidisciplinary field comprises of computer science, mathematics, biotechnology, pharmaceutics, molecular biology. The objective of bioinformatics is to develop a computational based solution for the storage, analysis of data being generated by the biological experiment. In the omics world there are Genomic, Protcomics, and Metabolomics, Genomics is a new field of science deals with the sequencing and to study the genetic makeup of the entire genome of an organism. The genomic sequences are used to study the function of a gene (functional genomics), compare the sequences with another organism for the evolutionary genetics' analysis (comparative genomics), structural genomics to compare or generate new structure of a protein. Proteomics deals with the task performed in the cell. In a cell the complete set of proteins is called proteomics and what is the function of every cell in the body. The major goal of proteomics is to understand how the structure and function of the protein. As scientists were gaining more and more insight into the human/animal body the molecular biology concerns between the cells, molecular synthesis, modification and interpretation of different mechanism in the body. If we talk about the central dogma of life then molecular biology deals with how the DNA is transcribed into the RNA and then to proteins. In the human body there a number of proteins are responsible for the different function via activation/overexpression, suppression/under expression of genes. As the technology emerges the more data is generated by researchers/scientist during the experiment and to analysis, store this information we need some kind of solutions as a result of algorithms, big data, Hadoop, databases and other kind of programming language to adhere this problem.

Data consists of computer-formatted numbers, words, measures, and observations. Big data means vast, structured, or unstructured sets of these data. For traditional data-processing software, the digital era presents some challenges: Information is available in such volumes, speeds, and veracity, that computations are centrally focused on people, researchers are not aware of that. There are five V's found in the data, we can describe it as big data: volume, variety, veracity, velocity, value. Volume refers to the scale of data available, velocity is the rate at which data is collected, variety refers to the various sources from which this data comes in the above three two other Vs are often added: The veracity refers to the coherence and certainty in the data obtained, while the value measures the utility of the data extracted from the data received. Good analytical data requires people with business skills, programming knowledge, extensive mathematical and analytical skills. But how can traditional techniques settle millions of credit card scores or billions of transactions that takes place at various banks? How to find, store and retrieve the genomic data of living organisms. How the protein structure predictions are taking place and how the proteins ligand data stores. How to study the protein–ligand interaction. These are some of the mysterious and complex questions that need to be answered. For to answer these problems machine learning, deep learning, and big data come into place for rescue scientists, researchers, and other data science developers, analytics engineers [3, 4]. With the use of big data researchers, scientists were able to solve the problem related to the storage and retrieval of biological experimental data. Here we will learn in detail about the different V's used in Big Data analytics.

The 5Vs of Big Data:

As we discuss the biology of the data here, we will understand the use of 5Vs in bioinformatics in this book chapter. The different researchers provide the following Big Data 5Vs: **Volume, Velocity, Variety, Veracity, Value**. The volume refers to how much data the experiments generate. Velocity discusses the frequency with which

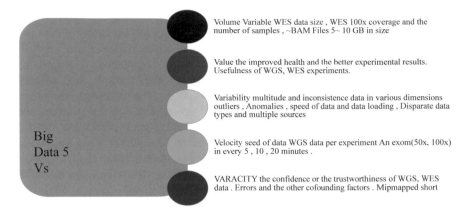

Volume Variable WES data size , WES 100x coverage and the number of samples , ~BAM Files 5~ 10 GB in size

Value the improved health and the better experimental results. Usefulness of WGS, WES experiments.

Variability multitude and inconsistence data in various dimensions outliers , Anomalies , speed of data and data loading , Disparate data types and multiple sources

Velocity seed of data WGS data per experiment An exom(50x, 100x) in every 5 , 10 , 20 minutes .

VARACITY the confidence or the trustworthiness of WGS, WES data . Errors and the other cofounding factors . Mipmapped short

Big Data 5 Vs

Fig. 1 5V's of Big Data used in big data analytics in bioinformatics

new data is created and stored efficiently. Since we have a variety of experiments, the heterogeneity of the data is different. Veracity, the "truthiness" or "disorderliness" of the biological experimental data. Value, the significance of data, and the outcome result that we got from here. Figure 1 shows the different types of V's are in Big Data analytics [5, 6].

2.1 Volume

In the world of big data, you are not, unless exabytes, petabytes, and more are the volume of data. Giants of big data technology such as Amazon, Google, Microsoft, and other e-commerce platforms are getting real-time, structured, and unstructured data from millions of customers around the world, especially smartphone users, lying each second between terabytes and zettabytes. They are doing almost real-time data processing and take decisions to deliver the best customer experience after running machine learning algorithms to analyse big data. In similar cases, a biological experiment like NGS of an organism generates a tremendous amount of data which is in sometimes GB, or terabytes. The data could be of different samples hence biological experiment generates a huge amount of data which refers to the volume in terms of big data.

Volume as a problem and what is the solution?

A typical human genome sequence is around 200 GB and supposes you want to sequence a genome of the human population of which is infected with the COVID-19 pandemic and we took a sample size of 100 people which is approximate ~10 TB. For simplicity just imagine is around least $300 is on a decent 10 TB hard drive. To manage a data petabyte of 100 × 300 USD = 30 thousand USD. You might receive a discount, but you are more than $10,000 in storage costs alone, even with

50% discount. Imagine just keeping a redundant disaster recovery version of the data. More disc space you would need. This makes the volume of data difficult to store on local storage devices if it goes beyond the normal limitations and becomes inefficient and costly way. Hence can imagine that volume is a great challenge and it is not restricted to only biological experimental data, there are other resources like drug design experiments or high throughput sequencing experiments which have the potential to generate enormous data which needs to be efficiently and effectively analysed. The appropriate solution for this problem is the use of popular options for storage is Amazon Redshift, a cloud-managed data warehouse service from AWS. It stores information distributed across several nodes that are disaster-resilient and quicker for calculations compared to relation-bases on the ground such as Postgres and MySQL. Data from relational databases to Redshift can also easily be replicated without downtimes.

2.2 Velocity

Velocity refers to the rate at which the data is being generated in biological experiments. Imagine a machine learning service that continually learns from a data stream for the cancer patients or any other NGS sequencing experiments platform with thousands of thousands of samples or long/short reads in the form of images or text (genomics sequence BAM, BAD file) 24 × 7 × 365 and upload it on the analytics server. In each second, millions of transactions occur, which means that petabytes of data are transmitted every interval of time to a data centre from one or more devices. This high-volume data inflow rate per second defines the data speed.

Velocity is a problem, what is the appropriate solution?

Speed*time = volume and volume lead to visions, and visions lead to more money. However, it is not without its costs that this path to increasing income. Many questions arise: how do you treat the maliciousness of every data packet coming through your firewall? How do you handle such structured and unstructured high-frequency fly data? In addition, if you have high data speeds, it almost always means that the amount of data processed every second will swing considerably. The solution for this problem comes in "streaming data" that have emerged which handle the data at the time of the beginning of the data. The Apache company has popular solutions such as Spark and Kafka, where Spark is excellent as well for batch processing as streaming, Kafka runs a publishing/subscription mechanism. Amazon Kinesis is also a solution that has several related APIs for streaming data processing. A further popular serverless function API is Google Cloud Functions (Google Firebase also includes). These are all excellent black-box solutions for the management of complex payload processing on the experimental data, however, they all require time and effort to construct data pipelines to process for analysis.

2.3   Variety

The real world is messy because of different types of data so it makes sense to deal with messy data as well. We have different type of biological data which is generated form the all the experiments are as text, images. Heterogeneity of data is frequently a source of stress in the construction of a data store. These heterogeneous data sets have a big challenge for the biological researcher, data scientist, and students in the analysis of large data.

Variety as a problem, what is the solution?

When a great volume of data is being consumed by the analyst, it can be used to massage the databases in a data warehouse with a variety (JSON, YAML, CSV ($x =$ C (comma), P (pie), T (ab), etc.) and a uniform type of data (XML). Data processing becomes even more painful when columns and keys, such as rebutting, introducing, and/or deprecating keys in the API, are not always guaranteed to exist. So not only one tries to uniformize a wide range of data types, the data types may differ from time to time. The solution for this problem is any transformation milestone applied to this data along the pipeline is recorded in a variety of data types. To begin with, store the raw data in a data storage cluster (a data cluster is a hyper-flexible repository of data collected and kept in its rawest form, like Amazon S3 file storage). Then transform the raw data with various types of data into a certain aggregate and sophisticated state which can be stored elsewhere in the data storage. Further, this is loaded to a relation data database or data warehouse.

2.4   Veracity

The data are so dynamic in the real world that it is difficult to know what is right. Truthfulness refers to the trustworthiness or confusion of data, and if the trustworthiness of the information is increased, the confusion is reduced. Veracity and value define the data quality together, which gives data scientists great insight.

Veracity is a problem, what is the solution?

In the case of NGS sequencing, all this data is very noisy, and the volume of data also increases with the lot of information, which may at times be exponential. Think about experiments of Sequences of different strains of SARS-COV2 in different countries and different populations with different sample sizes. In the NGS sequencing or case of drug design experiments like protein–ligand docking and, molecular modelling dynamics studies of the different simulations of protein and ligand, there is noisy data the noise reduces the overall quality of data that affects data management. In a typical ML, we have to consider so many feature vectors related to the model building which unnecessarily increases the computational complexity of the experiments in becoming difficult to build, train up the ML model. The solution for this problem is, if the data is not trustworthy enough, only high-value information must be extracted as the collection of all the data that you can do not always make sense because it is costly and requires further effort hence we can remove the data unwanted and select only those features which are use full in model building. Data processing pipeline filtering noises from the data while extracting them as soon as possible. This allows you only to process and load required and reliable data for the analysis of data.

2.5 Value

Value is very important in big data. It is useless until we can transform big data into something worthwhile. The costs of resources and the effort in big data collection and their value at the end of data processing are very important to understand. Value is very important since it has an impact on business decisions and a competitive edge. Biological database repository like NCBI (National Center for Biotechnology Information https://www.ncbi.nlm.nih.gov/) provide the information and access to the biological and genomic information, PDB (Protein Data Bank) https://www.rcsb.org/ is database which store the 3D annotation of all the protein, nucleic acid and complex assemblies that are generated by the X-ray crystallography and NMR (Nuclear Magnetic Resonance), Drug bank https://go.drugbank.com/ contains all the drug related molecules (chemical, pharmacological and pharmaceutical data with drug targets.) which is maintained by the University of Alberta. Consider NCBI, Drug bank, PDB, which collects data from various data sources, data visualisation, browsing patterns are extracted and transformed into the data processing pipeline only to generate high-value information which provides corresponding user interests in useful recommendations. In turn, this helps these biological servers to avoid the difficulty that of browsing the data to users and to the database. If the user was not satisfied, the resulting information could have been of low value. The value of big data, therefore, has an impact on many research-related decisions and a competitive advantage over others.

3 Proposed Model for Map-Reduce Architecture Hadoop Cluster

In Sect. 2, you have read about the Big data V's. Now we have experimental data. Have you thought about that how are these files were stored in the cluster of big data? The answer is a MapReduce architecture which is a data processing tool that is used in a distributed form to process the data in parallel. The paper, titled "MapReduce: Simplified Data Processing on Large Clusters," was developed in 2004 and published by Google [7, 8]. The MapReduce is a two-phase paradigm, the mapper and the reducer. The input is provided as a key-value pair in the Mapper. The Mapper's output is supplied to the input reducer. After the Mapper is over, the reducer only works. The reducer also uses key-value format input, and the reducer output is the final output. The reducer phase occurs after the mapper phase is complete, as the MapReduce name suggests. The first is the map job, which reads and processes an array of data for producing key-value pairs. The mapper/map job output (key-value pairs) is entered in the reducer. The reducer is supplied by a number of map jobs to the key-value pair. The reducers then add these intermediate tuples to a smaller set of tuples or key-value pairs, which is the final output. The reducer also aggregates them. Let's get more information about MapReduce and its components. MapReduce mainly has three categories. The map-reduce function in the Hadoop is work as follows:

- **Mapper Class**: Mapper Class is the first step in data processing with MapReduce. Here, Record Reader processes the respective key-value pair of every record input. This intermediate data is stored on the local disc in Hadoops Mapper.
- **Input Split**: It is the logical display of data. It is a work package with one map task in the MapReduce program
- **Record Reader**: It interrelates with the division input and converts the data collected into key-value pairs.
- **Reducer Class**: The transitional mapper output is supplied to the reducer, who processes it and produces the ultimate output, which is then saved in the HDFS.
- **Driver Class**: A driver class is the main component of a MapReduce job. He has to set up a MapReduce Job to run Hadoop. The Mapper and Reducer Class names with data types and their respective job names are specified.

Steps in Map Reduce

- The map takes pair-shaped data and returns a <key, value> pair list. In this case, the keys are not unique.
- The Hadoop architecture uses the output of Map to sort and shuffle. This type and shuffle will act on the <key, value> pair lists, and send unique keys and a list of values related to that unique <key, list (values),> key.
- The sorting and shuffling output are sent to the reducer stage. The reducer functions in a list of values for single keys, and the final output <key, value> is saved/shown (Fig. 2).

Features of MapReduce:

The two biggest advantages of MapReduce are:

- Parallel Processing
- Data Locality
- Reduced cost

Parallel Processing: The software split the job between multiple nodes in MapReduce and each node works with one part of the job. MapReduce is therefore based on a paradigm dividing and conquering, which helps us process data on various machines.

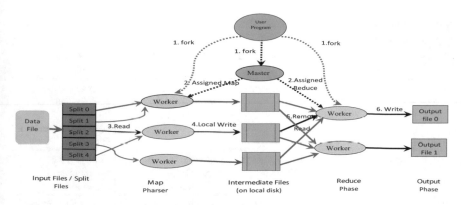

Fig. 2 The working of Map-Reduce architecture in big data analytics [7]

The time required for data processing is greatly reduced by multiple machines rather than one machine in parallel [9].

Data Locality: In the MapReduce Framework, the processing unit is shifted to the data rather than the processing unit. In the traditional SYSTEM, we used to bring data to the processing unit and process it [9]. However, as the data accumulated and became extremely vast, the following issues developed as a result of transporting such a massive volume of data to the processing unit:

Transferring enormous data to processing is expensive and degrades network performance.

The data is handled by one unit, which constitutes a bottleneck problem, processing the data and it takes time.

Reduced Cost: MapReduce now lets us overcome the problems mentioned above by bringing the processing unit to the information. As you can see in the picture above, data are distributed across multiple nodes where the part of the data that is on it is processed by each node. This allows us to benefit from the following: The transfer of the processing unit to the data is very economical. As all nodes work parallel with their part of the information, the processing time is reduced. Each node is processed by a part of the data, so there is no opportunity for an overloading node.

4   Machine Learning in Bioinformatics

At the base of machine learning are self-learning algorithms, which constantly evolve by improvements in their task. These algorithms will eventually result in pattern recognition and predictive modelling contexts when properly structured and feeding correct data. Data is like a supplement for machine-learning algorithms to perform better. Algorithms are designed and trained on a regular interval based on information (input data) as a coach train an Olympic athlete to improve their bodies and skills

Table 1 The difference between different ML algorithms used to train the model on the biological experimental data [10–12]

Sr. no	Criteria for measurement	Supervised learning	Unsupervised learning	Reinforcement learning
1	Definition	Used the labelled data	Unlabeled data without prior knowledge	Works with the environment integration needed—no defined data set
2	Types of problem	Carry regression, classification problems	Clustering and association algorithms	Works on the exploitation or exploration
3	Algorithms used	Use SVM, KNN, ANN, linear regression, logistic regression	Use the K- means, C-means, and Apriori algorithms	Use the Q-learning and SARASA algorithms
4	Training	Used for external supervision	No supervision is required	No predefined data
5	Approach to solve the problem	This maps the labelled inputs to the known outputs	Understand the patterns and discovers the output	It works on the trial and error method
6	Real-time application	Used for the risk evaluation, forecast the sales prediction in companies	Used for recommendation system(chatbots) and anomaly detection in medical science	Used for the self-driving vehicles and gaming and healthcare system

every day through training. Many languages, including Python, R, Java, JavaScript, and Scala, work with machine learning. Thanks to the Google TensorFlow library, which offers a complete ecosystem of machine learning tools, Python is the preferred choice for many developers. Machine learning is divided into various forms of supervised, unsupervised, semi-supervised, and reinforcement-based learning. The basic difference is the kind of data that is labelled or unlabeled by the algorithm used for classification [10–12]. The complete information about the difference between these different ML Model is shown herein in Table 1.

4.1 Artificial Neural Networks and Deep Neural Networks

A cosmic number of simple features called neurons consist of artificial neural networks (ANN), each taking simple decisions. Together, neurons can provide precise answers to complex problems such as the processing of natural languages, computer views, and AI. A typical neural network can have an input Neuron layer, just one/two/more "hidden layer" which processes inputs, or an output layer that provides the model's final output. An additional 2–8 layers of neural nets are usually

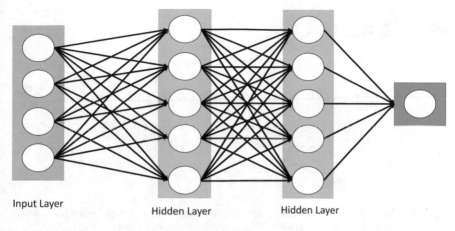

Input Layer   Hidden Layer   Hidden Layer

Fig. 3 The basic architecture of neural network (ANN/DNN)

found within a deep neural network (DNN) studies by Goodfellow, Bengio, and Courville, and other experts suggest that the number of hidden layers increases with neural networks. The neural network is composed of the input layer, hidden layer, and output layer [13, 14]. There are the activation functions used to determine the output of the neural network hence if someone wants to learn machine learning then most of the people confused while dealing with the activation function here, we will learn a detailed view about the activation functions (Fig. 3).

4.2 Activation Function

The activation function in a model is a mathematical equation that determines the neural network output. The function is attached to each neuron in the network and decides whether or not to activate it, depending on whether the input of each neuron is relevant for the prediction of the model. Activation functions also help to normalize the output of each neuron within ranges 1 to 0 or -1 to 1 [15, 16]. Another aspect of the activation function is the computational efficiency since thousands or even millions of neurons for each sample are calculated. Modern neural networks use a back-propagation technique to train the model, which puts a heavier computational pressure on the function and the derivative function. Due to the need for speed, new functions such as ReLu and Swish were developed which are further divided into linear and nonlinear functions. Here we will learn more about nonlinear and linear activation functions used in ANN.

Role of the Activation Function in a Neural Network Model

Numerical input data points are fed into neurons in the input layer in a neural network. Each neuron has a weight, which is transmitted to the next layer by multiplying the

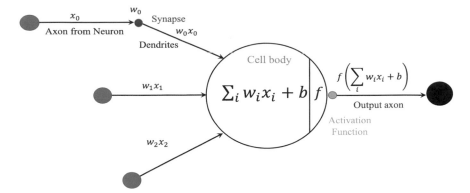

Fig. 4 The basic process carried out by a neuron in a neural network: input number (x^i, weight w^i), f (activation function), b (bias value) Biases are also assigned a weight

input number by weight. The activation function is a 'gate' between the input and the output of the current neuron to the next layer. Depending on a rule or threshold, it can be as simple as a step function that switches the neuron output on and off. It can also be a transformation that maps the input signals into output signals needed to function in the neural network. Neural networks increasingly use non-linear functions that can help the network learn complicated data, compute and learn nearly every question-related function, and provide accurate predictions [17] (Fig. 4).

Types of Activation Functions:

- Binary Step Function
- Linear Activation Function
- Non-Linear Activation Function

4.2.1 Binary Step Function

A binary step function is an activation function based on the threshold. If the input value is above or below a specified threshold, the neuron is enabled and sends the exact signal to the next layer. The problem with a step function is that multi-value outputs are not permitted for example the classification of inputs into one of several categories cannot be supported.

4.2.2 Linear Activation Function

A linear activation feature takes shape: $A = cx$ the inputs and weights of each neuron are multiplied and the output signal is proportionate to the input. A linear function is in one sense better than a step function because it allows several outputs, not just yes or no [18]. However, a linear activation function has two major problems:

1. Backpropagation (gradient descent) cannot be used for model training the function derivative is constant, it has no input relation, X. It cannot, therefore, be reversed and understood what weights in the input neurons can better forecast [18].

2. All neural network layers collapse into a single layer, the last layer will be the linear function of the first layer, regardless of how many layers of the neural network (because a linear combination of linear functions is still a linear function). The neural network becomes a single layer with a linear activation function. A linear activation network is nothing more than a linear regression model. The power and ability to handle varying input data parameters are limited [18].

4.2.3 Non-Linear Activation Functions

Modern neural network (NN) models use non-linear activation functions. The machine-learning model allows complex mapped inputs and outputs from the network that are necessary to learn and model complex data such as nonlinear, high-dimensional images, video, audio, and data sets. They are also necessary. Nearly any imaginable process can be seen as a functional computer in a neural network provided the activation function is nonlinear [19, 20].

Non-linear functions address the problems of a linear activation function:

In a Neural network they have a derivative function, they allow backpropagation, that is connected to a deep neural network, allowing several neuron layers to be 'stacked.' Multiple hidden layers of neurons are needed to achieve complex data with high accuracy. There are seven common nonlinear functions and a function can be chosen.

- **Sigmoid or Logistic Activation Functions**
- **Tanh or hyperbolic tangent Activation Function**
- **ReLu (Rectified Linear Unit) Activation Function**
- **Leaky ReLU**
- **SoftMax function**

Sigmoid or Logistic Activation Functions: The main reason for using a sigmoid is that it exists between two (0 to 1). It is thus particularly used for models where probability as an out must be predicted. Since there is only a likelihood of anything between 0 and 1, sigmoid is the right choice. The function can be distinguished. In other words, at any two points, we find the slope of the sigmoid curve. The function is monotonous but the derivative of the function is not. The sigmoid function of the logistics can cause a neural network to stay during the training. The SoftMax function has a more common, multi-class logistic activation function. The sigmoid function is the most used, but many more effective alternatives exist [21, 22].

$$s(x) = \frac{1}{1 + e^{-x}}$$

The main benefits of the sigmoid function are:

- The function can be differentiated. In other words, at any two points, we find the slope of the sigmoid curve.
- Monotonic function in nature.
- A smooth gradient will prevent the output values from "jumping".
- Output values bound from 0 to 1, standardizing the neuron output. It is thus particularly used for models where probability as output must be predicted. Since there is only a likelihood of anything between 0 and 1, sigmoid is the right choice.
- **Definite predictions**: The value Y (prediction) for X greater than 2 or less than − 2 tends to be at the edge of the curve, close to 1 or 0. This allows clear forecasts.

Disadvantages:

Vanishing gradients: If we look at the graph at the end of the function carefully, y values react to the changes in x very little. Let's consider what a problem it is! In these regions, the derivative values are very low and converge to 0. This is referred to as the extinction gradient and lessons are learned. If 0, no apprenticeship. When there is slow learning rate occurs, an optimization algorithm that minimizes errors can be attached to local minimum values and the artificial neural network model cannot achieve maximum efficiency.

- **Outputs are not zero cantered**
- **Computationally expensive**

Tanh or hyperbolic tangent Activation Function: Also, tanh is like sigmoid logistic, but better. The range of the tanh is from (−1 to 1). Tanh is sigmoidal as well (s-shaped). The advantage is that the adverse inputs are highly negative and that the zero inputs in the tanh graph are mapped to close to zero [21, 23, 24].

$$f(x) = \tanh(x) = \frac{(e^x - e^{-x})}{(e^x + e^{-x})}$$

The main advantage of Tanh functions are as follows:

- A function can be distinguished and differentiable in nature.
- It is monotonous, but its derivative is not monotonous.
- The tanh function is primarily used between two classes.
- The feed-forward networks use both tanh and logistical Sigmoid functions.
- It has a very Sigmoid-like structure. However, it is zero-centered when the function is defined (−1, +1).

The advantage compared to the sigmoid feature is that its derivative is steeper. So, because there is a wider range for faster learning and grading it will be more effective. However, it has similar disadvantages as the sigmoid function.

ReLu (Rectified Linear Unit) Activation Function: The ReLU is currently the world's most used activation function. Since then, it is used in nearly all neural networks or deep learning. The ReLU is half corrected, as you can see (from the bottom). f(z) is zero if z is below zero and f(z) is z when z is above or equals zero. The ReLu range from 0 to infinity in nature. The function is both monotonous and it is derivative. However, the problem consists of zero all the negative values, which decreases the model's ability to fit or train the data properly. This means that any negative input to the ReLU activation function changes the value into zero in a graph, which has an effect on the resultant graph because the negative values are not appropriately mapped [21, 25, 26].

$$f(x) = \begin{cases} 0 \text{ for } x \leq 0 \\ x \text{ for } x > 0 \end{cases}$$

Advantages

- **Computationally efficient**: ReLU is worth [0, +infinity] but what are the benefits and returns? Imagine a large size network of neurons with too many inputs, hidden layers. Almost all neurons were activated in the same manner by a sigmoid and hyperbolic tangent. As a result, the activation is quite intense. We want to have an effective computational charge since some of the neurons in the network are active and activation is infrequent. We got it with ReLU. The network will run faster with a 0 value on the negative axis. Because the computational charge is lower than the sigmoid and hyperbolic tangent functions, multi-layer networks have become more popular.
- **Non-linear**: At first sight, it seems that the linear function of the positives axis has the same characteristics. But ReLU is not linear, first and foremost. A good estimator or Any other function can also be converged by ReLU combinations.

Disadvantages

- **The Dying ReLU problem**: When entries are zero or negative, the function gradient becomes zero, the network can't reproduce and can't learn. To solve this problem, we have another type of non-linear activation function called the Leaky ReLu Network.

   Leaky ReLU: The leak helps the ReLU function to increase its range. The value of a usually is about 0.01. If an is not 0.01, then this is pronounced as randomized ReLU. That is why the Leaky ReLU range is available (−infinity to +infinity). The functions of Leaky and Randomized ReLU are both monotonous. Also, monotonic, their derivatives. The leak helps the ReLU function to increase its range. The value of a usually is about 0.01. If an is not 0.01, the randomized ReLU is called. That is why the Leaky ReLU range is available (−infinity to infinity). The functions of Leaky and Randomized ReLU are both monotonous in nature. Also, the monotonous nature of their derivatives [21, 27].

$$f(x) = \begin{cases} 0.01 \text{ for } x < 0 \\ x \quad \text{ for } x \geq 0 \end{cases}$$

Advantages

- **Prevents dying ReLU problem**: The variation of the ReLU in the negative field has a low positive incline, so even for negative input values, it allows backpropagation. The leaked value is given as 0.01 if the function name changes randomly as Leaky ReLU with a different value near zero. The leaky-ReLU definition range is still smaller. That is almost 0, but 0 with the value of the non-living gradients living in the RELU in the negative learning environment.

Disadvantages

- **Results not consistent**: For negative input values, leaky ReLU does not produce consistent predictions.

SoftMax function: The regression of SoftMax is a logical regression form that normalizes an input value into a value vector following the distribution of probability, whose total sum is up to 1. The output values are within the [0, 1] range, which is nice, as we can avoid binary classifying and accommodate as many classes or dimensions in our models. Therefore, SoftMax is often known as a logistic regression of multinomial. Aside, the Maximum Entropy (MaxEnt) Classifier is another name for SoftMax Regression. In discriminatory models such as Known cases of SoftMax regression are the feature is usually used to calculate losses expected when a data set is being trained. In discriminatory models such as Cross-Entropy and Noise, Contrast Estimation is known cases of SoftMax regression are used. These are just 2 of the different techniques that try to optimise the current training to increase the chance of the correct word or sentence being predicted. If you look from the beginning, the definition can sound trivial, but this regression functions h in the area of machine learning and NLP. If you look at it from the start, the definition may sound trivial but as a baseline comparator, it was useful in the field of machine learning and NLP. Investigators designing new solutions need to experiment with the results of SoftMax. However, SoftMax cannot be ideally used as an activation function such as Sigmoid or ReLU, but rather for multiple layers or one-size-fits-all layers. The feature is usually used to calculate losses expected when a data set is being trained [21].

$$f_i(\vec{x}) = \frac{e^{x_i}}{\sum_{j=1}^{J} e^{x_i}} \text{ for } i = 1, 2, \ldots, J$$

Advantages

- Capable of handling multiple classes in other functions only by one class normalises output between 0 and 1 for each class and divides by its sum, thus giving the input value a probability of being in a certain class.
- Useful for neurons: usually SoftMax is used for the output layer only, for neural networks requiring multi-categorization of input.

Neural Network Activation Functions in the Real World

The selection of active functions is critical when creating a model and training a neural network. Experimenting with various activation functions will aid you in achieving significantly better solutions for various problems. In a real-world neural network project, you can switch between activation functions using the profound deep learning framework of your choice. For instance, using Keras Library to activate the ReLU function. While it is easy to select and switch activation functions in profound education frameworks, it is difficult to manage multiple experiments and test different activation functions on large data sets. It can be hard to do:

- Keep track of experiment source code, parameters, and metrics across time as you attempt new activation functions for a model or variations of the same model.
- You're usually going to be required to run experiments on several machines that run multiple large-scale experiments, you have to supply and maintain those machines.
- Management training data management you need to test various sets of test data for several model variations on different machinery to achieve good results. It is difficult to move the training data whenever an experiment is needed, particularly if you process heavy inputs such as images or video. Missing Link can help you manage all these things by experimenting with your model's best activation feature. Find out more to see how easy it is to conduct profound learning experiments.

5 Conclusion Big Data Deep Learning Architecture and Challenges

As training datasets grow in size, machine-learning algorithms are more effective. Thus, we twice benefit from the combination of large data and machine learning: the algorithms help us keep pace with the continuous flux of data while the volume and variety of data feed and help algorithms to develop. Let's see how this process of integration works: We could expect to see defined and analyzed results such as hidden patterns and analytics that could support predictive modelling by adding big data to a machine-learning algorithm. These algorithms could automate previously human-centered processes for several companies. But most often, a company will review the findings of the algorithm and search for valuable insights that could guide business. This is where people are returning to the image. While AI and data analytics run on computers that exceed human beings with an enormous margin, they are not able to make certain decisions. Computers still have many human-inherent characteristics, including critical thinking, intention, and the ability to take holistic approaches [28, 29].

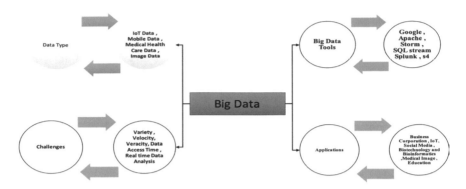

Fig. 5 Diagrammatic representation of Big Data (Application, Tools, Challenges, Data Types)

The architecture of Big Data:

See Fig. 5.

Challenges of Machine Learning/ Deep Learning:

Deep learning is still in its infancy and has a tremendous potential to help build value for scientists and researchers and practitioners face major challenges. Some of the problems are:

1. **Need enough quality of data for training the DL models**
2. **Security Concerns**
3. **Neural Network-based output (Black-box approach)**
4. **Real-time decision making**

- **Deep learning needs enough quality data**: As a rule, neural networks need additional data to make abstractions more powerful. The data are not always correct or of sufficiently high quality, while Big Data scenarios have abundant information to allow training. Small changes or unexpected input data features can throw away neural network models completely. Deep learning is not easy to change contexts, it is difficult to solve very similar problems, presented in another context, in a neural network model trained on certain problems. *For example*, when MRI images are represented, with different characteristics, deep learning systems which can effectively detect a set of images can be stumbled (grayscale vs. color, different resolution, etc.) [30, 31].
- **Security concerns**: Deep learning must be trained and retrained on massive, realistic datasets, and data must be transferred, stored, and securely handled during the development of an algorithm. In a mission-critical setting, when a deep learning algorithm is deployed, an attack or hacking may affect the output of the neural network by small, malicious input changes. This could alter experimental results, lead to misdiagnosis of the patient [30, 32].
- **Real-time decisions**: Many big data in the world are being streamed in real-time and data analytics are becoming more and more important in real-time.

Deep learning for real-time data analysis is hard to use because it is extremely computationally intensive. *For example*, for over 20 decades computer vision algorithms have been used for several generations to detect objects in live X-rays or MRI scans [30, 31].

- **Neural networks are black boxes**: Big data organizations need more than good answers. These answers must be justified and why they are correct. Deep learning algorithms rely on millions of parameters for decision-making, and "why" does the neural network choose one label over another often not explain. This opacity limits the ability to make use of deep knowledge for critical decisions like healthcare treatment or large financial investments [30, 33].

6 Deep Learning Framework for the Big Data Analysis in Bioinformatics

Deep learning in recent decades has demonstrated its ability to manage large numbers of data as a very powerful tool. Traditional methods, particularly in design recognition, have surpassed the interest in using hidden layers. Convolutional neural networks are one of the most popular profound neural networks. Researchers have struggled since 1950, the early days of AI, to build a system that can understand visual data. This field was named Computer Vision in the years that followed. Computer vision took a quantum leap in 2012 when an AI model that overcame best picture recognition algorithms and that too large was designed by a group of university scientists. AlexNet, named after Alex Krizhevsky, its main creator, won an incredible 85 percent accuracy in the ImageNet Vision Competition 2012 [34, 35]. A modest 74 percent was the result of the runner-up. The deep learning algorithms are extensively used in the various domains of biological life science, medical science, and pharmaceutics for drug discovery. In deep learning CNN, RNN is the method that is used to build up the DL models. Here we will learn in detail how these will work in the building of DL models [36].

- CNN (Convolutional Neural Network)
- RNN (Reinforcement Learning).

6.1 CNN (Convolutional Neural Network)

A monumental growth in artificial intelligence has witnessed bridging the gap between human and machine capacity. To achieve amazing things, researchers and enthusiasts both work on many aspects of the field. The Computer Vision domain is one of many such areas.

The agenda for this area is to enable machines to view the world, view it in the same way, and even use the knowledge for a wide variety of tasks such as image and video recognition, image analysis and classification, media recreation, recommendation, and natural language treatment. Deep Learning advances in computer vision are built and improved over time, mainly via one algorithm Convolution Neural Network. CNN was a special kind of neural network at the heart of AlexNet, which approximately imitates human vision. CNN has over the years become a major part of many applications for computer vision. A Convolutional neural network (CNN/ConvNet) is a deep learning algorithm that allows the input image to be taken, the emphasis on various aspects/objects (learnable weight and biases) to be attached, and differentiated between them. In comparison with other classification algorithms, the required pre-processing in ConvNet is much smaller. While primitive methods are hand-designed, ConvNets can learn these filters/characteristics with a sufficient level of training [35] (Fig. 6).

The ConvNet architecture is similar to that of Neurons' connectivity pattern in the human brain and has been inspired by the Visual Cortex organism. In the limited visual field region known as the receptive field, individual neurons only reply to stimulants. A collection of these fields covers the entire field of vision. Let us, therefore, examine the functioning of CNN.

The components of the CNN architecture are:

- Input image
- Pooling Layer
- FCC Layer (Fully Connected Layer)
- Output Layer
- Loss functions

Input Layer: Convolution neural networks consist of many artificial neuron layers. Artificial neurons are mathematical functions that calculate the weighted amount from multiple inputs and output to an activation value. A rough imitation of their biological counterparts. Each layer generates multiple Activation functions that are forwarded to the next layer when you enter a picture in ConvNet. Usually, the first layer extracts fundamental features like horizontal or diagonal borders. This

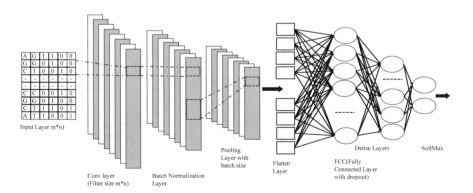

Fig. 6 Detailed CNN architecture [37]

output is transferred to the next layer, where complex properties, like corners or knit edges, are detected. When we move deeper into the network, even more, complex features like objects and faces can be identified. The classification level produces a set of confidence scoreboards (values from 0 to 1) which indicates how likely the image to be part of a "class" based on the final convolution layer activation map. For example, if you have ConvNets detecting diseased cells, cancerous cells and other cells, the output of the finished layer may be one of these disease cells entered in the image of the input layer [37, 38].

With the application of the appropriate filters, A ConvNets captures spatial and temporal dependencies in an image. Due to the reduced number of parameters involved and the reusability of weights, the architecture performs better for the image data set. The network can be trained in other words to better understand the sophistication of the image. The ConvNet's role is to reduce the images into a more easily processed form without losing the characteristics that are essential to a good prediction. This is important in designing an architecture that is not only good for learning but can be scaled to massive data sets (Figs. 7 and 8).

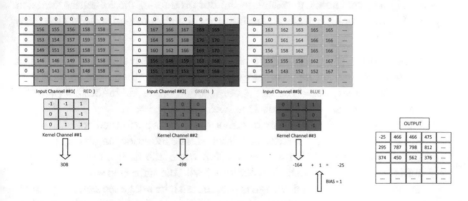

Fig. 7 The convolution operation (input feature, stride, and padding) [34]

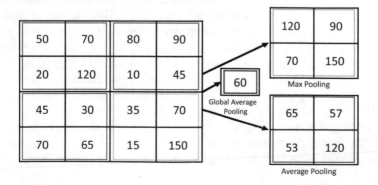

Fig. 8 The pooling layer Max pooling, average pooling, and global average pooling [39]

Kernel layer (Convolutional Layer): The most significant component in CNN architecture is the layer of convolution. It is made up of a set of filters (so-called kernels). The image of the input is expressed as N-dimensional CNN metrics for output characteristic map generation [38]. There are some of the operations that are going place at the time of convolutional which needs to understand and learned are as follows:

- **Kernel**: The kernel is described by a grid of discrete numbers or values. The kernel weight is known for each value. At the start of the CNN training process, random numbers are assigned as the weights of the kernel. Moreover, several different methods for initializing weights are used. The kernel then learns the extraction of significant features at every training period [38].
- **Convolutional operation in the CNN**: The input format of CNN is described in the beginning. The classical neural network's input is in the vector format, whereas CNN is input by the multi-channel image. *For example*, the single-channel format of a grey image is the 3-can image format of RGB. These are the convolutional operations that are performed in the CNN architecture, although it is a computationally expensive task to calculate the dot product of these metrics for each point. So, for simplicity lets us learn how this operation is performed in the 3 * 3 metrics [38].

  - **Input image**: The green section resembles our input picture $5 \times 5 \times 1$, I in the above demonstration. Kernel/Filter K, represented by the colour yellow, is the element that is involved in the convolution operation in the first part of a CNN layer. K has been chosen as a matrix for $3 \times 3 \times 1$.
  - **Stride Length**: Whenever a matrix multiplication is performed between K and part P of the image that hovers the kernel. Hence the stride length is depending upon the architecture. With a certain step value, the filter moves to the right until the full width is checked. It continues with the same step value and repeats the process until the whole image is passed (left) the image the same step value.
  - The kernel is the same depth as the input image for images with multiple channels (e.g. RGB). Matrix multiplication is made between K_n and in a stack ($[K_1, I_1]$; $[K_2, I_2]$; $[K_3, I_3]$) and all results have been added to the bias for a single-depth squashed feature output channel. The Convolution operation aims to extract from the input image high-level characteristics such as edges. Not just one convolutional layer needs to be limited to ConvNets. Conventionally, the first ConvLayer captures low-level characteristics such as edges, colours, gradients, etc. The architecture with additional layers also adapts to the features of high level, providing us with a network that understands the images in the data set in an entire way, similar to how we are.
  - **Padding**: There are two kinds of results for the operation one with a reduction of the size of the convolved function as compared to the input, the other with an increase of or remains the same dimension. In the case of former or latter,

this is done by using the same Valid Padding. When the $5 \times 5 \times 1$ picture is added to a $6 \times 6 \times 1$ picture and afterward the kernel is applied to it, it turns out that the convolved matrix is of $5 \times 5 \times 1$ dimensions. That's the name Same Padding. On the opposite side, we will be presented a matrix with dimensions of the kernel ($3 \times 3 \times 1$) itself if we do the same operation without padding, the Valid Padding.

- **Pooling Layer**: Similar to the Convolution layer, the spatial size of the Convolved feature decreases in the Pooling layer. This reduces the computing power necessary to process the data by reducing dimensionality. In addition, the process of effective model training is useful in the extraction of dominant features which are rotating and positional invariant. The pooling is of different types: Max pooling and Average and Global pooling. The maximum value from the image portion of the kernel is returned by Max Pooling. On the other hand, Average Pooling gives the average of all image values covered by the kernel in part. The input layer and the pooling layer, form a Convolution Neural Network in the ith layer. The number of such layers in a typical CNN model can be increased or decreased depending on the complexities of the data set in the images to capture even more low-level data but at the price of additional computing power. We have completed the above process, enabling the model to understand the characteristics. We will flatten the final output and supply it for classification to a regular neural network [34].
- **FCC Layer**: Adding a fully connected layer is (usually) an affordable way to learn nonlinear high-level combinations, represented by convolution layer output. The fully connected layer can be used in this space to learn a perhaps non-linear function. After our input image has been converted into an appropriate form for our multifunctional perceptron, the image will be flattened into a column vector. The flattened output is transmitted to a neural feed-forward network and back-propagation for each training iteration. The model can distinguish dominating features from certain low-level features in images over several periods and classify them via using the applied classification technique [36, 38].

Type	SoftMax loss function	Hinge loss function	Euclidean function
Definition	The CNN model performance measurement function is commonly used by SoftMax. The log loss function is also called. The probability [2] is its output. Moreover, it is usually used in multi-class classification problems as the substitution for the square error loss function. The SoftMax activations are used in the output layer to create the output in a probability distribution	The function is usually used in binary classification problems. This is mainly important for SVMs which use the hinge loss function when the optimizer attempts to maximise the margin around dual objective classes	The Euclidean function is widely employed in problems with regression. It is also the so-called medium square mistake
Mathematical equation	$H(q, x) =$ $$-\sum_i q_i \log(x_i) \text{ where } i \in [1, N]$$	$H(q, x) =$ $$\sum_{i=1}^{N} \max(0, m - (2q_i - 1)x_i)$$	$H(q, x) =$ $$\frac{1}{2N} \sum_{i=1}^{N} (q_i - x_i)^2$$
	$q_i = \frac{e^{a_i}}{\sum_{k=1}^{N} e_k^a}$ Here e^{a_i} is the unused output of the previous layer, while N stands for the number of neurons in the output layer. Finally, cross-entropy loss function mathematical representation	The margin m is usually set to 1. The predicted result is also marked as q_i while the desired result is marked as x	Euclidean function, mathematical representation represents q_i initial value, x_i represent value

Loss functions: Several layer types of CNN architecture were presented in the previous section. In addition, the final grade of the output layer, the last layer of the CNN architecture, is achieved. In the output layer, some loss functions are used to calculate, measure the predicted error created in the CNN model across training samples. This mistake reveals that the actual output differs from the predicted.

Through the CNN learning process, it will then be optimized. However, the loss function uses two parameters to compute the mistake. The first parameter is the estimated CNN output (called the prediction). The second parameter is the actual output (called the label) [40]. In different problem types, several types of loss functions are used. Some types of loss functions are explained concisely in the following.

- SoftMax Loss Functions (Cross-Entropy) [41]
- Hinge Loss function [42]
- Euclidean Loss function [43].

The advantages of using CNN

In the computer vision environment, the benefits of using CNNs are shown as follows over other conventional neural networks:

- CNN mainly considers the weight-sharing feature, reducing the number of parameters for the trainable network. In turn, it helps to improve generalisation and prevent overfitting.
- At the same time as learning the extraction layers and the classification layer, the output of the model is highly organised and relies heavily on the extracted features.
- CNN is much easier to implement large-scale networks than other neural networks.

6.2 RNN (Recurrent Neural Network)

RNNs in the discipline of DL are commonly used and familiar. The RNN is used primarily in language and NLP contexts. RNN uses sequential data in the network, in contrast to conventional networks. Since this function is essential for a range of applications, the integrated structure in the data sequence offers valuable information. To determine the meaning of a specific word in it, for example, it is important to understand the context of the sentence. Thus, the RNN can be considered as a short-term memory unit, where x is the input layer, y is the output layer, and is a (hidden) state layer. In Fig. 9, three different deep RNN techniques, namely "hidden-to-Hidden," "hidden-to-output," and "Input-to-hidden," are presented for a particular input sequence [44]. A typical unfolded RNN diagram is depicted in Fig. 9. There is a deep RNN that reduces the problem of deep network learning and brings benefits from a deeper RNN based on these three techniques [44].

However, one of the major issues with this approach is RNN's sensitivity to the gradient descent and other disappearing problems. Specifically, the reduplications of multiple big or small derivatives throughout the training phase may cause the

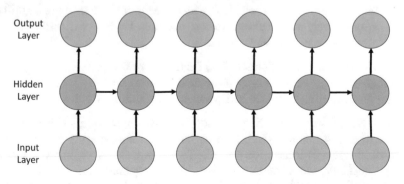

Fig. 9 Diagrammatic representation of RNN (recurrent neural network)

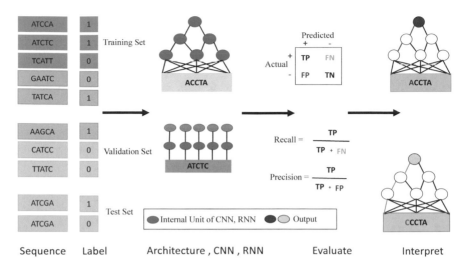

Fig. 10 The CNN and RNN architecture (an overview)

gradients to exponentially expand or decline in the reinforcement learning. With the entry of new inputs, the network ceases to think of the original; this sensitivity, therefore, decreases over time. In addition, LSTM can be used to handle the problem. This approach provides repetitive links to network memory blocks. Several memory cells are contained in each memory block that can store network temporal status [45, 46] (Fig. 10).

7 Application of Machine Learning for Big Data

The mechanical and geometrical properties play an important role in the compressive strength of FRP wrapped columns. Machine learning is a sub-branch of artificial which deals with the automation of task with the help of some learning algorithms. ML/DL is extensively used in almost various domains. Here we will discuss how the ML/DL algorithm is used in bioinformatics.

- **NGS (Next Generation Sequencing) Based Sequencing**
- **Biological Image Processing (Biomedical Informatics)**
- **Drug Discovery and Molecular Dynamics Simulation Studies**

NGS (Next Generation Sequencing) Based Sequencing: The sequencing methods of next-generation (NGS) are at the core of almost every area of biology and medical research. It has been essential for the processing and analysis methods in the produced data sets, addressing issues like metagenomic classification, variant calling, quantification of data, genomic characteristics detection, downstream analysis in larger biological or medical contexts, which have been steadily increasing in demand. Machine-learning (ML) techniques are frequently used for such tasks besides classical algorithmic approaches. Deep learning (DL), in particular, has gained significant traction for such applications, using artificial neural networks (ANNs) with multilayered methods for supervised, semi-supervised and unsupervised learning. The author has highlighted the major network topologies, application areas, and DL frameworks in the context of the NGS. The volume of the data generated in life sciences, biotechnology has been steadily increasing in recent years, driven in particular by rapid progress in the technology of high-performance sequencing such as Illumina systems. Low genome sequence costs (www.genome. gov/sequencingcosts) make major projects feasible, thus increasing the statistical power of larger sets of data. The wealth of genomic and transcriptomic sequencing data generated enables us to address a multitude of questions that could not previously be answered, both for accurate and personalized applications and drug discovery for *example*, more reliable studies of transcriptome can identify patterns of expression linked to diseases or differences of expression as adverse drug effects. Important to individual medicine can be improved variant calling and genome-wide associated studies (GWAS). Enhanced metagenomics can provide essential ecosystem information such as human intestinal microbiome, hospital bacterial populations, or viral patient progression. NGS methods are also an essential element in epigenetics research involving a wide range of diseases. It also helps to understand the roots and epidemiology of serious coronavirus (SARS-CoV2), acute respiratory syndrome. The huge data growth shifts life sciences from a model-driven science to a data-driven one continuously. As in other scientific fields, we are now seeing a growing use of statistical learning methods, which are not 'seeded' in mind for a biological model, but rather are trying to learn from data directly. DL methods based on multi-layered ANNs, in particular, have become more popular [47–49].

Application of Deep Learning in NGS:

- **Variant Calling**: The purpose of varied calls is to directly detect genomic variants from NGS data. Point mutations are the simplest variant type (single nucleotide variants; SNVs). Single nucleotide polymorphism (SNP) genotyping and the detection of SNVs in individuals with multiple tissue samples include typical SNV-calling applications. It is also important, but typically more difficult to identify more complex genetic changes from the NGS data, such as long or copy numbers (CNV). The calculation process is based on the initial harmonization of NGS reads to a reference genome sequence. In the prediction of the chance of variation in each position, *for example*: based on quality scores and all accounts of aligned readings in each position, traditional algorithms, such as a GTK apply various statistical models and manually tuned heuristics. Sequencing errors, however, generally depend on alignment positions and types of instruments. This makes it difficult to design accurate statistical modelling to differentiate between variants and sequencing errors or alignment devices. Use a controlled DL variant model. The need for hand-made heuristics can be substituted with a system that has data directly from the skill to learn feature patterns. The fundamental idea is to transform aligned readings focused on placement and application of the interest in the images (candidate variant). Computer vision solution CNN-based solution is there to handle the variant calling [50].

- **Metagenomics**: Metagenomics deals with environmental sample sequencing data. Taxonomic reassignment is an important processing step. Classical approaches, like Kraken or MetaCache, simply number accurate k-mer matches to a database to assign readings to a likely type of origin. This problem also can be defined as a monitoring learning task, taking into account the output categories for each species of interest and the input reads by NGS. This classification problem is more complex and suitable for RNN architectures due to the size of reference genomes (which are significantly longer than short reads) and a large number of classes (typically thousands for bacterial samples) [51].
- **Single-cell transcriptomics**: The single-cell RNA-seq protocols (scRNA-seq) is becoming even more prominent because they profile thousands or even millions of cells in a single test. The input to the corresponding analytical tasks is typically a gene expression matrix, which can be calculated through an estimate of the expression of each gene in each cell using a mapping to a reference's transcriptome. The clustering of supposed cell types is of central importance. The large dimensions, high dimensionalities, and droppings (due to low genes leading to many zero entries in the expression matrix) are challenging, which motivate the application of unattended ML methods to cluster, input, and decrease dimensionality DL techniques in this field are currently dominated by different types of AE because they can learn inherent distribution effectively by reducing dimension without supervision [52–55].

Biological Image Processing (Biomedical Informatics): Bioinformatics and Biomedical Imaging is an interdisciplinary subject that derives from comprehensive computer science, life sciences, and biology theories and methodologies that play an integral role in the diagnosis of diseases and treatment of diseases. In recent years, the fields of medical science, healthcare computing have progressed greatly and have led to in-depth analysis which is demanded by mass data collection in which the traditional analytical techniques, are no longer competent. Instead, algorithms have been improved significantly in bioinformatics and biomedical image analysis by Deep learning which is rapidly evolving (including convolutional neural networks (CNN), recurrent neural networks (RNN), auto-encoders, generative adversarial networks (GAN)). Thus, both academic and medical areas have emphasized the application of in-depth learning in bioinformatics, research, and biomedical images (radiology), to derive insight knowledge from data.

Due to today's rapid development of biotechnology in historical time, the biomedical data generated has increased exponentially in several fields of research and application ranging from molecular levels (gene functions, protein interactions (PPI), metabolic pathways, etc.), biological levels of tissue (brain neurons map, X-rayed image(s) (intensive care unit, electronic medical record, etc.) [56]. The unavoidable fact is that dealing with biomedical data with qualities like rising speed and heterogeneous structure is far more difficult than dealing with traditional data analysis approaches. More powerful theoretical methods and practical tools are therefore required to analyse and extract expressive meaning full information from the above complex biodata. This complex and heterogeneous data analysis is a typical complex problem in the system. The reliance, relation, or interaction between various data levels and their environment needs to be analysed. In this case, it is very difficult for us to model traditional methods due to the nonlinear, emergent, spontaneous order, adaptation, and feedback features of raw data. We can solve these problems only through deep learning. The author highlights in this section the latest developments in the application of advanced deep learning technologies in bioinformatics and bio-image analysis [12, 57].

Some of the areas where Machine Learning methods are used in biomedical image processing are:

- Profound genome sequence and single-cell sequence learning methods.
- Deep learning methods in regulatory analysis of epigenomics and genomics3D-genomics, functional genomics, and medical genomics methods of learning.
- Multi-omics integration profound learning methods.
- Biomedical signal/picture analysis Deep learning method.
- Deep methods of learning for the ultrasound/CT/MRI grading of tissues, lesions, and diseases.
- Computer-aided ultrasounds/CT/IRM deep learning methods.
- Deep image registration and/or ultrasound/CT/MRI fusion learning methods.
- Bioinformatics and Biomedical Image Artificial Intelligence and Algorithms.
- Artificial intelligent online databases and webservers, parallel bioinformatics, and biomedical imaging acceleration technologies.

Drug Discovery and Molecular Dynamics Simulation Studies: The process of drug discovery is very time-consuming and expensive. Computational modelling is an important tool for the conception and optimization of lead compounds in an attempt to accelerate the progress of a drug candidate through the discovery pipeline. Although physics-based modelling of (e.g. protein docking or molecular dynamics), in the computational chemistry community it has been standard de facto for many years, machine learning methods are now a powerful alternative to modelling. In particular, the deep learning paradigm can be viewed as a black box method because rules or regulations can be hard to extract from the trained model. The MD (Molecular Dynamics) is based on statistical mechanics. Setting up an MD run for complex systems can still be an easy task for continuous automation tools to enable greater use in academic and industrial environments. The contribution in this context discusses the implementation of a web server for the creation of hybrid molecular mechanics and simulations related to their protein–ligand molecular complex. The protein and ligand constitute the most important class of druggable goals, which is why it is important to have handy instruments to establish their systemic simulations. Big data and the Machine learning can be used in various ways in the field of drug discovery [58, 59]. Here are some of the areas where we can use the big data analytics and machine learning approach in drug discovery and drug design:

- **To understand the biological mechanism of system and disease in the body**: In most cases, a drug discovery program can only be initiated after scientists have come to understand a cause and a mechanism of action behind a particular disease, pathogens, or medical condition. Without exaggeration, biological systems are the most complex in the world and the only way to understand them is to follow a comprehensive approach, looking into multiple organizational "layers", starting from genes and to proteins, metabolites, and even external factors influencing inner "mechanics". The human genome decoding process began in 1990 by a group of scientists and took 13 years and cost $2.7 billion to finish the project. The genome reveals the kind of 'functions' for the organism to tell which proteins and why. With a full understanding of the genome, the understanding of our bodies will open up doors to a much more detailed understanding, what can be wrong with them, and under which conditions the genome is kind of 'instruction.' With full knowledge of the genome, our body can become much more understandable, what can go wrong and under what circumstances. Biological system research generates huge amounts of data to be stored, processed and, analyzed. The three billion chemical coding units which together constitute DNA would produce cannot be handled through the traditional storage system. The human proteome has been identified so far, with more than 30,000 distinct proteins. And there are more than 40,000 small body molecules, metabolites. Data mapping, derived from different experiments, associations, factors, and conditions combinations, generates billions of information points. Big data analysis and Machine Learning algorithms start shining here, which enable hidden data patterns to be derived, dependencies and associations previously unknown.
- **To find out the right drug molecule**: When scientists propose an appropriate biological target, it is time that molecules are searched that can interact selectively with the target and stimulate the desired effect of a "hit" of a molecule. To identify hit molecules several screening paradigms, exist. *For example,* a popular HTS high throughput sequencing) approach involves the direct screening of millions of chemical compounds against the drug target. This paradigm for screening involves the use of complex machinery, is expensive, and has a low success rate. But what is good is that it does not assume that the nature of chemicals that are likely to have activity with the target protein is known in advance. HTS, therefore, seems to be an experimental source of insights for further research, providing valuable "negative" results. Additional approaches include screening fragments and a more specialized, physiological screening approach. This technique on tissue is intended to provide an answer that is more in line with the ultimate in vivo effect than one specific medicine drug target. Computing scientists have taken advanced CADD (computer-aided drug discovery) with the help of protein ligand interaction between different protein responsible for the different disease in the human body (Auto Dock, Discovery Studio, Schrodinger etc.) approaches to the use of pharmacophore and molecular modelling to perform so-called "virtual" screens from compound libraries to reduce laboratory display cost, improve the efficiency and predictability of the above complex screens. Millions of compounds can be tested in silicon against a known 3D structure of a target protein (structure-based

approach) by using this approach; if the structure is unknown, drug candidates can be identified based on knowledge of other molecules known to activity in the direction of the target [60].

- **Personalized Medicines**: A drug usually interacts with multiple objectives, both on- and off-targets, and this interaction is highly influenced by the drug efficiency and the side effects. Different genetic, epigenetic, and environmental factors affect the disturbance of the individual biological system (e.g. a patient) by a drug molecule. Personalized medicine has been designed to meet individual patient characteristics to identify these hidden hierarchical data. Personalized medicines rely heavily on a scientific understanding that makes a patient vulnerable to disease and sensitive to a therapist with its unique characteristics, such as molecular and genetic profiles. Hundreds of genes were identified for their contribution to the human condition, driven by biomarker studies in the late 1990s, and patients' genetic variability was used for the distinguishing of individual responses to dozens of treatments. Although the studies have generated enormous data, like the Human Genome Project, computer modelling has become one of the key tools for personalized medicine. Metabolism modelling of the metabolic network and identification of population genetics patterns are several recent advances in this field that rely on computational modelling. The NIH Precision Medicine Initiative has led to a wide range of data generation, sharing, and computer modelling initiatives to help expand precision medicines. For example, the National Cancer Institute Genomic Data Commons program aims to provide a dataset that allows data sharing in the field of precision medicine over cancer genomic studies. Using this portal (https://gdc.cancer.gov/), case studies are submitted and shared so far. The approach to AI is widely used and many reviews on this popular bioinformatics are available [60, 61].

- **Pre-clinical trials**: One reason for a crisis and such a decline in research and development in the pharmaceutical industry is that animal testing by young drug candidates does not reflect the human outcome. Drugs fail in later stages and cost investors enormous money and businesses wasted time. More importantly, it costs patients' lives. There are now new algorithms of artificial intelligence and large-data approaches to simulate the activity of many drugs in many as in a "virtual" human [62].

8 Conclusion

book chapter comprises of different sections one section describes the importance of Deep Learning and the other section contains Big Data. A brief introduction to both fields was given in the book chapter. A DL Section deals with the available algorithms for the data analysis and what is the activation function needed in the DL model generation. The big data portion contains 5v's for data management along with the available HADOOP system and map-reduce functions. In the book chapter while writing the menu script of the book chapter author has found out the following observations which are as follows:

- Today, high-volume real-time data streams are constantly being generated from devices such as smartphones, IoT devices, computers, all these large data streams, and 5-V's are important features (the big data framework, if you wish) that help you to recognize all you need to study when the data stream is being scaled down.
- In many areas such as artificial intelligence (AI), business intelligence (BI), data science, and machine learning (ML), big data play an essential role, where data processing (extraction, transformation, and loading) results in fresh insights, innovation, and superior decision-making.
- The large data explosion gives those who conduct data analyses before making decisions about those who use traditional data to operate their business competitively. Solutions such as Amazon Redshift will surely be at the forefront of relation-based data warehousing and Spark and Kafka are promising warehouse data warehouse solutions.

Acknowledgements The author acknowledges Coursera course "Introduction to big data", "Deep Learning Specialization" which help to understand the basic concept of machine learning and big data analytics. Author sincere thanks and dedication of this book to all the reviewers who review the book chapter and helps in the improvement.

Author Contribution Both the author has the equal contribution in the preparation and design of manuscript of the concern book chapter. As Ajay Ph.D. Research Scholar (Dept of BT&BI) JUIT Solan has prepared the manuscript and all the revision part of the book chapter was done by Dr. Raj.

Conflict of Interest There is no conflict of interest.

References

1. Greene, C.S., et al.: Big data bioinformatics. J. Cell. Physiol. **229**(12), 1896–1900 (2014)
2. Li, Y., et al.: Deep learning in bioinformatics: Introduction, application, and perspective in the big data era. Methods **166**, 4–21 (2019)
3. Azmoodeh, A., Dehghantanha, A.: Big data and privacy: challenges and opportunities. In: Handbook of Big Data Privacy, p. 1–5. Springer (2020).
4. Wang, J., et al.: Big data service architecture: a survey. J. Internet Technol. **21**(2), 393–405 (2020)
5. Sagiroglu, S., Sinanc, D.: Big data: a review. In: 2013 International Conference on Collaboration Technologies and Systems (CTS). IEEE (2013)
6. Anuradha, J.: A brief introduction on Big Data 5Vs characteristics and Hadoop technology. Procedia Comput. Sci. **48**, 319–324 (2015)
7. Dean, J., Ghemawat, S.: MapReduce: simplified data processing on large clusters. Commun. ACM **51**(1), 107–113 (2008)
8. Dean, J., Ghemawat, S.: MapReduce: simplified data processing on large clusters (2004)
9. Liu, J., et al.: A novel configuration tuning method based on feature selection for hadoop MapReduce. IEEE Access **8**, 63862–63871 (2020)

10. Shastry, K.A., Sanjay, H.: Machine learning for bioinformatics. In: Statistical Modelling and Machine Learning Principles for Bioinformatics Techniques, Tools, and Applications, pp. 25–39. Springer (2020)
11. Li, H., et al.: Modern deep learning in bioinformatics. J. Mol. Cell Biol. (2020)
12. Srinivasa, K., Siddesh, G., Manisekhar, S.: Statistical Modelling and Machine Learning Principles for Bioinformatics Techniques, Tools, and Applications. Springer Nature (2020)
13. Pezeshki, M., et al.: Gradient Starvation: A Learning Proclivity in Neural Networks. arXiv preprint arXiv:2011.09468 (2020)
14. Havaei, M., et al.: Brain tumor segmentation with deep neural networks. Med. Image Anal. 35, 18–31 (2017)
15. Leshno, M., et al.: Multilayer feedforward networks with a nonpolynomial activation function can approximate any function. Neural Netw. 6(6), 861–867 (1993)
16. Ramachandran, P., Zoph, B., Le, Q.V.: Searching for Activation Functions. arXiv preprint arXiv:1710.05941 (2017)
17. Gomes, G.S.d.S., Ludermir, T.B., Lima, L.M.: Comparison of new activation functions in neural network for forecasting financial time series. Neural Comput. Appl. 20(3), 417–439 (2011)
18. Sharma, S., Sharma, S.: Activation functions in neural networks. Towards Data Sci. 6(12), 310–316 (2017)
19. Agostinelli, F., et al.: Learning Activation Functions to Improve Deep Neural Networks. arXiv preprint arXiv:1412.6830 (2014)
20. Obla, S., et al.: Effective activation functions for homomorphic evaluation of deep neural networks. IEEE Access 8, 153098–153112 (2020)
21. Goyal, M., et al.: Activation functions. In: Deep Learning: Algorithms and Applications, pp. 1–30. Springer (2020)
22. Zadeh, M.R., et al.: Daily outflow prediction by multi layer perceptron with logistic sigmoid and tangent sigmoid activation functions. Water Resour. Manag. 24(11), 2673–2688 (2010)
23. Gomar, S., Mirhassani, M., Ahmadi, M.: Precise digital implementations of hyperbolic tanh and sigmoid function. In: 2016 50th Asilomar Conference on Signals, Systems and Computers. IEEE (2016)
24. Kalman, B.L., Kwasny, S.C.: Why tanh: choosing a sigmoidal function. In: Proceedings 1992 IJCNN International Joint Conference on Neural Networks. IEEE (1992)
25. Xu, B., et al.: Empirical Evaluation of Rectified Activations in Convolutional Network. arXiv preprint arXiv:1505.00853 (2015)
26. Agarap, A.F.: Deep Learning Using Rectified Linear Units (RELU). arXiv preprint arXiv:1803.08375 (2018)
27. Zhang, X., Zou, Y., Shi, W.: Dilated convolution neural network with LeakyReLU for environmental sound classification. In: 2017 22nd International Conference on Digital Signal Processing (DSP). IEEE (2017)
28. Chan, J.O.: An architecture for big data analytics. Commun. IIMA 13(2), 1 (2013)
29. Boja, C., Pocovnicu, A., Batagan, L.: Distributed parallel architecture for "big data." Inf. Econ. 16(2), 116 (2012)
30. Hutter, F., Kotthoff, L., Vanschoren, J.: Automated Machine Learning: Methods, Systems, Challenges. Springer Nature (2019)
31. Li, H.: Deep learning for natural language processing: advantages and challenges. Natl. Sci. Rev. (2017)
32. Sharma, O.: Deep challenges associated with deep learning. In: 2019 International Conference on Machine Learning, Big Data, Cloud and Parallel Computing (COMITCon). IEEE (2019)
33. Angelov, P., Sperduti, A.: Challenges in deep learning. In: ESANN (2016)
34. Krizhevsky, A., Sutskever, I., Hinton, G.E.: Imagenet classification with deep convolutional neural networks. Adv. Neural. Inf. Process. Syst. 25, 1097–1105 (2012)
35. Deng, J., et al.: Imagenet: a large-scale hierarchical image database. In: 2009 IEEE Conference on Computer Vision and Pattern Recognition. IEEE (2009)
36. Albawi, S., Mohammed, T.A., Al-Zawi, S.: Understanding of a convolutional neural network. In: 2017 International Conference on Engineering and Technology (ICET). IEEE (2017)

37. Ma, N., et al.: Shufflenet v2: practical guidelines for efficient cnn architecture design. In: Proceedings of the European Conference on Computer Vision (ECCV) (2018)
38. Sun, Y., et al.: Completely automated CNN architecture design based on blocks. IEEE Trans. Neural Netw. Learn. Syst. **31**(4), 1242–1254 (2019)
39. Yu, D., et al.: Mixed pooling for convolutional neural networks. In: International Conference on Rough Sets and Knowledge Technology. Springer (2014)
40. Zhao, H., et al.: Loss functions for image restoration with neural networks. IEEE Trans. Comput. imaging **3**(1), 47–57 (2016)
41. Liu, W., et al.: Large-margin softmax loss for convolutional neural networks. In: ICML (2016)
42. Bartlett, P.L., Wegkamp, M.H.: Classification with a reject option using a hinge loss. J. Mach. Learn. Res. **9**(8) (2008)
43. De Soete, G., Carroll, J.D.: K-means clustering in a low-dimensional Euclidean space. In: New Approaches in Classification and Data analysis, pp. 212–219. Springer (1994)
44. Sherstinsky, A.: Fundamentals of recurrent neural network (RNN) and long short-term memory (LSTM) network. Phys. D: Nonlinear Phenom. **404**, 132306 (2020)
45. Yin, R., et al.: Tempel: time-series mutation prediction of influenza A viruses via attention-based recurrent neural networks. Bioinformatics **36**(9), 2697–2704 (2020)
46. Millham, R., Agbehadji, I.E., Yang, H.: Parameter tuning onto recurrent neural network and long short-term memory (RNN-LSTM) network for feature selection in classification of high-dimensional bioinformatics datasets. In: Bio-inspired Algorithms for Data Streaming and Visualization, Big Data Management, and Fog Computing, pp. 21–42 Springer (2021)
47. Gerratana, L., et al.: Abstract P5-01-10: next generation sequencing-based gene variant-oriented characterization in metastatic breast cancer: an innovative analysis using ctDNA. AACR (2020)
48. Schmidt, B., Hildebrandt, A.: Deep learning in next-generation sequencing. Drug Discov. Today (2020)
49. Naito, T., et al.: A deep learning method for HLA imputation and trans-ethnic MHC fine-mapping of type 1 diabetes. Nat. Commun. **12**(1), 1–14 (2021)
50. Cosgun, E., Oh, M.: Exploring the consistency of the quality scores with machine learning for next-generation sequencing experiments. BioMed. Res. Int. (2020)
51. Tonkovic, P., et al.: Literature on applied machine learning in metagenomic classification: a scoping review. Biology **9**(12), 453 (2020)
52. Huang, Y., Zhang, P.: Evaluation of machine learning approaches for cell-type identification from single-cell transcriptomics data. Brief. Bioinf. (2021)
53. Loher, P., Karathanasis, N.: Machine learning approaches identify genes containing spatial information from single-cell transcriptomics data. Front. Genet. **11**, 1743 (2021)
54. Oller-Moreno, S., et al.: Algorithmic advances in machine learning for single cell expression analysis. Current Opin. Syst. Biol. (2021)
55. Kupari, J., et al.: Single cell transcriptomics of primate sensory neurons identifies cell types associated with chronic pain. Nat. Commun. **12**(1), 1–15 (2021)
56. Zhu, W., et al.: The application of deep learning in cancer prognosis prediction. Cancers **12**(3), 603 (2020)
57. Wei, R., Mahmood, A.: Recent advances in variational autoen-coders with representation learning for biomedical informatics: a survey. IEEE Access (2020)
58. De Vivo, M., et al.: Role of molecular dynamics and related methods in drug discovery. J. Med. Chem. **59**(9), 4035–4061 (2016)
59. Lima, A.N., et al.: Use of machine learning approaches for novel drug discovery. Expert Opin. Drug Discov. **11**(3), 225–239 (2016)
60. Gertrudes, J., et al.: Machine learning techniques and drug design. Curr. Med. Chem. **19**(25), 4289–4297 (2012)
61. Klambauer, G.N., Hochreiter, S., Rarey, M.: Machine Learning in Drug Discovery. ACS Publications (2019)
62. Zhang, S., et al.: Learning for personalized medicine: a comprehensive review from a deep learning perspective. IEEE Rev. Biomed. Eng. **12**, 194–208 (2018)

Deep Learning for Medical Informatics and Public Health

K. Aditya Shastry⬤, H. A. Sanjay⬤, M. Lakshmi, and N. Preetham

Abstract The technology & healthcare intersect to constitute medical informatics (MI). MI enhances the outcomes of patients & healthcare through the skills of medical & computer sciences. The fusion of both these disciplines enables the related personnel to improve the patient care along with the research & clinical settings. Public health (PH) represents the science of safeguarding & improving community health. This outcome is attained by encouraging lifestyles that are healthy in nature, investigating new illnesses, preventing injuries and contagious illnesses. In general, PH focuses on safeguarding the health of entire communities. These communities can be small (local communities) or large (whole country/a global region). In this regard, Deep learning (DL) represents an interesting area of research. DL application has been observed in several domains due to the fast growth of both data & computational power. In recent years, the application of DL in the domain of MI has increased due to the probable advantages of DL applications in healthcare. DL can aid medical experts in diagnosing several illnesses, detecting sites of cancer, determining the impacts of medicines on each patient, comprehending the association among phenotypes & genotypes, discover novel phenotypes, & forecasting the outbreaks of contagious illnesses with higher precision. When compared to classical techniques, DL does not need data which is specific to a domain. DL is likely to transform the lives of humans. Regardless of these benefits, DL faces certain drawbacks related to data (higher number of features, dissimilar data, reliance on time,

K. Aditya Shastry (✉) · M. Lakshmi · N. Preetham
Department of Information Science & Engineering, Nitte Meenakshi Institute of Technology, Yelahanka, Bangalore, Karnataka 560064, India
e-mail: adityashastry.k@nmit.ac.in

M. Lakshmi
e-mail: lakshmi.m@nmit.ac.in

N. Preetham
e-mail: preetham.nagaraju@nmit.ac.in

H. A. Sanjay
M S Ramaiah Institute of Technology, Bangalore 560054, India
e-mail: sanjay.ha@msrit.edu

© The Author(s), under exclusive license to Springer Nature Switzerland AG 2022
K. R. Ahmed and H. Hexmoor (eds.), *Blockchain and Deep Learning*,
Studies in Big Data 105, https://doi.org/10.1007/978-3-030-95419-2_13

unsupervised data, etc.) & model (dependability, understandable, likelihood, scalable, safety) for real world applications. This chapter emphasizes on DL techniques applied in MI & PH, recent case studies related to the application of DL in MI & PH, and certain critical research questions.

Keywords Big-data analytics · Agriculture · Prediction · Analysis

1 Introduction

For analysis of data layered algorithmic architecture such as DL or deep structured learning is employed. Hierarchical learning is another name given to DL which is a type of machine learning (ML). The models in DL filters data through numerous hidden layers. Each successive layer utilizes the output from previous layer to predict its results. Accuracy can be achieved by processing more data and refining the previous results that can lead to more correlations and connections. The idea for DL is roughly centered on the basic connection of biological neurons with one another in the brains of animals to process information. It's similar to how electrical signals travel across cells in living organisms. The nodes in the subsequent layer get activated when they receive a stimulus from their neighboring neurons [1].

Figure 1 demonstrates the fundamental design of a DL architecture.

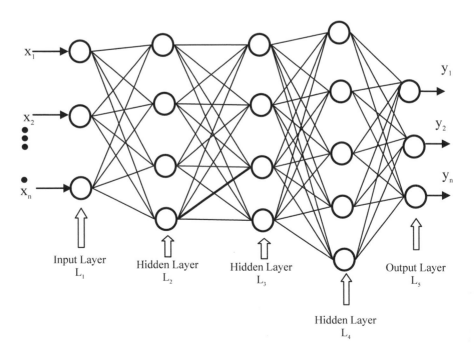

Fig. 1 DL architecture [2]

As demonstrated in Fig. 1, the network model comprises of many hidden layers with each layer containing several neurons. When each layer is allocated a portion of a transformation task, the layers processes the data multiple times to further refine and optimize the final output. Mathematical translation tasks are used by hidden layers to turn a raw input to a more significant output [2, 3].

Pre-processing of data is comparatively less in DL techniques. The filtering and normalization tasks in several other ML techniques needs to be completed by a human programmer whereas in DL, it is handled by the network only. As specified in article Nature the conventional ML techniques have limited capacity for processing raw data.

For decades, substantial domain proficiency & cautious engineering was needed for creating an ML system for devising an attribute extractor capable of converting raw data (like values of image pixels) into an appropriate vector of attributes. This transformed data is given to the learning subsystem, usually a classifier, that could identify or classify patterns. DL networks have the capability to determine the patterns needed for automatic classification which reduces the need for supervision and which in turn helps speeding up the procedure of mining the essential understandings from datasets that have not been as extensively organized. DL models utilize advanced mathematical models for development of networks. Several variations of networks exist that control different sub-strategies within this field. The developing discipline of DL is very quickly powering the advanced computing capabilities in the world, bridging every industry and adding substantial value to user experiences and viable decision-making [1].

MI & systems which are related to supporting decisions form the basic components for DL in health Informatics. The researchers train the models by feeding the data obtained from the clinical information provided by the clinicians by analyzing the patient condition. Data may include reading clinical images, outcome prediction, determining the relationships among genotypes and phenotypes or between the disease and phenotype, studying the response to the treatment, tracking an abrasion/change in structure. Predicting results of a disease based on analysis of risks can be extended to develop a system that detects early symptoms of a disease. Recognizing correlations and patterns can be drawn-out to global pattern research and population healthcare, providing predictive treatment for the total population [2, 3].

DL in MI can be trained without a prior knowledge, which results in lack of labelled data and is a burden on clinicians. For instance, the target points that were overlapped & 3D/4D medical images, dealt with complexity of data. Scientists delivered improved & understandable results by employing supervised/semi-supervised learning, transfer learning, augmentation of information & architectures related to multi-modality [4–8]. Additionally, the research assisted the scientists in determining non-linear associations among variables which in turn assisted the medical experts in understanding the illness & its related solutions. Since decisions are based on data provided, no models or human interference can divide the category into subgroups according to their clinical data. Sequences of RNA / DNA in bioinformatics, were researched to recognize alleles of genes along with ecological aspects that caused the illnesses. Secondly, it assisted in examining the interactions between proteins, comprehend processes at a higher level, devise therapies that were customized, determine resemblances among 2 phenotypes, etc. [9].

DL algorithms were applied to predict the interweaving movement of exons, the characteristics of the binding proteins of RNA/DNA, & DNA methylation [10–12]. Thirdly, it validates its effectiveness, particularly when predicting rapidly developing diseases such as acute renal failure. Instead of regular medical visits, research determined that it is adequate to employ novel phenotypes for forecasting in real-time [13–21]. Fourth, it is anticipated to be extensively utilized for transferred patients, for 1st time inpatients, outpatients lacking the information of charts and weak healthcare infrastructure patients [22, 23].

For instance, neuro signals such as EEG, PPG, etc. are utilized to forecast the freezing of Parkinson's illness. Also, information from accelerometer & mobile apps are employed for screening health status, arthritis, diabetes, heart related illnesses, and chronic diseases for offering information related to health before the patients undergo hospitalization in emergency situations [24, 25]. Also, analysis of X-ray images captured by a mobile phone helped for treating marginalized communities and resource-poor people [26]. Clinical notes together with discharge notes summarization are intended to determine how summarization records express accurate, effective and reliable information for comparing the information with medical records. Finally, social behavior, disease outbreaks, research related to medicines and analysis of treatments have proven to avoid illnesses, lengthen life along with the monitoring of epidemics [2, 27–33].

Several hospitals are making use of the EHR systems. As per [34, 35], around 70% of clinicians & 90% of USA hospitals are utilizing EHR system for improving their competence levels & efficiency. Patient information can be recorded by employing technologies related to imaging, genomes, and sensors that are wearable. DL architectures with advanced computation power support GPUs have a major impact on the practical acceptance of DL. Hence, several experimental works have applied DL models for MI which is mostly being utilized by several clinical experts. Yet, the DL application to health informatics leaves us with a number of challenges that must be resolved, including lack of data (missing values, expensive labelling, class imbalance), data interpretation (heterogeneity, high dimensionality, multi-modality), reliability of information & integrity, reliability and model interpretability (convergence issues and tracking along with overfitting), feasibility, scalability and security [2].

The chapter remainder has 4 sections. The 2nd section emphasizes the different DL methods in MI & PH. The 3rd section demonstrates how DL can be applied in the field of MI & PH with different real-world case studies. The fourth section analyses the various research issues related to DL in MI & PH that can be determined by various stakeholders in healthcare. The summary of the chapter is given at the end.

2 Deep Learning Techniques in Medical Informatics and Public Health [36]

In this segment, we elaborate the operational functions of 5 DL models in which each model has various forms. When an input is given, an expected output is produced according to the fundamental principle of approximating a function. The diverse models are better suited to manage diverse challenges & diverse data types for performing the expected tasks [36].

The speech or time series classification model is comparatively different from that of image classification. The data dimensionality is reduced by applying a pre-processing phase in the models. Primarily the model structure consists of several interconnected neurons to each other & hidden layer that links the input to the output. Consequently, an activation sequence is generated via the weighted connection from the neurons. This process is known as feedforward [37]. Comparison between DL and NN models puts forward the fact that DL model includes more of hidden layers than NN. NN model can be prepared only for supervised learning jobs whereas DL is used for training both unsupervised & supervised learning jobs. Output layer results are compared with the actual value at the end of the feedforward process. An error value is computed by calculating the difference in the values of predicted & actual targets. As the process of feedforward concludes, the weights of the ANN model are updated in order to reduce this error and make the predicted and actual values closer. This technique is known as backpropagation [38, 39].

2.1 Autoencoders (AE)

AE is intended for extraction of features by utilizing information driven learning. AE is subjected to unsupervised training since the training takes place for recreating the input vector instead of assigning the class label. Similar number of input & output neurons are present in AE with the neurons being fully connected with those of the next layers. The input neuron number is typically lesser than the hidden neuron number. The reason for this design is to code information in reduced feature space & to accomplish the mining of attributes. Whenever data dimensionality is high, several AEs are stacked together for the creation of a deep AE system. Several variants of AE have been devised over the previous decade for handling diverse patterns of data and for executing particular functions. For instance, Vincent et. al [40] proposed the denoising AE. Its purpose was to strengthen the traditional AE model. This technique reconstructs the input by inducing certain disturbances to the patterns, consequently compelling the design to grasp the input design.

Sparse AE represents another variant of the classic AE model. In this variant, sparse representation is utilized for making the data more separable [41]. Another variant of AE known as the convolutional AE [42] incorporates shared weights among nodes for processing 2-D patterns & preserving spatial locality. Contractive AE is

Fig. 2 Autoencoders [36]

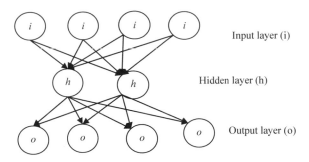

Input layer (i)

Hidden layer (h)

Output layer (o)

like denoising AE, yet as opposed to inducing noise to disrupt the training dataset, it alters the function to determine error by including analytical contractive cost [43, 44]. The learning procedure for Autoencoder is depicted as limiting a cost function C with the end goal that C (k, m(t(u))). t(u) is a capacity that maps u to h & the capacity k maps h to the output that is a reproduction of the input. r denotes the weight interfacing the layers. Figure 2 demonstrates the simple architecture of AE.

Figure 2 shows the design of a simple AE demonstrating the i, h and o layers. The direction of the arrows depicts the interconnection between the neurons.

2.2   Recurrent Neural Network (RNN)

RNN is a category of DL that comprises of associations with neurons in the 'h' layer to create a series of directed graph. This component provides it a progressive unique mechanism. This is significant in situations where the output relies upon the past calculations, for example, the examination of texts, noises, DNA groupings, and persistent electrical body signs. The RNN training is done on information having dependencies for maintaining info about the preceding interval. Performance outcome at time 't-1' impact the selection at time t. It deliberates the preceding output (O) & the current input (I) & a number in the range of 0 to 1 is generated from the cell state M. Here, 1 denotes saving the value while 0 signifies discarding the value.

Sigmoid layer (Gate layer) makes this choice. Hence, the RNN principle defines the recurring function with respect to time steps that are done utilizing the equation: $M = f(M \times W \times I \times W)$, where M signifies time state k, M denotes previous gate output, I represents input at time k, and W is the network weight parameters. The RNN can be regarded as a state that comprises of the feedback loop [45]. Feedforward RNN implementation is demonstrated in Fig. 3.

The outcome of the output layer is looped back as input part to the input layer. The I_t & O_t denote the input & output at time t & O_{t-1} represents the output for the preceding input at time t-1. Apart from the structural difference, RNN utilizes similar weights across all layers whereas other DL employ diverse weights. This prominently reduces the total features that the system must study. In spite of the

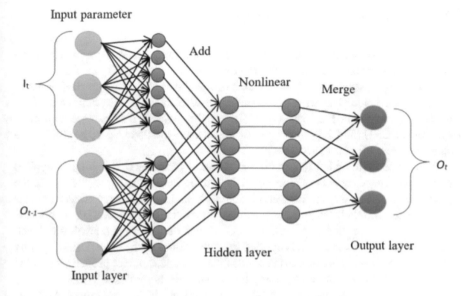

Fig. 3 Feedforward RNN [36]

application of the effective model, the disappearing slope by extensive input order & detonating slope issues form the major hindrance as depicted in [46]. To deal with this constraint, LSTM was invented by [47]. In particular, LSTM shown in Fig. 4 is mainly appropriate for applications possessing lengthy time lags having undefined sizes among vital events.

To accomplish this, LSTMs employ novel information sources for storing, writing & reading to/from node at every step. At the training time, errors in classification are reduced by permitting either reading/writing [48]. One more form of RNN is the gated repetitive component, which denotes enhanced form of LSTM that has performs comparable to LSTM [49].

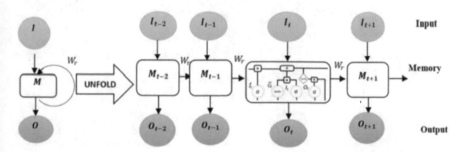

Fig. 4 LSTM [36]

2.3   Convolutional Neural Network (CNN)

CNN was motivated by natural procedures of the human cerebrum in which the connections of patterns is similar to the brain of humans [50, 51]. A classic CNN consists of input, several hidden layers, & an output layer. Following are the important elements of the hidden layers in CNN: - layers of normalization, fully connected (FC), pooling, convolution. For instance, analyzing the imagery data represents a good example of CNN [52]. Figure 5 demonstrates a CNN architecture for recognizing a character from a 3 × 3 matrix image.

The model is intended to perceive X, O and / characters. The filter is applied upon input image by convolution layer. Operation is done using 2 filters over the input image possessing similar weight. Hence, 8 parameters are generated. For this example, the simplicity is preferred over bias. Frequently, an activation (nonlinear) layer called the rectified linear unit (ReLU) is included after the convolution layer. The function g(l) = max (0, l) is applied to all the values of the convolution layer by the activation layer. This process enhances the non-linear model properties and the network as a whole without impacting the convolution layer fields.

In this manner, the vanishing problem that occurs in the traditional ANNs is solved. The neuron cluster output present in the convolution layer is combined into a solitary neuron by the pooling layer [48]. The average, max or sum pooling are used to achieve this [53].

In Fig. 6, the general CNN model is introduced, that depict the layers of input, output, FC, and convolution + pooling.

The attribute mining is performed by conv+pool layers. In order to characterize the output, the FC layer plays the role of a classifier above the features and allots a likelihood score for the information image. In this layer, m × m × r image is the input (m→image height, w→image width and r→number of channels). The size of the filter (a privately associated structure) can be formed with the measurements

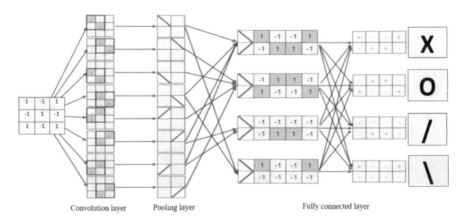

Convolution layer   Pooling layer   Fully connected layer

Fig. 5 CNN architecture [36]

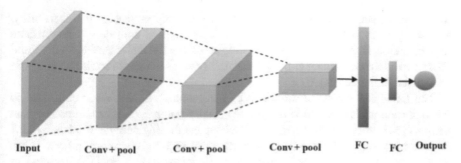

Fig. 6 Framework of CNN [36]

of k having m–n + 1. Each attribute map is applied with sigmoidal nonlinearity & additive bias. The attribute vector of the information which is a complex & combined data from the conv+pool layers is represented by the FC layer. The input image is predicted by utilizing the final feature vector [36].

2.4   Deep Boltzmann Machine (DBM)

A system of evenly coupled stochastic visible and hidden units is known as a Boltzmann machine (BM) [54]. The design of BM is outlined in the first diagram of Fig. 7. To assess the data dependent & independent prospects in a pair of connected binary features, BM algorithm needs Markov chains that are arbitrarily initialized to attain symmetric distributions [55]. Using this framework, learning strategy is extremely slow in real-time [54]. Restricted Boltzmann machine (RBM) was devised to accomplish effective learning. RBM has no connections among hidden units [56].

A simple design of RBM with associations between neurons is as depicted in the second outline of Fig. 7. An RBM feature which is valuable is the hidden unit

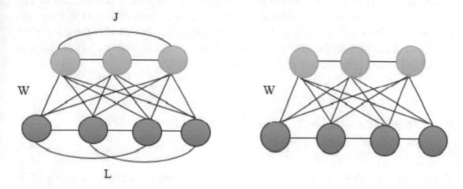

Fig. 7   Structure of BM [36]

distribution given the units that are visible. By directly increasing the probability, a cluster of marginal posterior distributions is obtained by considering the attribute representation of the RBM where the inferences are tractable. Besides, DBM and DBN are the two fundamental Deep Learning systems in this class which have been introduced in literatures [57].

The DBM NN has more variables & layers that are hidden when compared to RBM. Inside all layers, DBM design has totally undirected associations between neurons [54]. In Fig. 7, the image on RHS depicts the engineering of a basic DBM NN for one visible & hidden layer. Nevertheless, amid the neurons in a layer, it has connections that are undirected between all layers of the system. For DBM training, in order to boost the lower bound of the likelihood, a stochastic most extreme likelihood-based calculation is applied. This is on the grounds that computing the posterior hidden neurons distribution, given the neurons that are visible, can't be accomplished by directly maximizing the probability due to the collaborations among the neurons that are hidden.

Also, in instances of semi supervised learning, high level representation could be worked via restricted labeled information and huge contribution of un-tagged sources of input would then be able to be utilized to tweak the model for specific tasks. Feedback in top-down fashion can be incorporated to facilitate the propagation of uncertainty and therefore handle ambiguous inputs in a more robust fashion.

2.5 Deep Belief Network (DBN)

DBN signifies another RBM variation in which several hidden layers learn by feeding one RBM output as input to another RBM layer [55, 56]. The connections among its 2 topmost layers are undirected while the next layers are connected directly. The technique of training follows the greedy approach which is done during unsupervised training of the DBN. Based on the output expected fine tuning of parameters is done.

The figure to the left of Fig. 8 demonstrates the DBN design comprising of a 3-layer configuration depicting the connections which are symmetric in nature. The design consists of numerous layers of hidden neurons that are trained utilizing BP algorithm [58, 59]. As demonstrated in Fig. 8, the DBN design comprises of connection units among neurons present in different layers. Nevertheless, in contrast to RBM, intra-connections between neurons of a layer does not exist.

3 Applications of Deep Learning in Medical Informatics and Public Health [60–62]

This section describes the different case studies of DL that is applied for doing MI & PH.

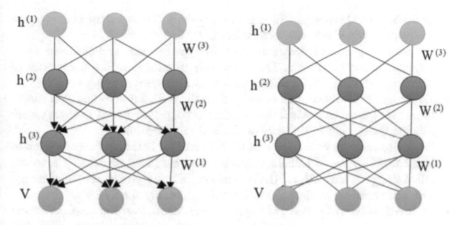

Fig. 8 DBN architecture [36]

DL methods utilize data gathered in the records of EHR for addressing several issues related to healthcare such as reduction of the misdiagnosis rate & forecasting of the procedure results. ANNs are able to assist medical experts in analyzing vast data & identify numerous factors such as:

- Analysis of samples of blood
- Tracking of the levels of glucose in patients suffering from diabetes
- Identifying problems related to heart
- Detection of tumors by applying image analysis
- Cancer diagnosis by detecting cells that are cancerous
- Early diagnosis of osteoarthritis from a MRI scan.

3.1   The Use of DL for Cancer Diagnosis

For many years, oncologists have been employing medical imaging approaches such as Computed Tomography (CT), MRI & X-ray for diagnosing cancer. Though these methods are effective for detecting several cancer types, certain cancers still exist that are not diagnosable accurately. CNN shows a great potential for detecting such cancer types. ANNs are able to diagnose the cancer early with a reduced rate of misdiagnosis on similar medical images thus delivering improved results for the patients. Following are some studies in which scientists have effectively utilized DL models for diagnosing diverse cancer types with a higher accuracy:

- The rate of breast cancer misdiagnosis was reduced by around 85% by employing a DL model as per the Nvidia study.
- Hossam Haick designed a device to treat cancer motivated by the fact that his roommate was detected with leukemia. A team of scientists trained an ANN

model based on his design. This model recognized 17 diverse cases based on the breath smell of patients with an accuracy of 86%.

- Effective lung cancer detection from CT images was demonstrated by a researcher's team at Enlitic. They were able to achieve a 50% higher rate of detection on test data when compared to a team of expert radiologists.
- Training of CNN model was done by Haenslle et al. for detecting skin cancer by assessing whether a skin lesion displayed in digital imaging can cause cancer with the same or higher accuracy of a skilled skin specialist.
- Google researchers have devised a CNN model for detecting cancer related to breast from images of pathology quicker with enhanced accuracy. The model called LYmph Node Assistant (LYNA) attained a very high success rate of around 99% in comparison to physicians who had an accuracy of only 38%.
- DL based analysis of radiology images is being done in over 100 hospitals around the globe which has assisted radiologists in treating millions of patients [60–62].

3.2 DL in Disease Prediction and Treatment

The costs of hospitalization in US was in billions during 2006. All these hospitalizations could have been prevented. Majority of the patients who were hospitalized suffered from either heart related ailments or diabetes. DL can be utilized for improving the rate of diagnosis and forecast more accurately which in turn can reduce the number of hospitalizations leading to reduced cost.

Some research teams have already applied their solutions to this problem. Researcher team from the University of Boston teamed up with hospitals in the Boston locality. They were able to achieve an accuracy of around 82% for predicting which patients suffering from heart ailments & diabetes would require hospitalization during the subsequent year. The EHR records were analyzed using DL model for forecasting heart failures for the subsequent 9 months before the medical experts could predict [60–62].

- **Diabetic Retinopathy (DR)**: DR is a form of blindness caused due to poor control of diabetes. Around 415 million people in developing nations suffer from DR. DL has been found very effecting in preventing DR. For instance, the CNN technique analyzes the retinal image data for recognizing internal bleedings in the eye, initial symptoms, and any characteristics indicating DR. The major cause of DR is due to the extreme fluctuations in the levels of blood glucose. Regular monitoring of the glucose levels of diabetic patients can be a good way of controlling diabetes and hence preventing DR. Hence, the DL model can be very effective in analyzing the glucose level data for forecasting sharp surges or sharp lows. This in turn can aid the patients in controlling their diet. For instance, if low sugar occurs then they can eat sugar related food or if high sugar occurs, they can inject insulin [60–62].
- **Human Immunodeficiency Virus (HIV)**: It has been found that around 36 million people around the globe are affected by HIV. Regular antiretroviral dosages are required by HIV affected patients. Since HIV can quickly mutate,

the drugs given to the patients need to be periodically changed. A form of DL model called Reinforcement Learning (RL) can assist in this activity. For treating the HIV patients continuously, the RL technique can keep track of several indicators in the environment with each administration of the drug. This assists in providing the superlative action to modify the sequence of drugs for continuous treatment. Toronto University devised a CNN model called DeepBind which took genomic data & forecasted the DNA & RNA sequence binding proteins. It can detect changes in the DNA sequence that can lead to the development of improved tools for medical diagnosis & medications [60–62].

- **Drug Discovery**: The discovery & development of drugs is significantly assisted by DL. The medical history of patient is analyzed for delivering the better treatment for them. Besides, DL is acquiring insights from the symptoms of patients & tests. The domain of drug discovery adds significant economic value to ML application developers and other stakeholders like physicians, CEOs, patients, nurses, insurance companies, etc. as it involves large incentives and profits. The pharmaceutical or drug companies are very lucrative industries that will become potential customers for the drug development process. IBM has started the concept of drug discovery from very early days. Nowadays, Google has also been involved in the process of drug discovery using DL along with other major organizations [60–62].
- **Medical imaging**: Several dangerous illnesses like cancer, tumors related to brain, diseases associated with heart can be diagnosed using medical imaging methods like scans of MRI, CT, and ECG. In this regard, DL assists medical experts for analysis of illness.

  The computer vision represents one of the noteworthy innovations that came up due to the ML and DL. It is being widely applied in medical informatics. For instance, an initiative called InnerEve from Microsoft that was found in 2010 is currently developing tools for image diagnostic. In this regard, they have published several videos describing their progress. DL plays a primary role in the application development related to diagnostic. This is very much possible as more data sources comprising of rich & diverse forms of medical imagery are being made available & accessible. Nevertheless, DL represents a "black box" in which their predictions cannot be explained even if they are accurate. This type of "black box" problem becomes even more challenging in medical related applications where it is a matter of life and death. This is because the medical experts require a clear understanding on how the DL application arrived at its outcome even if these forecasts were estimated to be precise earlier [60–62].
- **Insurance fraud**: DL is effectively utilized to identify deceitful claims by analyzing medical insurance policies. Future occurrence of frauds can also be detected using the DL techniques. Additionally, DL assists the insurance companies in targeting potential customers/patients and send them discounts to lure them [60–62].
- **Alzheimer's disease**: The early detection of this illness in humans has been a major challenge faced by the medical organizations. Currently, this illness can only be identified once it occurs in a person. Using traditional methods, early

diagnosis of Alzheimer's disease is not possible. However, DL is being employed for identifying this illness during early stages [60–62].

- **Genome**: DL methods are being employed to comprehend genomes & assist patients in understanding the illness that may impact them. The future of DL in the area of genomics & insurance industry is promising. For example, Entilic employs DL methods to assist doctors in making diagnosis faster & accurate. Cellscope utilizes DL strategies for assisting parents in monitoring their children's health via a smart device. This reduces the frequent visits to doctors. DL has the ability to deliver remarkable applications in public health & medical informatics that can aid doctors in providing improved medical treatments [60–62].

- **Treatment Queries and Suggestions**: Disease diagnosis is a very difficult process and comprises of several varying factors that can vary from the person's color to what food he/she consumes. Currently, machines are not fully capable to match the diagnosis made by human doctors. Nevertheless, in future DL methods can assist the medical experts in the process of diagnosis & treatments, just by delivering an extension of medical knowledge. For example, the department of Oncology of the Memorial Sloan Kettering (MSK) has a partnership with IBM Watson. The MSK possesses huge amount of medical data on cancer patients & the related treatments utilized over several years. Using this information, it can advise various treatment choices to physicians when they encounter future cancer cases by extracting useful information from the historic data. This tool is currently is in preliminary use [60 62].

- **Scaled Up/Crowdsourced Medical Data Collection**: Nowadays, there is a lot of focus on collecting & analyzing live mobile data for medical informatics & public health. For instance, ResearchKit from Apple is focusing on the treatment of Parkinson's disease & Asperger's syndrome by permitting user access to interactive apps (such as ML application for recognition of faces). It is able to evaluate the conditions of the patient over a period of time. This app use allows data to be fed continuously creating a huge pool of data for experimentation and future predictions. Real time health data related to insulin & diabetes is being collected by IBM in partnership with Medtronic for extracting useful diabetic knowledge. IBM also bought a Truven which is an analytics company related to healthcare for 2.6 billion dollars. Even though, tremendous medical data collection has taken place, the IT industry is still struggling to analyze this data and suggest treatments in real time. However, as the pooling of patient data is progressing, in future, researchers may come up with better innovations in handling diverse diseases [60–62].

- **Robotic Surgery**: Recently, surgery performed by robots are gaining significance. In this domain, the da Vinci robot permits surgeons to make the limbs of robots nimbler that aids in precise surgeries with reduced tremors when compared to hands of humans. Although, all the surgical operations performed by robots don't employ DL, certain systems utilize computer vision which is assisted by DL in recognizing specific body parts (such as detecting follicles of hair during hair transplantation operation). Furthermore, DL can be employed to stabilize robotic motions when getting instructions from humans controlling the robot [60–62].

3.3 Future Applications

The list of DL applications in healthcare that are rapidly progressing are summarized below:

- **Personalized Medicine (PM)**

It refers to the process of automatically prescribing drugs based on personalized recommendations. For instance, when a wisdom tooth is pulled from a child, usually a drug called Vicodin. is given. Similarly, a person suffering infection of the urinary track usually gets a medicine named Bactrim. In the near future, it is expected that certain patients may get the same drug dose. Knowing the history and genetics of patients, certain patients may be prescribed with the same drugs. PM has the potential to make the medical recommendations & disease treatments of each person customized based on their genetics, levels of stress, past medical conditions, diet, etc. It is applicable to minor medical circumstances such as giving reduced Bactrim dosage for UTI as well as major medical conditions like deciding whether chemotherapy should be performed on a person based on his age, gender, genetics, etc [60–62].

- **Automatic Treatment or Recommendation**

The autonomous treatment refers to the process of treating patients automatically without disturbing his daily life. For instance, a machine may be developed that automatically adjusts the patient's drug dose of antibiotics by keeping track of their blood level, diet, levels of stress, and sleep. In such scenarios, humans need not remember how many medicines should be taken as an ML agent like Alexa may remind him about what pills to be taken, how much dose to be taken, what time to take or call a doctor if condition worsens. For instance, Hooman Hakami developed by Medtronic is able to autonomously monitor the levels of blood glucose and inject insulins as required without disturbing the daily activities of the patient. But there are several legal issues that need to be addressed before giving such power to machines especially in the medical domain since even small mistakes can cause loss of lives. So, lengthy trials need to be done before autonomous treatments can be introduced into real life [60–62].

- **Enhancing Performance (Beyond Amelioration)**

Recently, IBM & Orreco have partnered for developing a software that could enhance the performance of athletes. A similar partnership exists between IBM & Armor. Although, the key objective of western medicine is on the treatment and cure of disease, greater focus must be given to prevention of diseases as well. In this regard, IoT devices are being employed. Apart from prevention of diseases and performance improvement of athletes, the DL based healthcare apps can be employed to track performance of workers, reduce levels of job stress or seek optimistic enhancements in groups that are at a higher health risk. However, before these applications become feasible, lot of ethical issues need to be resolved [60–62].

- **Autonomous Robotic Surgery**

Currently, robots mostly act as an extension to the human surgeon's abilities such as precision. For mastering surgeries, DL can be employed for integrating visual data & motor patterns within robots. In certain kinds of painting & visual art, machines have adapted the skill to go beyond the expertise of humans. For example, if a machine can simulate the creative ability of a Picaso, then with sufficient training, the machine can perform surgeries better than humans [60–62].

4 Open Issues Concerning DL in Medical Informatics and Public Health [63]

Most of the modern statistical methods, DL are offering novel prospects to put into operation the previously available and fast increasing data sources for the advantages of patients. In spite of progressive research being performed recently, particularly in imaging, the literature still lacks' transparency, lags in exploration for potential ethical worries, clarity in reporting to encourage duplicity & does not effectively demonstrate. It may be beneficial for health care if interdisciplinary groups are combined to perform research and develop projects that involve DL which explicitly addresses a series of queries related to reproducibility, effectiveness, transparency and ethics (TREE). The 9 critical questions suggested here serve as a background for researchers to highlight the reporting, design and conduct, for editors & peer reviewers for evaluating contributions towards literature; and critically review new findings by policy makers and clinicians for benefit of patients. Emphases on some of the Critical Questions are shown below:

(Q.1)   *What is the safety concern for the value of the patient?*

Majority of prediction models published are not practically applied by clinicians [64]. One of the factors is the nonexistence of a medical process for making decisions from which the model can learn meaningful information/optimize data. This has assisted in transitory invention from the previous years in DL for health [65]. Hence it is strongly advised by researchers to see the context in a wider organizational perspective. Researchers need to be sensible during developing & implementing the parts of proposed research in healthcare data science. What is crucial is that this requirement is expressed up front, just like the principles on which research registration is based.

(Q.2)   *What proofs exist to show that best clinical research practices and the design of epidemiological studies influenced the creation of the algorithm?*

Similar patterns to those of past problems with medical investigation have started to emerge in DL-based study, like utilizing the result attributes as forecasters, not considering the underlying mechanisms, inadequate explanations of the DL outputs, and recording precisely the type of patients that were allowed in the research. The PECO design rules of epidemiology (i.e. identifying sample residents, contacts utilized, & medical results) play a significant part in certain problems that emerged in medical science. These rules constitute as effective guidelines for validity assessment and significance of research proof [66].

(Q.3)   *What is the Patient involvement in the gathering, review, delivery & utilization of data?*

With the rising utilization of regularly gathered patient information, frequently having alternate lawful justification (i.e., valid interests) for specific permission, it is now further significant that patient participation is viewed as complimenting the healthcare investigation. Exempting researchers from obtaining separate permission doesn't mean that they are excused from involving the patients & the community completely. Consequently (if applicable) healthcare DL plans must comprise of a clear strategy for assessing the suitability of the devised technique & results for certain data-gathering entities, consumers (i.e., clinicians) and those affected (i.e. those who use the model to notify clinical management).

Several existing frameworks [67] explain how a research project could include patients and the public. It is strongly recommended for the researchers to decide the phases of their project, if any, which is appropriate for public & patient involvement (at the outset). For example, to define the need for a predictive modeling approach, to support the improvement of the algorithm, & to evaluate the algorithm's suitability.

(Q.4)   *Do the data gathered answer the medical queries-i.e. are they heteroge-nous in nature as they exist in the real-world & is the quality of the data sufficient?*

The main problem here is to verify if the data available can address the clinical query. For instance, a dataset that does not contain the (known) appropriate or signif-icant forecasters of a result is improbable to respond reasonably to queries about it. Without sufficient data, no DL strategy can produce accurate results. To assist in explaining certain possible problems that are present in deciding if the information is of appropriate value; Following are two key areas where researchers often find it difficult to apply DL techniques to data relating to healthcare:

- **Inherent sample features**: When data are available but of low quality or are not appropriate, it is impossible to produce a successful DL application. The correct-ness of data gathering approaches, participant sampling, eligibility requirements & missing information should be taken into consideration for evaluating the potential for the development of effective & generalizable DL strategies [68].
- **Task significance**: Because of the probability of failure while working beyond the training data range, techniques are frequently incapable of reaching the accu-racy observed during training. For example, when you first encounter a cyclist at night, the DL system for identifying the image/driverless car may fail. The data—including time scale, heterogeneity must therefore be consistent.

(Q.5) *Does the methodology for validation represent the practical restrictions &*
       operating processes related with information gathering & storage?

DL work is gradually using regularly collected data, including health-care data (e.g., genomic information, and EHR), municipal organizational information (e.g., educational achievement and death records), and portable & wearable device data [69]. Despite the real-world limitations, DL algorithms are mostly tested on past data, providing definite results if the process of producing information doesn't alter. These conventions are frequently not followed in practice & results in DL techniques performing badly when installed as opposed to the performance observed in development [70].

This problem can be considered by researchers as 2 distinct, but associated problems. The first concerns ensuring the development of a fool-proof authentication system. For instance, approaches considering time and producing momentarily disorderly training & test data can be considered when collecting & processing the data [71, 72]. The issue number 2 is to avoid redundancy of a suitable result due to drift in institution data gathering, or methods of storage. However, developers and researchers can do little to prove their work in the future, apart from utilizing best reproducibility practices to decrease the volume of effort essential to reinstall the appropriate service.

(Q.6) *Which computational & technological resources are needed for the*
       mission, & are the existing resources enough to deal with this issue?

Working with several factors is prevalent in predictive modeling associated with health like DL based image & computational genetics [73]. It's therefore a normal routine to determine the data complication & the available resources for computation, since these resources form the restrictive feature (with conventional statistical models) that determine the analysis to be performed [74]. In certain instances, additional resources for computation might permit training of improved models. For instance, the utilization of prototypes centered on complicated ANNs can be absurdly tough without sufficient computer resources, particularly if these strategies need extra, complicated functions for preventing overfitting (e.g., regularization) [75]. Preferably, investigation shouldn't be constrained by the amount of computation resources. However, scientists must appreciate the limitations that they work in, so that any study can be customized to the needs. Comparable issues may occur when utilizing protected computing scenarios, like reserves of data, wherever the applicable software systems might not be accessible and will therefore require implementation from scratch onwards [76, 77]. Therefore, understanding the repercussions of using particular software's is also critical, as the underlying license may have comprehensive effects on the market prospective & additional traits of the future of the procedure [78].

(Q.7) *Are the metrics of performance stated applicable to the clinical context*
       that the model should be used in?

Selecting metrics of performance is required for converting the good performance during training into good performance when the system is deployed in the real clinical environment. This disparity in the performance of the method may occur for many reasons; the most important of which is that assessment measures are not good alternatives for patients to demonstrate better results (E.g., mis-categorization error for Unbalanced Class screening application). Another common error is to select a success measure that is strongly connected to, nonetheless not demonstrative of better medical results for patients [79]. Nonetheless, published studies explaining WFO do not disclose performance indicators appropriate for statistics (e.g. calibration, discrimination) and clinical based (e.g., type of net benefit). Rather, they concentrate on consistency (true positive rate where physician offers the ground truth. To prevent these pitfalls, the following guidelines for researchers are presented: [80–82]

- Consult with all concerned stakeholders like patients, researchers, doctors, etc. to assess the suitable devising of the statistical objective, like forecasting an event's absolute risk, or determining a rank order or trend detection or classification.
- Choose the correct performance metrics. That target comprises of specific criteria and doing the statistical aim clear can assist scientists to decide what the appropriate prognostic success indicators are for every particular scenario. For instance, if forecasting is the objective, then standardization & discrimination form the basic reporting needs. In addition, the correct scoring rules (or at least side-byside histograms) should be used to compare two versions [83].
- Results to report. Even though outcomes of training may not be enough, to prove the model's usefulness, they provide valuable insights into the characteristics of samples & any out-of-sample outcomes that are also being produced. However, the most relevant to mention are reliable estimates (i.e., the ones that have been modified correctly for overfitting) [84].

5 Conclusion

DL has achieved a significant position in recent years in the domain of MI and PH. ML is slowly impacting the manner in which treatment related to health care & nursing is being done. This is due to DL. In comparison with ML & feature engineering, DL can analyze big data in an effective manner. Several variants of DL methods have been developed across several domains such as processing of health records, images of biomedical, sensors & physiological processing of signals, human motion & analysis of emotions, etc. This chapter discusses the different DL techniques like RNN, DBN, AE, CNN, and DBM which are being used in MI and PH. Different case studies in real world like diagnosis of cancer, prediction of DR, HIV & its related treatment along with future applications have also been presented. Certain open research issues associated to the usage of DL in MI & PH are also discussed.

When compared to other domains, healthcare requires interactions between various disciplines, dedicated processes & information repositories. Hence, if large models are generated without completely comprehending the outcome would be practically disastrous in healthcare. When compared to other domains in which researchers can work alone without much interaction, healthcare domain requires active collaboration between the medical experts, data scientists, & experts related to informatics. Comprehensive knowledge of present medical workflow is required so that the models can extract data from appropriate sources of data and apply them in real-world. In conclusion, DL is being prominently utilized in the domain of MI & PH.

References

1. Lauzon, F.Q. An introduction to deep learning. In:11th International Conference on Information Science, Signal Processing and their Applications (ISSPA), Montreal, QC, pp. 1438–1439 (2012)
2. Hyunjung Kwak, G., Hui, P. DeepHealth: review and challenges of artificial intelligence in health informatics, 42 p (2019) (In press)
3. Ravi, D., Wong, C., Deligianni, F., Berthelot, M., Andreu-Perez, J., Lo, B., Yang, G.Z. Deep learning for health informatics. IEEE J. Biomed. Health Inf. **21**(1), 4-21 (2017)
4. Chang, H., Han, J., Zhong, C., Snijders, A.M., Mao, J.H. Unsupervised transfer learning via multi-scale convolutional sparse coding for biomedical applications. IEEE Trans. Pattern Anal. Machine Intell. **40**(5), 1182-1194 (2017)
5. Nie, D., Zhang, H., Adeli, E., Liu, L., Shen, D. 3D deep learning for multi-modal imaging guided survival time prediction of brain tumour patients. In: International Conference on Medical Image Computing and Computer-Assisted Intervention, pp. 212--220. Springer (2016)
6. Samala, R.K., Chan, H.P., Hadjiiski, L., Helvie, M.A., Wei, J., Cha, K. Mass detection in digital breast tomosynthesis: deep convolutional neural network with transfer learning from mammography. Med. Phys. **43**(12), 6654–6666 (2016)
7. Xu, T., Zhang, H., Huang, X., Zhang, S., Metaxas, D.N. Multimodal deep learning for cervical dysplasia diagnosis. In: International Conference on Medical Image Computing and Computer-Assisted Intervention, pp.115–123, Springer (2016)
8. Yan, Z., Zhan, Y., Peng, Z., Liao, S., Shinagawa, Y., Zhang, S., Metaxas, D.N., Zhou, X.S. Multi-instance deep learning: Discover discriminative local anatomies for bodypart recognition. IEEE Trans. Med. Imag. **35**(5), 1332–1343 (2016)
9. Cho, K., Van Merriënboer, B., Gulcehre, C., Bahdanau, D., Bougares, F., Schwenk, H., Bengio, Y. Learning phrase representations using RNN encoder-decoder for statistical machine translation. In: Proceedings of the 2014 Conference on Empirical Methods in Natural Language Processing (EMNLP), pp. 1724–1734. Doha, Qatar (2014)
10. Alipanahi, B., Delong, A., Weirauch, M.T., Frey. B.J. Predicting the sequence specificities of DNA-and RNA-binding proteins by deep learning. Nature Biotechnol. **33**(8), 831–838 (2015)
11. Angermueller, C., Lee, H.J., Reik, W., Stegle, O. Accurate prediction of single-cell DNA methylation states using deep learning. Genome Biol. **18**(1), 1-13 (2016)
12. Xiong, H.Y., Alipanahi, B., Lee, L.J., Bretschneider, H., Merico, D., Yuen, R.K., Hua, Y., Gueroussov, S., Najafabadi, H.S., Hughes, T.R., Morris, Q., Barash, Y., Krainer, A.R., Jojic, N., Scherer, S.W., Blencowe, B.J., Frey, B.J.: The human splicing code reveals new insights into the genetic determinants of disease. Science **347**, 6218 (2015)
13. Bamgbola, O.: Review of vancomycin-induced renal toxicity: an update. Therapeutic advances in endocrinology and metabolism. **7**(3), 136–147 (2016)
14. Davis, S.E., Lasko, T.A., Chen, G., Siew, E.D., Matheny, M.E.: Calibration drift in regression and machine learning models for acute kidney injury. J. Am. Med. Inform. Assoc. **24**(6), 1052–1061 (2017)
15. Goldstein, S.L. Nephrotoxicities. F1000Research. **6**, 55 (2017)
16. Hoste, E.A.J., Kashani, K., Gibney, N., Perry Wilson, F., Ronco, C., Goldstein, S.L., Kellum, J.A., Bagshaw, S.M. Impact of electronic alerting of acute kidney injury: workgroup statements from the 15th ADQI Consensus Conference. Can. J. Kidney Health Disease **3**(1), 1--9 (2016)

17. Knaus, W.A., Marks, R.D. New phenotypes for sepsis. JAMA **321**(20) (2019)
18. Prendecki, M., Blacker, E., Sadeghi-Alavijeh, O., Edwards, R., Montgomery, H., Gillis, S., Harber, M.: Improving outcomes in patients with Acute Kidney Injury: the impact of hospital based automated AKI alerts. Postgrad. Med. J. **92**(1083), 9–13 (2016)
19. Seymour, C.W., Kennedy, J.N., Wang, S., Chang, C.C., Elliott, C.F., Zhongying, X., Berry, S., Clermont, G., Cooper, G., Gomez, H., Huang, D.T., Kellum, J.A., Mi, Q., Opal, S.M., Talisa, V., Poll van der, T., Visweswaran, S., Vodovotz, Y. Weiss, J.C., Yealy, D.M., Yende, S., Angus, D.C. Derivation, Validation, and potential treatment implications of novel clinical phenotypes for sepsis. JAMA **321**(20), 2003-2017 (2019)
20. Tomašev, N., Glorot, X., Rae, J.W., Zielinski, M., Askham, H., Saraiva, A., Mottram, A., Meyer, C., Ravuri, S., Protsyuk, I., Connell, A., Hughes, C.O., Karthikesalingam, A., Cornebise, J., Montgomery, H., Rees, G., Laing, C., Baker, C.R., Peterson, K., Reeves, R., Hassabis, D., King, D., Suleyman, M., Back, T., Nielson, C., Ledsam, J.R., Mohamed, S.: A clinically applicable approach to continuous prediction of future acute kidney injury. Nature **572**, 116–119 (2019)
21. Wang, L., Zhang, W., He, X., Zha, H. Supervised reinforcement learning with recurrent neural network for dynamic treatment recommendation. In: Proceedings of the 24th ACM SIGKDD International Conference on Knowledge Discovery & Data Mining, pp. 2447-2456. ACM (2018)
22. Barth, J., Klucken, J., Kugler, P., Kammerer, T., Steidl, R., Winkler, J., Hornegger, J., Eskofier, B. Biometric and mobile gait analysis for early diagnosis and therapy monitoring in Parkinson's disease. In: Annual International Conference of the IEEE Engineering in Medicine and Biology Society, pp. 868--871. IEEE (2011)
23. Wilson, S., Ruscoe, W., Chapman, M., Miller, R.: General practitioner-hospital communications: A review of discharge summaries. J. Qual. Clin. Pract. **21**, 104–108 (2002)
24. Jindal, V., Birjandtalab, J., Baran Pouyan, M., Nourani, M. An adaptive deep learning approach for PPG-based identification. In: 38th Annual international conference of the IEEE engineering in medicine and biology society (EMBC), pp. 6401-6404. IEEE (2016)
25. Nurse, E., Mashford, B.S., Yepes, A.J., Kiral-Kornek, I., Harrer, S., Freestone, D.R. Decoding EEG and LFP signals using deep learning: heading TrueNorth. In: Proceedings of the ACM International Conference on Computing Frontiers, pp. 259-266. ACM (2016)
26. Cao, Y., Liu, C., Liu, B., Brunette, M.J., Zhang, N., Sun, T., Zhang, P., Peinado, J., Garavito, E.S., Garcia, L.L. et al. Improving tuberculosis diagnostics using deep learning and mobile health technologies among resource-poor and marginalized communities. In: IEEE First International Conference on Connected Health: Applications, Systems and Engineering Technologies (CHASE), pp. 274-281. IEEE (2016)
27. Alimova, I., Tutubalina, E., Alferova, J., Gafiyatullina, G. A machine learning approach to classification of drug reviews in Russian. In: Ivannikov ISPRAS Open Conference (ISPRAS), pp. 64–69. IEEE, Moscow (2017)
28. Bodnar, T., Barclay, V.C., Ram, N., Tucker, C.S., Salathé, T. On the ground validation of online diagnosis with Twitter and medical records. In: Proceedings of the 23rd International Conference on World Wide Web, pp. 651–656. ACM (2014)
29. Chae, S., Kwon, S., Lee, D.: Predicting infectious disease using deep learning and big data. Int. J. Environ. Res. Public Health **15**(8), 1–20 (2018)
30. de Quincey, E., Kyriacou, T., Pantin, T. # hayfever; A Longitudinal Study into Hay Fever Related Tweets in the UK. In: Proceedings of the 6th International Conference on Digital Health Conference, pp.85-89. ACM (2016)
31. Garimella, V.R.K., Alfayad, A., Weber, I. Social media image analysis for public health. In: Proceedings of the CHI Conference on Human Factors in Computing Systems, pp. 5543-5547. ACM (2016)
32. Phan, N.H., Dou, D., Piniewski, B., Kil, D. Social restricted Boltzmann machine: human behavior prediction in health social networks. In: Proceedings of the 2015 IEEE/ACM International Conference on Advances in Social Networks Analysis and Mining, pp. 424-431. ACM (2015)

33. Tuarob, S., Tucker, C.S., Salathe, M., Ram, N. An ensemble heterogeneous classification methodology for discovering health-related knowledge in social media messages. J. Biomed. Infor. **49**, 255--268 (2014)
34. Birkhead, G.S., Klompas, M., Shah, N.R. Uses of electronic health records for public health surveillance to advance public health. Annual Rev. Public Health **36**, 345--359 (2015)
35. Henry, J., Pylypchuk, Y., Searcy, T., Patel, V. Adoption of electronic health record systems among US non-federal acute care hospitals: 2008–2015. ONC Data Brief. **35**, 1--9 (2016)
36. Tobore, I., Li, J., Yuhang, L., Al-Handarish, Y., Kandwal, A., Nie, Z., Wang, L. Deep learning intervention for health care challenges: some biomedical domain considerations. JMIR Mhealth Uhealth **7**(8) (2019)
37. Schmidhuber, J.: Deep learning in neural networks: an overview. Neural Netw. **61**, 85–117 (2015)
38. Yousoff, S.N., Baharin, A., Abdullah, A. A review on optimization algorithm for deep learning method in bioinformatics field. In: Proceedings of the Conference on Biomedical Engineering and Sciences, pp. 707–711.. Kuala Lumpur, Malaysia, IEEE (2016)
39. LeCun, Y., Bengio, Y., Hinton, G.: Deep learning. Nature **521**(7553), 436–444 (2015)
40. Vincent, P., Larochelle, H., Bengio, Y., Manzagol, P.A. Extracting and composing robust features with denoising autoencoders. In: Proceedings of the 25th International Conference on Machine learning; ICML'08; July 5–9 2008, pp. 1096–103. Helsinki, Finland (2008)
41. Ranzato, M.A., Poultney, C., Chopra, S., LeCun, Y. Efficient learning of sparse representations with an energy-based model. In: Proceedings of the 19th International Conference on Neural Information Processing Systems; NIPS'06; 4–7 December 2006, pp. 1137–44. Vancouver, British Columbia, Canada (2006)
42. Masci, J., Meier, U., Cirean, D., Schmidhuber, J. Stacked convolutional auto-encoders for hierarchical feature extraction. In: Proceedings of the Artificial Neural Networks and Machine Learning; ICANN'11; 14–17 June 2011, pp. 52–9. Espoo, Finland (2011)
43. Ororbia, L.A., Kifer, D., Giles, C.L.: Unifying adversarial training algorithms with data gradient regularization. Neural Comput. **29**(4), 867–887 (2017). https://doi.org/10.1162/NECO_a_00928
44. Rifai, S., Vincent, P., Muller, X., Glorot, X., Bengio, Y. Contractive auto-encoders: explicit invariance during feature extraction. In: Proceedings of the 28th International Conference on Machine Learning; ICML'11; June 28-July 2, 2011, pp. 833–40. Bellevue, Washington, USA (2011)
45. Schmidhuber, J. WebCite. [Demonstrates Credit Assignment Across the Equivalent of 1,200 Layers in an Unfolded RNN] http://www.webcitation.org/71i6G4Jawwebcite (1993)
46. Bengio, Y., Simard, P., Frasconi, P.: Learning long-term dependencies with gradient descent is difficult. IEEE Trans Neural Netw. **5**(2), 157–166 (1994). https://doi.org/10.1109/72.279181
47. Hochreiter, S., Schmidhuber, J.: Long short-term memory. Neural Comput. **9**(8), 1735–1780 (1997). https://doi.org/10.1162/neco.1997.9.8.1735
48. Krizhevsky, A., Sutskever, I., Hinton, G.E.: ImageNet classification with deep convolutional neural networks. Commun. ACM **60**(6), 84–90 (2017). https://doi.org/10.1145/3065386
49. Cho, K., van Merriënboer, B., Gulcehre, C., Bahdanau, D., Bougares, F., Schwenk, H., Bengio, Y. Learning phrase representations using RNN encoder-decoder for statistical machine translation. In: Proceedings of the Conference on Empirical Methods in Natural Language Processing; EMNLP'14; 25–29 October 2014, pp. 1724–34. Doha, Qatar (2014)
50. Matsugu, M., Mori, K., Mitari, Y., Kaneda, Y.: Subject independent facial expression recognition with robust face detection using a convolutional neural network. Neural Netw. **16**(5–6), 555–559 (2003). https://doi.org/10.1016/S0893-6080(03)00115-1
51. Hubel, D.H., Wiesel, T.N.: Receptive fields, binocular interaction and functional architecture in the cat's visual cortex. J. Physiol. **160**(1), 106–154 (1962). https://doi.org/10.1113/jphysiol.1962.sp006837
52. Lecun, Y., Bottou, L., Bengio, Y., Haffner, P.: Gradient-based learning applied to document recognition. Proc. IEEE **86**(11), 2278–2324 (1998). https://doi.org/10.1109/5.726791

53. Cirean, D., Meier, U., Schmidhuber, J. Multi-Column deep neural networks for image classification. In: Proceedings of the Conference on Computer Vision and Pattern Recognition; IEEE'12; June 16–21 2012, pp. 3642–9. Providence, RI, USA (2012)
54. Salakhutdinov, R., Larochelle, H. Efficient learning of deep boltzmann machines. In: Proceedings of the Thirteenth International Conference on Artificial Intelligence and Statistics; AISTATS'10; May 13–15, 2010, pp. 693–700. Sardinia, Italy (2010).
55. Hinton, G.E., Osindero, S., Teh, Y.W.: A fast learning algorithm for deep belief nets. Neural Comput. **18**(7), 1527–1554 (2006). https://doi.org/10.1162/neco.2006.18.7.1527
56. Hinton, G.E., Salakhutdinov, R.R.: Reducing the dimensionality of data with neural networks. Science **313**(5786), 504–507 (2006). https://doi.org/10.1126/science.1127647
57. Ravi, D., Wong, C., Deligianni, F., Berthelot, M., Andreu-Perez, J., Lo, B., Yang, G.Z.: Deep learning for health informatics. IEEE J. Biomed. Health Inform. **21**(1), 4–21 (2017). https://doi.org/10.1109/JBHI.2016.2636665
58. Ryu, S., Noh, J., Kim, H.: Deep neural network-based demand side short term load forecasting. Energies **10**(1), 3 (2016). https://doi.org/10.3390/en10010003
59. Goodfellow, I., Bengio, Y., Courville, A.: Deep Learning (Adaptive Computation and Machine Learning Series) Cambridge. MIT Press, Massachusetts (2016)
60. Esteva, A., Robicquet, A., Ramsundar, B., Kuleshov, V., DePristo, M., Chou, K., Cui, C., Corrado, G., Thrun, S., Dean, J.: A guide to deep learning in healthcare. Nat Med **25**, 24–29 (2019). https://doi.org/10.1038/s41591-018-0316-z
61. https://www.allerin.com/blog/top-5-applications-of-deep-learning-in-healthcare.
62. https://emerj.com/ai-sector-overviews/machine-learning-healthcare-applications/.
63. Vollmer, S., Mateen, B.A., Bohner, G., Király, F.J., Ghani, R., Jonsson, P., Cumbers, S., Jonas, A., McAllister, K.S.L., Myles, P., Granger, D., Birse, M., Branson, R., Moons, K.G.M., Collins, G.S., Ioannidis, J.P.A., Holmes, C., Hemingway, H. Machine learning and artificial intelligence research for patient benefit: 20 critical questions on transparency, replicability, ethics, and effectiveness. BMJ. **368**, l6927 (2020). https://doi.org/10.1136/bmj.l6927
64. Steyerberg, E.W., Moons, K.G., van der Windt, D.A. et al. PROGRESS Group. Prognosis Research Strategy (PROGRESS) 3: prognostic model research. PLoS Med. **10**, e1001381 (2013). https://doi.org/10.1371/journal.pmed.1001381
65. Snooks, H., Bailey-Jones, K., Burge-Jones, D., et al.: Effects and costs of implementing predictive risk stratification in primary care: a randomised stepped wedge trial. BMJ Qual. Saf. **28**, 697–705 (2019). https://doi.org/10.1136/bmjqs-2018-007976
66. Avati, A., Jung, K., Harman, S., Downing, L., Ng, A., Shah, N.H.: Improving palliative care with deep learning. BMC Med. Inform. Decis. Mak. **18**(Suppl 4), 122 (2018). https://doi.org/10.1186/s12911-018-0677-8
67. UK Standards for Public Involvement in Research. Homepage. https://sites.google.com/nihr.ac.uk/pi-standards/home (2018)
68. Cortes, C., Jackel, L.D., Chiang, W.P. Limits on learning machine accuracy imposed by data quality. In: Advances in Neural Information Processing Systems, pp. 239–46 (1995)
69. Willetts, M., Hollowell, S., Aslett, L., Holmes, C., Doherty, A.: Statistical machine learning of sleep and physical activity phenotypes from sensor data in 96,220 UK Biobank participants. Sci. Rep. **8**, 7961 (2018). https://doi.org/10.1038/s41598-018-26174-1
70. Siontis, G.C., Tzoulaki, I., Castaldi, P.J., Ioannidis, J.P.: External validation of new risk prediction models is infrequent and reveals worse prognostic discrimination. J. Clin. Epidemiol. **68**, 25–34 (2015). https://doi.org/10.1016/j.jclinepi.2014.09.007
71. Hyndman, R.J., Athanasopoulos, G. Forecasting: principles and practice. OTexts. https://otexts.com/fpp2/ (2018)
72. Lyddon S, Walker S, Holmes CC. Nonparametric learning from Bayesian models with randomized objective functions. In: Advances in neural information processing systems, 2018:2072–82.
73. Simonyan, K., Zisserman, A. Very deep convolutional networks for large-scale image recognition. https://arxiv.org/abs/1409.1556 (2014)

74. Inouye, M., Abraham, G., Nelson, C.P. et al. UK Biobank cardiometabolic consortium chd working group. Genomic risk prediction of coronary artery disease in 480,000 adults: implications for primary prevention. J. Am. Coll. Cardiol. **72**, 1883–93 (2018). https://doi.org/10.1016/j.jacc.2018.07.079.
75. Canziani A, Paszke A, Culurciello E. An analysis of deep neural network models for practical applications. arXiv [Preprint] 2016 May 24. https://arxiv.org/abs/1605.07678.
76. Hinton, G.E., Srivastava, N., Krizhevsky, A., Sutskever, I., Salakhutdinov, R.R. Improving neural networks by preventing co-adaptation of feature detectors (2012) https://arxiv.org/abs/1207.0580
77. Collins, G.S., Moons, K.G.M. Comparing risk prediction models. BMJ **344**, e3186 (2012)
78. Morin, A., Urban, J., Sliz, P.: A quick guide to software licensing for the scientist-programmer. PLoS Comput. Biol. **8**, e1002598 (2012). https://doi.org/10.1371/journal.pcbi.1002598
79. Epstein, A.S., Zauderer, M.G., Gucalp, A., et al.: Next steps for IBM Watson Oncology: scalability to additional malignancies. J. Clin. Oncol. **32**(suppl), 6618 (2014). https://doi.org/10.1200/jco.2014.32.15_suppl.6618
80. Suwanvecho, S., Suwanrusme, H., Sangtian, M., Norden, A., Urman, A., Hicks, A. et al. Concordance assessment of a cognitive computing system in Thailand. J. Clin. Oncol. **35**(15), 6589 (2017)
81. Somashekhar, S., Kumarc, R., Rauthan, A., Arun, K., Patil, P., Ramya, Y.: Double blinded validation study to assess performance of IBM artificial intelligence platform, Watson for oncology in comparison with Manipal multidisciplinary tumour board–First study of 638 breast cancer cases. Cancer Res. **77**(4 suppl), S6-07 (2017)
82. Baek, J., Ahn, S., Urman, A., et al. Use of a cognitive computing system for treatment of colon and gastric cancer in South Korea. J. Clin. Oncol. **35**(15), e18204 (2017)
83. Moons, K.G.M., Altman, D.G., Reitsma, J.B., et al.: Transparent reporting of a multivariable prediction model for individual prognosis or diagnosis (TRIPOD): explanation and elaboration. Ann Intern Med **162**, W1-73 (2015). https://doi.org/10.7326/M14-0698
84. Rieke, N., Hancox, J., Li, W. et al. The future of digital health with federated learning. NPJ Digit. Med. **3**, 119 (2020). https://doi.org/10.1038/s41746-020-00323-1

Transiting Exoplanet Hunting Using Convolutional Neural Networks

Dhruv Kaliraman, Gopal Joshi, and Suchitra Khoje

Abstract The human race has spent and invested considerable resources to understand the processes of your solar system. However, since the discovery of exoplanets we have the means to determine whether or not we actually understand these processes. The most compelling reason to find exoplanets is that it opens the door for us to look for other habitable planets as well as understand our own solar system better. For years, scientists have been utilizing data from NASA's Kepler Space Telescope to look for and identify thousands of transiting exoplanets. Thanks to new and better telescopes, astronomical data is rapidly increasing. Traditional human judgment-based prediction and classification methods are inefficient and vulnerable to vary depending on the expert doing the study. The widely used methodology for exoplanet discovery, the Box-fitting Least Squares technique (BLS), for example, creates a large number of false positives that must be manually checked in the event of noisy data. As a result, an automated and unbiased approach for detecting exoplanets while removing false-positive signals imitating transiting planet signals is required. A new convolutional neural network-based mechanism for finding exoplanets is introduced using the transit technique. Since the dataset is large and highly imbalanced, SMOTE is used to resample the data, while the exponential decay approach along with dropout and early stopping techniques are used to reduce model overfitting. In addition, the model employs the Grid-SearchCV approach to fine-tune hyperparameters. Finally, for a robust and full model, the model is evaluated using k fold cross-validation. Performance criteria such as accuracy, precision, recall, f1 score, sensitivity, and specificity are used in the study. After analyzing the data, the research concluded that the convolutional neural network produced a maximum accuracy of 99.6% on the testing data.

Keywords Convolutional neural network · GridSearchCV · K fold cross validation · Machine learning · SMOTE · Exoplanet detection · Transit method

D. Kaliraman (✉) · G. Joshi
School of Computer Science, MIT WPU, Survey No. 124, Paud Rd, Pune, Maharashtra, India

S. Khoje
School of Electronics and Communication Engineering, MIT WPU, Survey No. 124, Paud Rd, Pune, Maharashtra, India
e-mail: suchitra.khoje@mitwpu.edu.in

© The Author(s), under exclusive license to Springer Nature Switzerland AG 2022
K. R. Ahmed and H. Hexmoor (eds.), *Blockchain and Deep Learning*,
Studies in Big Data 105, https://doi.org/10.1007/978-3-030-95419-2_14

1 Introduction

Exoplanets are those planets that are far away from our planetary system. The majority of exoplanets orbit other stars, while rogue planets, also known as free-floating exoplanets, orbit the galactic center and are unattached to any star.

Because modern telescopes can only view up to that point, the great majority of exoplanets identified so far are limited to a restricted region of our Milky Way star cluster. The Kepler Space Telescope of NASA claims that in the galaxy there are more planets than stars [1].

Thousands of extrasolar planets have been found out and verified around other stars. Van Maanen, who found the first novel sub-planet in 1917, found out the first traces of exoplanets. It would take until the 1990s for it to be confirmed. The discovery of extrasolar planets increased at a remarkable rate following the introduction of the Kepler Space Telescope.

The main aim behind the launch of the Kepler space telescope was to find out planets that were similar to Earth and which lie in the habitable zone or close to it (The region around the sun where planets could have liquid water on their surface, also known as the "Goldilocks zone") and determine the association of stars with planets. Despite deactivating the Kepler spacecraft in 2018, the details collected by it are still used to identify exoplanets.

Spitzer Space Telescope with its infrared capabilities helped in TRAPPIST-1 system's significant exploration. In order to find more exoplanets orbiting around the stars, The Transiting Exoplanet Survey Satellite (TESS) was introduced in 2018 as the next replacement to Kepler. Exoplanet missions such as NASA's James Webb Space Telescope and the Nancy Grace Roman Space Telescope hold a lot of promise in terms of the areas we may be able to learn from. The spectroscopic approach (a method to measure light fingerprints) is employed by astronomers to understand more about planet atmospheres and conditions [2].

The motivation for hunting for exoplanets stems from one of humanity's most deep and thought-provoking questions: "Are we alone in the universe?" For thousands of years, philosophers and curious people have pondered this topic, but this is the first generation to have the tools to begin answering the scientific observations. Not only are there thousands of planets orbiting other stars but also the presence of hundreds of billions more are inferred, diverse ranges of planets are being discovered, some of which are extremely different from those from our planetary system. With the discovery of a new star system, the question of how the universe works is answered. This helps in figuring out how our solar system fits in this universe.

The newer methods are being leveraged to improve on the current algorithm-based ways for detecting exoplanets in astrophysics. Neural Networks is a relatively young field that deals with algorithms based on the structure and function of the human brain (NN). A subset of machine learning known as deep learning is being concentrated on. It has networks that can learn data from both labeled and unlabeled input. With the use of the data gathered from photos provided by the Kepler telescope, the topic of exoplanet discovery using deep learning has been addressed in this paper.

The following is an overview of the paper. A summary of prior attempts to discover exoplanets using machine learning is presented in Sect. 2. The basic principle for our paper is outlined in Sect. 3. Our findings and analyses are presented in Sect. 4. In Sect. 5, we present and elaborate on our deep learning model's final architecture and hyperparameters. Our conclusion is presented in the last portion of the paper.

2 Literature Survey

Aisha et al. [1] propose an approach for an automated meteorite detection system that uses an unmanned aerial vehicle (UAV). Machine learning is used to program it to detect and find meteorites. This was accomplished utilizing several architectures such as MobileNetV2 and InceptionV3. The parameters in a CNN differ with different layers of convolution (with InceptionV3 employing the most remarkable parameter count). The machine-learning algorithm was implemented using Python, Keras, and Tensorflow.

The SBC (single board computer) performs onboard processing by examining a single frame of the video and applying it to the deep learning model. Transfer Learning on Convolutional Neural Networks is used to train this model using photos of several sorts of commonly encountered meteorites (CNN). The outcomes were encouraging, with an accuracy rate of around 90%. Abhishek et al. [2] present a way for employing the transit method to provide a novel machine learning-based tool for detecting exoplanets. They employed these techniques to improve on the current algorithm-based methods for detecting exoplanets in astrophysics. They analyzed light curves with the time-series analysis package TSFresh, extracting 789 features from every curve that extracted data about the qualities of the curve. They then utilized the machine learning application lightgbm to build a gradient boosting classifier utilizing these characteristics. This method was put to the test on simulated data, and it proved to be more successful than the standard box least-squares fitting (BLS) method. They also discovered that their strategy achieved results that were equivalent to existing deep learning models at the same time were significantly more efficient in terms of computing and requiring no folding as well. This approach predicted a planet with an accuracy of 0.948 using Kepler data, meaning that 94.8% of actual planet signals are rated higher than non-planet signals. The ultimate recall value is 0.96. On the Transiting Exoplanet Survey Satellite (TESS) dataset their resultant metrics were: an accuracy of 0.98, recall of 0.82, and a precision of 0.63.

Erqian et al. introduce a Galaxy Detection and Classification Tool (GalaDC), which efficiently, accurately discovers, and sorts out galaxies. GalaDC is an easy-to-use statistical analysis tool, which allows batch processing, and can handle pictures with several galaxies. GalaDC uses a convolutional neural network and a variety of computer vision techniques in this study. 90% of the data is utilized to train the network, with the remaining 10% being utilized to assess the network's generalization and overfitting. The testing accuracy was 91.0%, while the training accuracy was 91.5%.

For identifying exoplanet transits, Pattana et al. offer a machine-learning methodology based on a two-dimensional convolutional neural network. Five distinct types of deep-learning models with and without k fold cross-validation are developed and analyzed to test this novel strategy (MLP, 1D-CNN-folding-0, 1D-CNN-folding-1, 2D-CNN-folding-0, 2D-CNN-folding-1). All models with k fold cross validation have accuracy above 98% when S/N is less than 10. These models use the light curves from Kepler Data Release 25 as their input. The correctness, reliability, and completeness of the data are measured and compared.

Support Vector Machines and deep learning models are used in combination with NASA's Kepler dataset, Mohd Aijaj et al. offer a novel methodology for finding comparable planets. They have used the PCA (Principal Component Analysis) methodology to reduce the total attributes in the dataset as a preventive step. This model achieved a maximum precision of 99.1%. Four 1-D layers are created which is a routine filter and the data length is extended by a factor of 4 through pooling. In the end, 2 dense layers are utilized. Batch normalized layers are used to increase the rate of convergence exponentially (Table 1).

3   Methodology

The data for this analysis has been cleaned and comes from NASA's Kepler space telescope observations. Convolutional neural networks were utilized to build a method for automatically categorizing Kepler Transiting planet candidates. This study uses SMOTE to resample the data, as well as an exponential decay and early stopping methods to reduce model overfitting. In addition, the model employs the GridSearchCV approach to fine-tune hyperparameters. Finally, for a robust and full model, the model is tested using k fold cross-validation. Performance criteria such as accuracy, precision, recall, f1 score, sensitivity, and specificity are used in the study. The network model is represented in Fig. 1.

3.1   Dataset Analysis

What is the transit method?

The transit method measures the light intensity of an exoplanet. In this method when a planet passes between an observer (Space Telescope) and the stars around which it revolves, that star genuinely dims for a brief amount of time. It's a minute difference, yet it's sufficient to tip astrophysicists off to the discovery of an extrasolar planet orbiting a faraway star.

The astronomers refer to the 'light curve,' as depicted in Fig. 2. It's a graph showing the star's light output. Over time the brightness of the stars fades away for a brief amount of time when a planet passes in front of it and blocks a part of its light,

Table 1 Literature survey

Name of paper	Author	Year of study	Dataset used	Method used	Accuracy (%)	Research gap
Discovering Exoplanets in Deep Space using Deep Learning Algorithms	M. A. Khan, M. Dixit	2020	NASA's Kepler space telescope dataset of flux intensity of distant stars. (Kaggle)	Deep neural networks and support vector machine	99.1%	This model fails to classify the true positive exoplanets
Exoplanet Detection using Machine Learning	Abhishek Malik, Benjamin P. Moster and Christian Obermeier	2020	Kepler dataset, TESS dataset	Time series analysis library TSFresh, to train features a gradient boosting classifier using the machine learning tool lightgbm was used	Kepler data – 94.8% TESS data – 98%	When compared to the other two data sets, their model performed worse with the TESS data set. The severe class imbalance in TESS data was the key explanation behind this

(continued)

Table 1 (continued)

Name of paper	Author	Year of study	Dataset used	Method used	Accuracy (%)	Research gap
Detecting Exoplanet Transits through Machine- learning Techniques with Convolution Neural Networks	Pattana Chintarungr uangchai and Ing-Guey Jiang	2019	Keplar dataset	MLP, 1D CNN folding 0, 1D-CNN-folding-1, 2D-CNN-folding-0, 2D-CNN-folding	98%	All models with folding can have accuracy above 98% even when S/N is less than 10. The accuracy of models without folding can become about 85% when S/N is less than 10. The precision and recall have a similar trend
Meteorite Hunting Using Deep Learning and UAVs	Aisha AlOwais, Safa Naseem, Takwa Dawdi, Mariam Abdisalam, Yusra Elkalyoubi, Anas Adwan, Khawla Hassan and Ilias	2019	Google images	CNN, transfer learning	90%	Fewer images in dataset led to less accuracy
GalaDC: Galaxy Detection and Classification Tool	Erqian Cai	2020	"Galaxy Zoo–The Galaxy Challenge" on Kaggle	CNN open CV Pyqt5	91%	This model doesn't perform well with different redshifts band of images

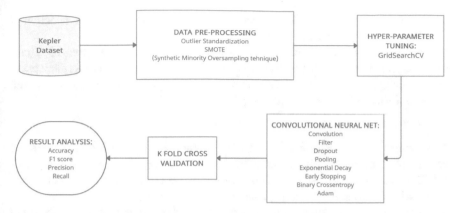

Fig. 1 Model architecture

Fig. 2 Transit method
depiction

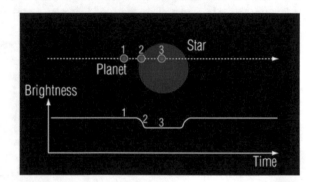

eventually, the light intensity is restored as the planet moves away from the observer
and the star [3].

The duration and magnitude of the dip in light intensity reveal a lot about the
target planet. Larger planets block lighter, resulting in more pronounced light curves.
Furthermore, the farther a planet is from its star, the more time it takes to revolve
around it. As a result of which the characteristics of the light curve can tell exactly
how distant the star is from the planet.

Apart from exploring exoplanets, the technique is also useful in studying the
planet's atmosphere and temperature. Some of the starlight's travel through an
exoplanet's atmosphere which is then detected in the transit light curve. This spectrum of light is then studied by astrophysicists to understand more about the elements
that can be found in the planet's atmosphere. With this technology, scientists have
discovered elements ranging from methane to water (Fig. 2).

Secondary eclipses (in this the planet passes behind the star as a result of which the
observer cannot see the light originating from the planet) can also reveal data about
the temperature and radiation of a planet. Astronomers measure the photometric
intensity of the star at this moment, then deduct it from the readings taken before

the second eclipse. This enables temperature readings in addition to the detection of cloud forms in the planet's ozone layer [4].

Finding new exoplanets has been quite effective using the transit approach. Thousands of probable exoplanet discoveries were discovered by NASA's Kepler spacecraft, which searched for exoplanets using this technique, providing scientists with crucial information on the galaxy's exoplanet distribution.

Dataset:

This dataset was obtained from Kaggle and comprises the flux intensity variation of thousands of stars. A target label of 1 or 2 is assigned to each data point or star. 2 denotes the presence of at least one exoplanet circling the target star; a few observations are due to the multi-planetary systems' nature [5].

The planets do not glow on their own, as stated in the preceding sections, but they do obstruct the parent star's light through which they are transiting. There is a consistent diminishing of the light flux intensity if the set star is monitored over a lengthy period of time. This is proof that a planet is revolving around a star and hence qualifies as a candidate exoplanet.

A blue planet revolves around a target star. in the dataset utilized here. Given the telescopic location, the star's flux intensity declines at $t = 1$ because it is hidden by the revolving exoplanet. At $t = 2$, The star's light reverts to its former value.

Trainset:

- 5087 rows or observations.
- 3198 columns or features.
- Column 1 is the label vector. Columns 2 - 3198 are the flux values over time. Testset: 3198 columns or features. Column 1 is the label vector. Columns 2 - 3198 are the flux values over time. 5 confirmed exoplanet-stars and 565 non-exoplanet-stars.

Figure 3 displays the flux intensity time series data for instances with exoplanets. Figure 4 displays the flux intensity time series data for instances without exoplanets.

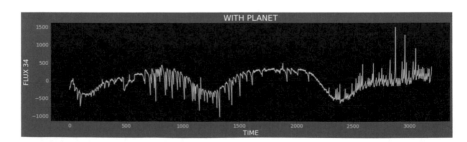

Fig. 3 Flux Intensity of a planet with exoplanet

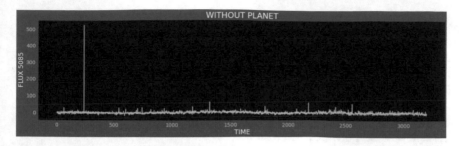

Fig. 4 Flux intensity without a planet

3.2 Data Pre-processing (Outliers)

Our model doesn't employ normalization procedures since scaling the data reduces the contrast between stars with and without exoplanets. Outliers abound in the statistics, therefore this fact must be taken into account. Because we haven't applied to scale, determining a truncation threshold is a sensitive operation. After several testing, it's a good idea to replace each star's maximum FLUX value with its mean. A plot comparison of data before, in Fig. 5, and after outlier correction, Fig. 6, is shown.

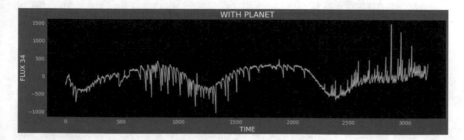

Fig. 5 Instance with outlier

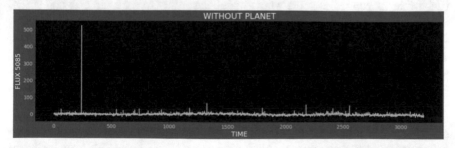

Fig. 6 Instance without outlier

3.3   SMOTE

A difficulty with unbalanced classification is that there are too few samples of the minority class for a model to learn the decision threshold successfully. SMOTE (Synthetic Minority Oversampling Technique) solves this problem [6].

SMOTE works by picking instances in the feature space that are close together, creating a line between the instances in the feature space, and generating a fresh example at a location along that line.

To be more specific, a random case from the minority class is picked initially. Then, for that example, k of the closest neighbors is found (usually k = 5). A randomly determined neighbor is picked, and a synthetic example is constructed at a randomly determined location in feature space between the two instances.

Random under sampling is used to reduce the instances that lie in the majority class, and subsequently. To balance the class distribution, SMOTE is utilized to oversample the minority class.

The method works because novel synthetic instances of the minority class are constructed that are realistic, that is, In terms of features, they're identical to current minority class instances.

The methodology has the drawback of creating synthetic instances without considering the predominant class, which might lead to ambiguous instances in the case that there is significant overlap between classes.

Since our dataset was highly imbalanced, smote was used for data augmentation of the minority class.

3.4   CNN

A convolutional neural network (CNN) is a form of artificial neural network that is specially intended to analyze pixel input and is used in image recognition and processing.

A local connection is accounted for via a Convolutional Neural Network (CNN) (Each filter is panned over the whole image according to a certain size and stride, allowing it to discover and match patterns regardless of where they are in the image). The weights are smaller and shared. They are thus less wasteful and easier to teach than Multilayer Perceptron's. It's also possible to go deeper. Rather than being entirely linked, layers are sparsely linked. It accepts both matrices and vectors as inputs. Rather than being entirely linked, the layers are sparsely or partly linked. Every node in the network is not linked to every other node.

CNN's filter-panning feature effectively allows parameter and weight sharing, allowing the filter to detect a particular pattern while being location invariant—able to recognize the patterns that are present anywhere in an image. This is really useful for detecting objects. Patterns may be found in a variety of spots across the images.

Layer (type)	Output Shape	Param #
reshape_11 (Reshape)	(None, 3197, 1)	0
conv1d_11 (Conv1D)	(None, 3196, 10)	30
max_pooling1d_11 (MaxPooling	(None, 1598, 10)	0
dropout_22 (Dropout)	(None, 1598, 10)	0
flatten_11 (Flatten)	(None, 15980)	0
dense_33 (Dense)	(None, 44)	703164
dropout_23 (Dropout)	(None, 44)	0
dense_34 (Dense)	(None, 20)	900
dense_35 (Dense)	(None, 1)	21

Fig. 7 Neural network architecture

1D Max pooling:

The 1D Max pooling technique reduces the amount of data, the number of parameters, and the amount of computation necessary while also preventing overfitting. The 1D max-pooling block sweeps a pool (window) of a certain size through the incoming data with a predetermined stride, computing the maximum in each window.

After one or more convolutional layers, max-pooling layers are implemented to enable inner convolutional layers to get input from a larger fraction of the original vector. If we consider convolutional layers to be detectors of a single feature, max-pooling preserves just the feature's "strongest" value within the pooling rectangle. Each channel (and hence each feature) is given its own treatment.

Figure 7 represents the convolutional neural network architecture used.

Stride: The number of cells that will be moved along the vector when pooling. 2 is the default.

Padding: The same results are obtained by adding the input to make the output the same length as the original input. "No padding" is what "valid" signifies.

3.5 Optimization of CNN Architecture

Dropout:

Dropout is a training approach in which randomly chosen neurons are rejected. They "disappeared" at random. This means that the contribution that the neurons have on the forward pass to the lower layer neurons is stopped temporarily, and similarly, on the backward side, no weight changes are applied [7].

It is routine that the weights of neurons settle in the neighborhood as the neural network keeps on learning. All neurons have weights that are configured for a particular specialization or a characteristic. It is hence logical that the neurons in the same neighborhood start to depend on these specific characteristics that make the model too rigid.

A solid solution to this problem is that neurons can be taken out of the training model randomly so as to make the neighboring neurons more independent and not rely on the missing neurons. As it turns out, this approach works and makes these neighboring neurons learn various characteristics that they would have learned had the neurons not been removed randomly.

This results in a more robust model that is independent of the weights of individual neurons and the model and network are able to make better generalizations which further results in the model being less exposed to overfitting.

To reduce overfitting our model has two separate layers of dropout.

Early Stopping and Exponential Decay:

It is difficult to predict or guess the optimum number of neurons that are required to train the neural network. Overfitting the training dataset can come from using too many epochs, whereas using fewer epochs can impact the model adversely as well by having an underfit model.

To avoid the above scenario a technique called early stopping allows users to give the model a randomly large number of epochs to train on and then quit training the moment the validation dataset performance metrics stop increasing [8].

This necessitates passing a validation split to the fit() function, as well as an EarlyStopping callback to define the performance metric that will be watched on the validation split.

This research employs the early stopping method to avoid overfitting.

It is common to drop the learning rate as the training advances while developing deep neural networks. Pre-determined learning rate schedules or adjustable learning rate approaches can be used to accomplish this. Learning rate schedules aim to reduce the learning rate according to a pre-determined timeline during training.

To avoid overfitting, our study uses the exponential decay approach as well.

3.6   Parameters Optimization

Grid Search:

GridSearchCV is a sklearn library function included in the model selection package. GridSearchCV is a methodology for finding the optimal parameter values from a given set of parameters in a grid. It's essentially a cross-validation technique. The model as well as the parameters must be entered. After extracting the best parameter values, predictions are made [9].

Deciding the hyperparameters that are to be experimented with is the next phase. It is dependent on the assessor that is used. Simply construct a dictionary with the hyperparameters as input keys and a list through which the loop will iterate with the settings that are to be experimented on.

All that is left to be done is to make a GridSearchCV object. These named parameters have to be declared here:

Estimator: an object is created that acts as an estimator.

Params_grid: this is the object of type dictionary that contains the hyperparameters
Scoring: evaluation metric that is to be used.

Cv: the value of k that is the input for k fold cross-validation

N_jobs: this represents the total processors that will execute in a parallel manner
This research employs the gridsearchcv method to find the optimum number of hidden neurons in the two hidden layers and the optimum dropout rate [10].

Figure 8 represents the optimum number of neurons that are used in the hidden layers.

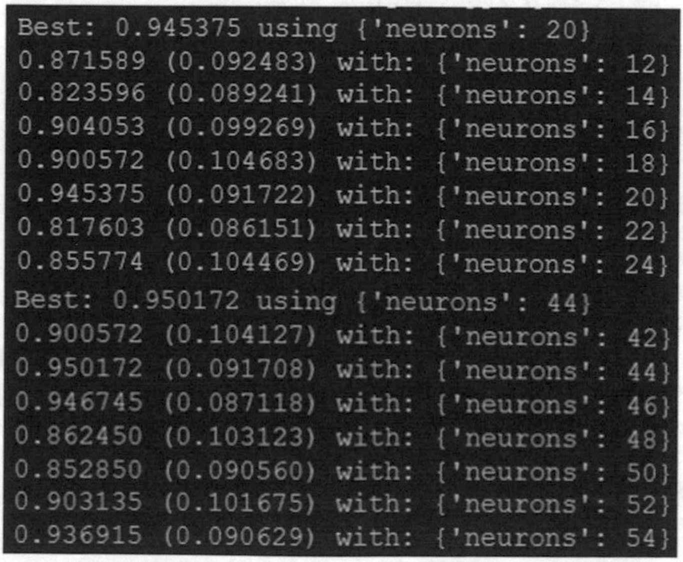

```
Best: 0.945375 using {'neurons': 20}
0.871589 (0.092483) with: {'neurons': 12}
0.823596 (0.089241) with: {'neurons': 14}
0.904053 (0.099269) with: {'neurons': 16}
0.900572 (0.104683) with: {'neurons': 18}
0.945375 (0.091722) with: {'neurons': 20}
0.817603 (0.086151) with: {'neurons': 22}
0.855774 (0.104469) with: {'neurons': 24}
Best: 0.950172 using {'neurons': 44}
0.900572 (0.104127) with: {'neurons': 42}
0.950172 (0.091708) with: {'neurons': 44}
0.946745 (0.087118) with: {'neurons': 46}
0.862450 (0.103123) with: {'neurons': 48}
0.852850 (0.090560) with: {'neurons': 50}
0.903135 (0.101675) with: {'neurons': 52}
0.936915 (0.090629) with: {'neurons': 54}
```

Fig. 8 GridsearchCV for first and second hidden layer neuron

```
Best: 0.945375 using {'neurons': 20}
0.871589 (0.092483) with: {'neurons': 12}
0.823596 (0.089241) with: {'neurons': 14}
0.904053 (0.099269) with: {'neurons': 16}
0.900572 (0.104683) with: {'neurons': 18}
0.945375 (0.091722) with: {'neurons': 20}
0.817603 (0.086151) with: {'neurons': 22}
0.855774 (0.104469) with: {'neurons': 24}
Best: 0.950172 using {'neurons': 44}
0.900572 (0.104127) with: {'neurons': 42}
0.950172 (0.091708) with: {'neurons': 44}
0.946745 (0.087118) with: {'neurons': 46}
0.862450 (0.103123) with: {'neurons': 48}
0.852850 (0.090560) with: {'neurons': 50}
0.903135 (0.101675) with: {'neurons': 52}
0.936915 (0.090629) with: {'neurons': 54}
```

Fig. 9 GridsearchCV for first and second dropout

Figure 9 represents the optimum dropout rate used in the convolutional neural network.

3.7   Proposed Method for model's Validation

K Fold Cross-Validation:

Cross-validation is a resampling approach used to test predictive models on a restricted set of data. The algorithm has only one variable, k, that determines how many groups a given data sample will be split into. Therefore, this entire process is known as k fold cross-validation [11].

A small sample size is used to assess how and why the model would perform in aggregate and it is used to generate predictions on instances that were not utilized while training.

This is a popular strategy since it is straightforward to grasp and results in a lower biassed or optimistic approximation of model competence than other approaches, like a straightforward training and testing data split.

To improve the accuracy and generalization of the model, k-fold cross-validation is implemented.

Table 2
GridsearchCV-dropout

Dropout	Accuracy for 1st dropout layer (%)	Accuracy for 2nd dropout layer (%)
0.1	87.9	91.1
0.2	80.9	90.9
0.3	91.1	89
0.4	76.3	97.2
0.5	95.7	97.7
0.6	84.1	74.1
0.7	78.3	59
0.8	94.4	58.8
0.9	65.5	58.8

4 Results

4.1 Experimental Results and Discussions

After applying the gridsearchcv method we have attained the optimum hidden neuron values and the optimum dropout values for both hidden layers and both dropout layers as represented in Table 2.

We can concur from the tables above that the optimum dropout values for both dropout layers are 0.5 and the optimum values for the two hidden layer neurons are 20 and 44 respectively as represented in Table 3.

After applying the k fold cross-validation method we have attained the optimum accuracy and precision [12].

We can conclude from the observations above that the fivefold cross-validation achieves the highest scores in all the metrics the model is evaluated on as represented in Table 4. Furthermore, the accuracy is a flawed metric as the basis for evaluating the model's performance as the number of exoplanets that are present in the dataset is just 5. Therefore, precision is the metric the model must be evaluated on as it gives

Table 3 GridsearchCV-hidden layer neurons

Number of neurons for hidden layer 1	Accuracy for hidden layer 1 (in %)	Number of neurons for hidden layer 2	Accuracy for hidden layer 2 (in %)
12	87.1	42	90
14	82.3	44	95
16	90.4	46	94.6
18	90	48	86.2
20	94.5	50	85.2
22	81.7	52	90.3
24	85.5	54	93.6

Table 4 K-fold cross validation

	3-fold cross validation	5-fold cross validation	10-fold cross validation
Accuracy	0.995	0.996	0.995
Precision	0.667	0.8	0.667
Recall	0.8	0.8	0.8
F1 score	0.727	0.8	0.727
Sensitivity	0.8	0.8	0.8
Specificity	0.996	0.998	0.996

Fig. 10 Epochs versus loss

Fig. 11 Epochs versus accuracy

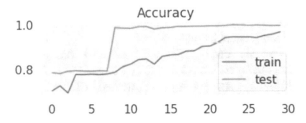

you a clear indication of the true positives. Precision is highest for fivefold cross-validation. The following plots Figs. 10 and 11, represent epochs versus accuracy and epochs versus loss to demonstrate that 30 was the optimum number of epochs for training the model [13].

Figure 12 represents the confusion matrix which shows why our model is unique. Previous research has not been able to achieve high precision whereas our model achieves a precision of 80% [14].

Final Model

NETWORK ARCHITECTURE:

Input layer;

1D convolutional layer, consisting of 10 2 × 2 filters, L2 regularization and RELU activation function;

1D max pooling layer, window size: 2 × 2, stride: 2;

Dropout with 50% probability;

Fully connected layer with 44 neurons and RELU activation function;

Dropout with 50% probability;

Fully connected layer with 20 neurons and RELU activation function;

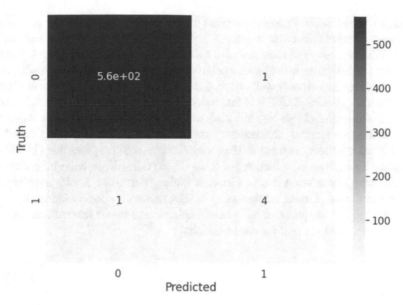

Fig. 12 Confusion matrix

Output layer with sigmoid function.

The model comprises 6 layers in effect, which includes a convolution layer, a max- pooling layer, two fully linked layers, and ultimately an output layer. The Adam optimization technique is used in the training, 30 epochs, 0.01 initial learning rate, and a batch size of 64. The training data set is split into 2 halves [15]. The data has been split into a 4:1 ratio which is used for training and validating the dataset respectively. The testing accuracy was 99.6%, while the overall training accuracy was 99.3%. We have attained the maximum accuracy using fivefold cross-validation. Our model was unable to predict all true positives and failed in one case as can be seen from the confusion matrix.

5   Conclusion

Deep learning was used to construct a method for automatically categorizing Kepler Transiting planet candidates. The dataset did not contain many exoplanets and therefore the true positives were resampled using SMOTE. The value of dropout and number of neurons in hidden layers was determined using GridsearchCV. 3 fold, 5 fold, and 10 fold models were constructed and tested on the Kepler dataset, and an accuracy of 99.5%, 99.6%, and 99.5% was achieved respectively. Previous papers have succeeded in achieving high accuracy, but the accuracy is not a reliable metric for assessing the model as there are very few true positives in the data. This research is unique because it addresses that issue and is successful in achieving a high precision which is the real metric through which the model can be judged. Despite performing resampling and k fold cross validation the model still cannot classify all of the exoplanets accurately and more research needs to be carried out on the same. Furthermore, the methods used in our paper works very well on imbalanced datasets but more research is required on how well our model will work on balanced datasets. Finally, one drawback of using SMOTE is that it doesn't take into account the neighbouring examples from other classes and this leads to additional noise and increase in overlapping classes and therefore more research can be done on similar methods. Current and future space missions, such as the Transiting Exoplanet Survey Satellite (TESS) and Planetary Transits and Oscillations of Stars (PLATO), will generate large datasets. One future space mission that is aimed at finding Earth-like Rocky exoplanets is the Roman Space Telescope Mission by NASA that will probably launch by 2025. Deep-learning techniques that have been implemented in our research can be used as an efficient analysis tool for such missions.

References

1. https://exoplanets.nasa.gov/what-is-an-exoplanet/overview/
2. https://exoplanets.nasa.gov/what-is-an-exoplanet/in-depth/
3. https://exoplanets.nasa.gov/faq/31/whats-a-transit/
4. https://www.universetoday.com/137480/what-is-the-transit-method/
5. https://www.kaggle.com/keplersmachines/kepler-labelled-time-series-data
6. Chawla, N.V., et al.: SMOTE: synthetic minority over-sampling technique. J. Artif. Intell. Res. **16**, 321–357 (2002)
7. Park, S., Kwak, N.: Analysis on the dropout effect in convolutional neural networks. In: Asian Conference on Computer Vision. Springer, Cham (2016)
8. Hinton, G.E., et al.: Improving neural networks by preventing co-adaptation of feature detectors. arXiv preprint arXiv:1207.0580 (2012)
9. González, R.E., Munoz, R.P., Hernández, C.A.: Galaxy detection and identification using deep learning and data augmentation. Astron. Comput. **25**, 103–109 (2018)
10. Shallue, C.J., Vanderburg, A.: Identifying exoplanets with deep learning: a five-planet resonant chain around kepler-80 and an eighth planet around kepler-90. Astron. J. **155**(2), 94 (2018)
11. Dattilo, A., et al.: Identifying exoplanets with deep learning. II. Two new super-earths uncovered by a neural network in k2 data. Astron. J. **157**(5), 169 (2019)
12. Pearson, K.A., Palafox, L., Griffith, C.A.: Searching for exoplanets using artificial intelligence. Mon. Not. R. Astron. Soc. **474**(1), 478–491 (2018). https://doi.org/10.1093/mnras/stx2761
13. Akeson, R.L., et al.: The NASA exoplanet archive: data and tools for exoplanet research. Publ. Astron. Soc. Pacific **125**(930), 989 (2013)
14. Williams, D.M., Pollard, D.: Earth-like worlds on eccentric orbits: excursions beyond the habitable zone. Int. J. Astrobiol. **1**(1), 61–69 (2002)
15. Linde, A., Linde, D., Mezhlumian, A.: From the Big Bang theory to the theory of a stationary universe. Phys. Rev. D **49**(4), 1783 (1994)

Emerging Technologies

The Future of Blockchain Technology, Recent Advancement and Challenges

Ajay Sharma, Vandana Guleria, and Varun Jaiswal

Abstract Blockchain is now in the news, and a lot of companies are working on blockchain technology. The first time the blockchain come into existence in the case of cryptocurrency where the decentralized architecture is followed for mining and storage. The trend of cryptocurrency is increasing day by day and, in other fields, the blockchain is in development. In this book, the chapter author has discussed the future and recent trends, history of blockchain technology. How to make a block and what is the initial requirement to build up the blockchain. The Book chapter gives an overview of the challenges associated with this technology. Later in the section, you will learn about how the blockchain work and the basic functionality of the block with some case study like smart farming, credit fraud detection, etc.

Keywords Blockchain · Healthcare · Bitcoin · Smart contract · Database · Cryptocurrency · Data mining · Blockchain architecture

1 Introduction

A Blockchain appears complicated, and it certainly can be, but its fundamental theory that is moderately simple. The concept of blockchain was first given by Satoshi Nakamoto. The date is still unknown and the person is still anonymous, many personal claims to be Nakamoto but still, it is a mystery [1]. A blockchain is a decentralized

A. Sharma
Department of Biotechnology and Bioinformatics, Jaypee University of Information Technology (JUIT), Waknaghat Solan 173234, Himachal Pradesh, India

A. Sharma · V. Guleria
School of Electrical and Computer Science, Department of Biotechnology, Shoolini University, Solan 173212, Himachal Pradesh, India

V. Guleria
Department of Biotechnology, Shoolini University, Solan 173212, Himachal Pradesh, India

V. Jaiswal (✉)
Department of Food and Nutrition, College of BioNano Technology, Gachon University, Seongnam-si 13120, Gyeonggi-do, Korea

© The Author(s), under exclusive license to Springer Nature Switzerland AG 2022
K. R. Ahmed and H. Hexmoor (eds.), *Blockchain and Deep Learning*,
Studies in Big Data 105, https://doi.org/10.1007/978-3-030-95419-2_15

type of database that stores data. To be able to understand blockchain you have to understand the database, database is a collection of data that are electronically stored at the different or same location on a computer system. Information or data is usually structured in table format in databases, to make it easier to search and filter for information. This book chapter is focused on what is the difference between someone using a spreadsheet to store information rather than a database or blockchain technology. A blockchain is designed to store limited quantities of information for one person or a small group of people. In contrast, a database has been developed to contain significantly more information, to which several users can quickly and easily access, filter, and handle data at once and in easy steps. This is achieved through large databases by accommodating data on servers made of powerful computers. These servers can sometimes be built with hundreds or thousands of computers, so many users can simultaneously access the database with the computational power and storage capacity necessary [2]. While a device or database is available to a number of individuals, it is often owned by a company and administered by an individual who has a complete mechanism over how it functions and the data within it. Efficiency, security, and consensus mechanisms are the main features of each blockchain network (Table 1).

To accomplish this one can, deliberate blockchain as a bunch of interrelated components which give explicit highlights to the framework, as outlined in (Fig. 1). At the most reduced level of this foundation; we've been signed exchanges between peers. These exchanges signify an understanding between two members, which may include the exchange of physical or advanced resources, the completion of an assignment, etc. In any event one member signs this exchange, and it is dispersed to its neighbors. Typically, any substance which interfaces with the blockchain is called a node. However, nodes that confirm all the blockchain guidelines are called full nodes. These nodes bunch the transactions into a block and they are responsible to

Table 1 Classification and main characteristics of blockchain networks

Property	Public	Private	Federated
Consensus Mechanism	Costly All miners	Light, Centralized organization	Light Leader node-set
Identity Anonymity	Anonymous Malicious	Identified users Trusted	Identified users Trusted
Protocol efficiency and consumption	Low efficiency High energy	High efficiency Low energy	High efficiency Low energy
Immutability	Almost impossible	Collusion attacks	Collusion attacks
Ownership and management	Public Permission less	Centralized Permissioned whitelist	Semi-centralized Permission nodes
Transaction approval	Oder of minutes	The Oder of milliseconds	The Oder of milliseconds

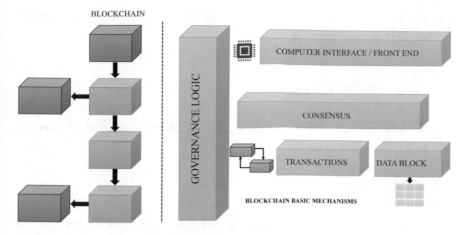

Fig. 1 An overview of blockchain architecture

decide if the transactions are substantial, and should to be retained in the blockchain, and which are not [3, 4].

1.1 History of Blockchain

A person, or a group of individuals, named Satoshi Nakamoto in 2008, conceptualized the first blockchain. Nakamoto significantly upgraded the plan by using a Hash-cash strategy for time signature blocks without expecting a trustworthy party to sign them and know a trouble limit with the balance rate at which the blocks are added into the chain. The plan was carried out the next year by Nakamoto as a center segment of the digital money bitcoin, where it fills in as the public record for all exchanges on the interconnected network [1, 5].

In August 2014, the bitcoin blockchain document size, containing accounts of all exchanges that have happened on the network, arrived at 20 GB (gigabytes). In January 2015, the size had developed to very nearly 30 GB, and from January 2016 to January 2017, the bitcoin blockchain developed from 50 to 100 GB in size. The record size had surpassed 200 GB by mid-2020. The words block and chain remained utilized independently in Satoshi Nakamoto's unique paper, yet remained in the long run advocated as a single word, blockchain, by 2016. As per Accenture, utilization of the diffusion of invitations theory recommends that blockchains accomplished a 13.5% reception rate with financial services in 2016, along these lines arriving at the early adopter's phase. Industry exchange groups joined to make the Worldwide Blockchain Discussion in 2016, a drive of the chamber of digital commerce [6, 7].

In May 2018, Gartner noted only 1% of CIOs showed any kind of blockchain appropriation within their partnerships, with only 8% of CIOs "arranging or looking at the dynamic blockchain experimentations". For the year 2019 Gartner disclosed

5% percent of the blockchain technology accepted by the CIO as a 'game changer's in the field of information technology [7, 8].

1.2 Blockchain Features

The blockchain concept has three pillars. Some of the key features of a blockchain are given below.

- Peer-to-peer Network
- Decentralized approach
- Incorruptible and Immutable

Peer-to-peer Network: In a blockchain network, exchanges have occurred directly between two nodes in an organization. There is no need for an outsider mediator. *For example*, if two nodes in a blockchain financial exchange are occurring, they can directly work with the exchange deprived of doing it through a bank. This is otherwise called a point (P2P) network where exchange happens between two nodes which get confirmed by the wide range of various nodes in a blockchain network. Subsequently, members of a blockchain organization can make immediate and secure exchanges within seconds [8].

Decentralized approach: The whole blockchain framework is a decentralized and appropriated one. This means, there is no central substance that controls and deals with the blockchain however every node in the organization has an equivalent position and has access to the archives, records in data block, node. Every computing node has a duplicate of the record and has the privilege to confirm an exchange or conduct a transaction. It ensures a secure and transparent blockchain network [8, 9].

Incorruptible and Immutable: Because of its highlight point and decentralized nature, there is no center man in the exchanges and everybody in the organization has a duplicate of records with them. That makes the information is sealed and permanent in the blockchain. Furthermore, it is impossible for someone to hack and change records because blockchains use cryptography or hash methods for getting a square (sets of standards) [8]. Every square has its own unique hash and is connected to the hash of the past block. If one tries to mess up with the records in a blockchain, it will change the haze capacity and disturb the whole hash chain. This makes it easy to interfere with records and place interference (Fig. 2).

1.3 Types of Blockchain

Two types of blockchains are mainly present; public and private blockchains. However, several varieties, similar to consortiums and hybrid blockchains, are also available. Let us first see what similarities they share before we dive into details of

Fig. 2 Features of
blockchain

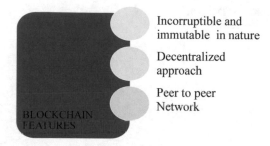

Incorruptible and
immutable in nature

Decentralized
approach

Peer to peer
Network

BLOCKCHAIN
FEATURES

the various types of blockchain. Each blockchain includes a set of nodes that work
on a distributed network (P2P) framework. A copy of the common record is available
to each node in your network, which is easy to refresh. Each node can monitor, begin
or exchange transactions and make blocks [10].

- Public blockchains
- Private blockchains
- Consortium blockchains
- Hybrid blockchains

Public blockchains: A public blockchain has not at all any entrance limitations.
Anybody through a computer network can send, conduct interactions to it just as
become a validator (i.e., partake in the execution of an agreement protocol) [11].
Generally, such organizations suggest financial motivating forces for the individuals
who secure them and use some sort of a Proof of Stake or confirmation of work
algorithm [12]. **Examples**: Bitcoin, Ethereum, Litecoin.

Private blockchains: A private blockchain is allowed [11]. You cannot do that
unless you are greeted by the executives of the organization. Access to Member
and Validator is limited. The wording Conveyed Record (DLT) is typically used for
private blockchains to recognize open blockchains and other decentralised peer-to-
peer database applications that do not open improper registration. Private networks
for voting, supply network management, digital identity, and asset ownership are
deployed [13]. **Examples:** Multichain and Hyperledger projects (Fabric, Sawtooth),
Corda, etc.

Consortium Blockchain: A blockchain consortium is a semi-decentralized type,
which deals with a blockchain network in several associations. This is in contrast to
what we found in a private blockchain, run by just one association. Blockchain in this
type use trade information or mining, not only is an organization able to act as a node.
Blockchain consortiums are frequently used by banks, government associations [14,
15], etc. **Example**: Energy Web Foundation, R3.

Hybrid Blockchain: A blockchain hybrid is a private blockchain mixture. It uses
and highlights the feature of mutual blockchains, which are a privately-owned framework as well as a lesser framework for public consent. Clients can handle who is

Fig. 3 Diagram types of blockchain

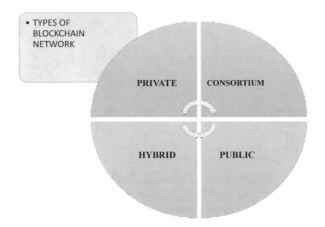

allowed to receive which information in a blockchain by a particular crossover organization. Only a selected part of the blockchain information or logs can be opened to the world and keep the rest secret in a private network. The blockchain hybrid system is adaptable so customers can join a private blockchain with various public blockchains without much deformation. In this organization, an exchange in a private blockchain hybrid network is usually checked. The public blockchains grow the haze and include additional control nodes. This improves blockchain network security and transparency [16, 17] (Fig. 3). **Example**: Dragonchain.

1.4 Benefits of Blockchain Technology

Blockchain technology is properly accepted and implemented, Blockchain technology is an innovative technology. The way to conduct financial transactions or replace files in a network is directed and secure (Fig. 4). The information available in the form of digital freedom and decentralization, anonymity and privacy are unchangeable and secure [18, 19]. Here are different major benefits of the blockchain device, making it so unique and popular:

Application of Blockchain Technology: Blockchain technology is an emerging field and a proportion of research, advancement is going on in this field and hence it is using in a wide range of different systems. The Application of blockchain technology is as follows.

Cloud Storage: Blockchain gives a decentralized framework to putting away records on the cloud. It resembles a Bury Planetary Document Framework (IPFS) in the idea where sharing and putting away records occur in an appropriated web framework instead of a customer worker web which we as of now have. Every website can go about as a node that has a P2P move of documents with different nodes directly, rather than mentioning a central server [20].

Fig. 4 Benefits of blockchain technology

Cryptocurrency: Blockchain innovation goes about as the foundation of cryptographic money frameworks like Bitcoin. Cryptographic money is scrambled advanced cash that everybody can use as a mechanism of trade in exchanges. Such exchanges happen through the blockchain network. The blocks in a blockchain store the records identified with age and exchange of cryptographic money between two nodes for all time. Nowadays, there even exist digital money wallets where you can send and get cryptographic forms of money or trade them for different currencies [21].

Healthcare: With the assistance of block chaining, we can store data about patients and medications in a data set safely. Specialists can get to patient records and history to investigate a case better at an offered highlight guarantee appropriate treatment. Also, associations can screen and deal with drug forging in the clinical store network. This is because grounds that they can store the supply network information as permanent blocks in the blockchain [21].

Smart Contracts: It is preferably a paper-based agreement signed by the parties if two parties agree to or make an arrangement. But this strategy is dependent on the danger of extortion and degradation for arrangements and contracts. A smart agreement idea appeared with blockchain as the basis of innovation. Brilliant agreements are signed and put in a blockchain, which is self-executable. A customized smart contract deal shall include each of the terms and conditions of a two-party agreement and will therefore be applied as soon as each of these conditions is satisfied [22, 23]. This gives 100% assurance and deals from any fraud (Fig. 5).

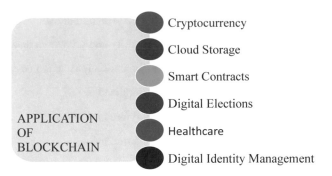

Cryptocurrency

Cloud Storage

Smart Contracts

Digital Elections

APPLICATION
OF   Healthcare
BLOCKCHAIN
Digital Identity Management

Fig. 5 Applications of blockchain

2 Advantage and Disadvantage of Blockchain Technology

2.1 The Advantages of the Blockchain

Decentralized: The decentralized framework of Blockchain innovation is and is the fundamental benefit of this innovation. Why is it important to our lives? The answer to this question is very simple—working with the outside organisation or the focal supervisor is not important. It means that the framework works without intermediation and all the Blockchain members make the choices. Each system has a base of information, and this data set is essential. The information base security process can take a long time and spend a lot of money. It also means that the transactions can be independently confirmed and processed [24].

Protection: Each activity is recorded on the Blockchain and data is available and cannot be changed or deleted by every member in this Blockchain. The results of this recording give transparency, durability, and confidence to the Blockchain [25].

Trust Factor: At least two members who don't know each other depend on their acceptance of the trust of Blockchain. The main thought is that these unknown individuals are genuine and not uselessly transacted. Confidence can be further expanded as more cycles and records can be shared [25].

Economically feasibility: Since no contribution is made by third parties, the cost is automatically compact. In contrast to others, the account holders in Blockchain can go for the peer-to-peer transaction openly by avoiding all additional cost cuts [25].

Faster Transaction: Transactions duplicate measurements achieve transparency of the Blockchain. Every transaction in the Blockchain organization is duplicated as it was composed. It also means that every activity is shown to members of the Blockchain by each member to look at all transactions. No one can insensibly do anything. The Blockchain plans such that it can show any issues and right them if it is essential. This benefit makes Blockchain innovation traceability [25].

Fig. 6 The advantages of
the blockchain [26]

Decentralization

Protection

Economically Feasible

Faster Transaction

All Time Accessibility

ADVANTAGES
OF
BLOCKCHAIN
TECHNOLOGY

All-Time Accessibility: The last advantage is the quicker treatment. Generally, preparing and starting up a bank association takes a long time to complete the transaction. The use of Blockchain innovation can reduce the time for operation and startup to several times—from approximately three days to several minutes or even seconds [25].

Easy Sharing Database for Business to Business Arrangement: Various organizations depend on different PCs for various information bases. Furthermore, it becomes difficult when one needs to impart this data set to another business. The basic arrangement is Blockchain innovation as it keeps a single shared ledger. And it is not difficult to share this record with some other businesses (Fig. 6).

2.2   The Blockchain Disadvantages

Energy Consumption: High energy consumption is the main disadvantage of the Blockchain network. Electricity consumption and computing power are required to keep running a system that keeps a record. Every time the new node is created and it talks to each other at a similar time. In this manner, transparency is established. The organization's miners are attempting to solve a lot of arrangements each second to approve transactions. They are utilizing generous measures of personal computing power [26].

Signature Performance: Signature verification is the test of the Blockchain because the large computing power is important to the estimation of interaction to the sign, each transaction should be signed in cryptography. This is one reason for the high consumption of energy [26].

Slow Performance: At the point when this decentralized data set is contrasted with a centralized information base then you will become to knowing it's a lot slower than the later [26].

Redundant Data: The amount of computation needed in Blockchain system is much more than in a concentrated data set. Each node running on this shared organization needs to go through a similar interaction independently [25].

Energy Consumption

Signature Verification Of Blockchain

Slow Performance

Redundant Data

DISADVANTAGE OF BLOCKCHAIN

The High Cost Of Operation

High Expertise To Handle The Data

Fig. 7 Blockchain disadvantages [25]

The High Cost: The Blockchain's high costs are a big and major disadvantage. The normal transaction costs between $75 and $160, and the majority of the transaction covers energy consumption. One reason was described above for the present situation. The second reason is Blockchain's high initial capital costs (Fig. 7).

3 The Architecture of Blockchain Technology

The generally accepted definition of a *blockchain* is "a distributed ledger with smart contracts". A ledger is important because it records the transactions that capture all changes to a set of business objects. In a distributed ledger system, organizations collaborate to maintain a consistent copy of a replicated ledger in a process that is called *consensus*. A smart contract is important because it defines the rules and conditions for querying a ledger and generating new transactions that are recorded on it. Smart contracts describe, in code, the lifecycle of one or more business objects that are stored on the ledger. It describes how they are created, updated, and queried [3, 27]. *For example*: a supply chain of manufacturers, distributors, suppliers, and retailers might use a set of smart contracts and a distributed ledger to maintain a record of transactions. The record details the movement of goods and services through the supply chain. The logical architecture of blockchain is divided into three important concepts: blocks, nodes, and miners (Fig. 8). The logical architecture of blockchain and each block is shown in the Fig. 8 and 10 respectively.

Nodes

Decentralization is one of the most significant concepts in blockchain technology. The chain cannot be maintained by any computer or organization. Instead, it is a distributed book through the linked nodes. Nodes can be an electronic device of any kind that keeps duplicates of the blockchain and the network working. Every node has its own blockchain and a newly mined block to update, rely on, and verify the chain must algorithmically be approved in the network. Decentralization is one of the key ideas for blockchain innovation. The chain cannot be maintained by any

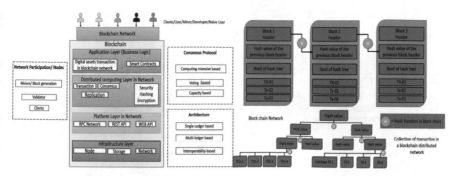

Fig. 8 The detailed physical architecture of blockchain (left side), Logical architecture of blockchain nodes, blocks, hash-function architecture (right side) [3, 27]

personal computer system or a group of minor, associations. Being all equal, it's a record conveyed through the chain-related nodes. Nodes can be an electronic gadget of any kind that keeps blockchain duplicates up and keeps the organization. Each node has its own blockchain duplicate and the company must support an algorithm for a refreshing, trusting, and validating a new mining block [28, 29].

Block: Block includes block header information and transactions. Blocks are data structures designed to bundle and replicate all Network nodes. Miners are creating new blocks and add in the blockchain. Mining is the process of creating a valid block that the rest of the network accepts. Nodes take pending transactions, check their encryption accuracy and package it into blocks to be stored on the blockchain. Block header is the metadata used to verify a block's validity [30]. The block metadata content is shown in (Figs. 9 and 10) respectively. There is a transaction in the rest of a block. Depending on the choice of the miner, there may be several transactions grouped into blocks [3, 28, 29]. Each chain comprises of different blocks and each block has three fundamental components:

Types of Blocks: There are multiple blocks present in each chain and each block has three elements are followed as a data block, Nonce, and Hash function.

- **Data Block**: The data in the block where all the data related to the blockchain type is stored, data related to healthcare, smart farming, etc.
- **Nonce**: A nonce was called an all-number 32-bit. The nonce is randomly generated by generating a block and generating a hazardous block header.
- **Hash Function**: The hash is a 256-bit number. It must start with a huge number of integers (i.e., be extremely small). The cryptographic hash is generated when you create the first block of a chain. The block data is deemed to be signed and always linked to the nonce, unless they are mined.

The current main blockchain is simply enlarged by many blocks and is also the longest network chain. These blocks are referred to a "**main branch blocks**". The other longest blockchain parent block. These blocks are referred to as "**side branches block**". Some blocks refer to a parent block that the node that processes the block is not known. These are referred to as the "**orphan blocks**". Side branch blocks may

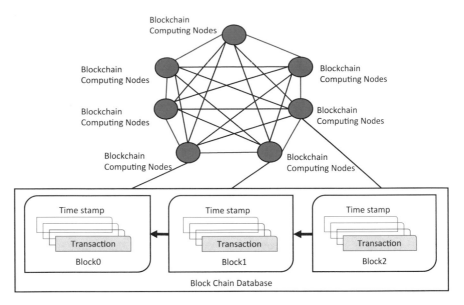

Fig. 9 The interconnection of blockchain computing nodes and the blockchain diagram [28]

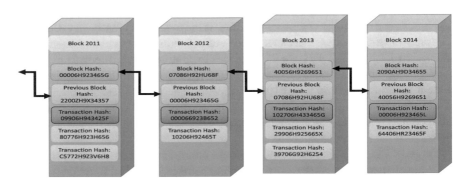

Fig. 10 Architecture of block in blockchain network

not be part of the main branch at present, but if more blocks have been created that refers to the parent, a particular side branch may be restructured in the main branch. This leads to a forking concept in blockchain networks [3].

Miners

Through a process called mining, miners create new blocks on the chain. Each block has its own unique nonce and hash function in a blockchain, but it also mentions to the hash function of the previous block, therefore that it's not easy to replace a block, specifically on large chains. Miners use special software to solve the incredibly complex mathematics problem of finding an accepted hash nonce. Because the nonce is only 32 bits and the hash is 256, it is necessary to extract about 4 billion possible nonce-hash combinations before finding the right. Once that happens, miners have the "golden nonce" found and their block has been added to the chain. Changing to every block in the chain earlier requires remaining not only the block with the change but all the following blocks. Therefore, manipulation of blockchain technology is extremely difficult. Think of it as "reliability in math," because it takes an enormous amount of time and computing power to find golden nonce. If a data block is successfully mined, all nodes in the grid accept the change, and the miner is financially recompensed [28–30].

Decentralization is one of the most significant concepts in blockchain technology. The chain of blocks cannot be solely maintained by any computer or organization. It is instead a distributed ledger through the chain-linked nodes. Nodes can be an electronic device of any kind that keeps copies of the blockchain and the network working. Every computing node has a duplicate copy of its own blockchain and a newly mined block for updating and trusting the chain must be algorithmically approved by the network. Each accomplishment transaction in the ledger can be certainly verified, checked, and seen as the blockchain is transparent. A single alphanumeric ID number showing the transactions of each participant is provided. The combination of public information with a checks and balances system helps the blockchain to uphold, regulate integrity and build user trust. Blockchains can be regarded as a technological trust framework in the real-time environment [3].

Smart contract

A *smart contract* is a code that defines the lifecycle of one or more business objects. It is used by an application to generate transactions that record the changes to those objects. It can also be used to query the current value and transaction history of those objects. Think of a smart contract as providing controlled access to the public, private, and confidential information that is on a distributed ledger [22, 24].

Consider the example of a vehicle lifecycle. It is manufactured, distributed, sold, insured, driven, repaired, resold, and eventually scrapped. A smart contract can be used to generate transactions that capture this entire lifecycle. These vehicle transactions are immutably recorded in a set of ledger instances that are owned by each member of a blockchain network. These organizations can query their copy of the ledger to determine the current state of a vehicle or any of its previous transactions and to generate new transactions [22].

A smart contract can also define the business conditions and rules that affect the generation of a transaction. For example, a condition of sale might be that only the owner of a vehicle can sell it. Or the rule that whenever a vehicle changes ownership, an event is generated to notify the appropriate government agency [22, 24].

Because a smart contract generates a multi-party transaction, it must be installed in every organization that is required to sign the transaction. For a transaction to be considered valid, it must be signed by the required organizations according to a network-agreed endorsement policy.

Consensus describes the process by which a multi-party transaction is generated, distributed, and immutably recorded in each organization's copy of the ledger. All multi-party transactions contribute to the ledger history, but only valid transactions update their current value [22, 24].

Ledger

The most common type of ledger that you experience is your bank account. It's a good example of a ledger because it contains a current value and a transaction history. For a bank account, the current value is the balance of the account. In a ledger that contains transactions that are related to a set of business objects, the current value is the most up-to-date state of those objects. The current value of the ledger is usually the most relevant value. In your bank account, your balance determines how much money is available to spend. In the example of your bank account, the transaction history contains all debits and credits to the account. In general, a transaction history contains all transactions that are related to a set of business objects. A ledger continuously increases in size as it appends transactions to it. A ledger's transaction history is more important than its current value because the history can be used to regenerate the current value by replaying all its transactions in order [3, 30].

Blockchain ledgers are special in two ways:

They're distributed. A blockchain network comprises a set of actors, each of whom holds an identical copy of the ledger. A blockchain ledger is decentralized because no single actor owns it. Instead, all replicated copies are kept synchronized with each other in a process that is called *consensus*.

They're immutable. It's impossible to tamper with the transactions in a blockchain ledger because every block of transactions is cryptographically hashed and linked to the previous block. It's impossible to change any transaction in this chain without rendering the hashes invalid. As the blockchain are divided into the different categories depending on the ledger and the network on which they are hosted. Table 2 shows some of the properties and the difference public, privated and Consortium base blockchain networks with their respective ledgers.

The decentralization and immutability provide a blockchain ledger with significantly increased transparency, trust, and resilience compared to a centrally managed ledger (Table 2).

3.1 Summary of Blockchain Architecture

In this section, the author has illuminated the quick summary of the blockchain as a conclusion. Here are the most important takeovers:

- A blockchain is a distributed booklet that is open to everybody. After it is recorded within a blockchain it is very difficult to change the data.
- An increasing list of ordered records known as blocks is supported by the blockchain architecture. A timestamp and a link to the previous block are kept for each block.

Table 2 The difference and properties between the public, consortium blockchain, and private blockchain network

Sr. No	Property	Public blockchain	Consortium blockchain	Private blockchain
1	Consensus determination	All miners	The selected set of nodes	Within one organization
2	Read permission	Public	Public or restricted	Public or restricted
3	Immutability level	Almost impossible to tamper	Could be tampered	Could be tampered
4	Efficiency (use of resources)	Low	High	High
5	Centralization	No	Partial	Yes
6	Consensus process	Permission less	Needs permission	Needs permission
7	Property	Public blockchain	Consortium blockchain	Private blockchain

- In smart contracts, crowdfunding, supply chain management, KYS and AML processes, medical records, transparent management, and decentralized economic sharing projects, you can use a blockchain architecture.
- To work with Blockchain architectures, Blockchains developers need a strong experience, background in CS (computer science), or technology. Moreover, there is high demand for knowledge of consensus methods, data structures, decentralized headlines, cryptography, and data security.
- Developing an architecture in blockchain calls for a functional specification, user interface design, and architecture plan. It is essential to define the application functionality and user roles as well as to take account of system flow and user interaction.

4 Services Provided by the Blockchain

Several Blockchain provide various services in the various domain of engineering and biotechnology, banking/finance transactions. Some of the services which are provided by the blockchain are as follows:

- **Confidentiality**: The confidentiality principle agrees that it should be possible just to access the content of a message from the sender and the intended beneficiary. The block generated through blockchain mining is highly confidential. The data in these are highly confidential and follows the decentralized approach to store the data at different location [31].
- **Authentication**: Mechanisms of authentication help to establish identity evidence. This process confirms the correct identification of the message source.

The block is highly authenticated with the 256-bit hash function and 32 bit nuance function out of which on which one combination is a golden nonce [32].

- **Integrity**: The integrity mechanism confirms that when the recipient is received by the sender, the subject of the message is the same. The block is integrated to form a blockchain and hence the larger blockchain is carried out [33].
- **Non-repudiation**: Non-repudiation will send a message to refute the claim that the message is not transmitted. There is no repudiation of blocks in the blockchain. Every block is distinctively unique in nature and contains data that is highly secured [34].
- **Access Control**: Checks who can access what Access Control specifies that the authenticated user will have the access to the data within the block. The data is secured and encrypted with the help of various cryptographic algorithms (SHA, SSH), etc. [34].
- **Availability**: The accessibility principle stipulates that resources should always be accessible to authorized parties. Only the authorized person can have the access to the data and data is available in the secure encrypted format [34].

5   Case Studies of Blockchain Use in Various Domains

Blockchain uses in the Banking sector, Crowd-sourcing, Insurance sector, Licensing and payment system sector, Medical treatment and healthcare sector, Copyright management sector, Electronic government, public sector, Eco-friendly management system sector, and Sharing economy sector, etc.

- **Banking sector**: Bank services, such as stores and loans may consider an increase in their reliability quality as dispersed records and brilliant agreements accessible with blockchain advances are utilized. In addition, similar technologies can be applied to unfamiliar trade organizations. International settlements and issuances of letters of credit and the executives in trade can continue at lower costs and higher rates. *For instance*, Ripple coin (XRP) impacts inside blockchain technology are considered to be substantial. Ripple is a continuous gross settlement system as well as a money trade and settlement network made by Ripple Labs, a US-based technology company [35].
- **Crowdsourcing**: Blockchain technology can also be applied to coordinating with administrations that associate ordering and getting business as well as occupation searchers and customers on the Internet to solve different issues. These problems incorporate precision of coordinating, speed of results, or high commission charges. Moreover, the use of blockchain technology in crowd financing is increasing. The capacity of people or organizations to create public financial help for an item or service is a significant benefit of group crowdfunding. This interaction permits financial contributions through little financial commitments that worry about little concern. As such, publicly supporting is a construction where people and organizations are directly associated without the use of intermediaries, and this dovetails well with the qualities of blockchains [36].

- **Insurance sector**: With blockchains, another kind of security management system can easily recognize and use the enormous amount of data that insurance contracts contain. In this process, the blockchain incorporates real information and smart agreements. Different projects are currently investigating these advancements. In one prominent case, an insurer at American Worldwide Group has applied blockchains to the protection business in participation with IBM [37].
- **Licensing and payment system sector**: Blockchain-based digital forms of money are utilized in a variety of installment systems. Also, these digital forms of money can be broadly utilized in regions where licenses are granted to get certain substances (for example music, films) for a set number of times or a set duration use [38].
- **Medical treatment and healthcare sector**: Clinics and hospitals have difficulties managing, keep up, and fix medical treatment systems, bringing about the distribution of massive financial and human resources. It is assessed that the presentation of blockchains can surprisingly reduce the costs that medical foundations cause. One significant contribution would be the recording of medical records on a blockchain so that patient's unique clinical records can be easily accessed from any place inside the industry [39].
- **Copyright management sector**: The utilization of blockchain technology will encourage extraordinary change in copyright management as it tracks ownership and prevents forgery. For instance, blockchain technology can digitize show passes to prevent scalping, the unapproved resale of tickets. Similarly, it can prevent the illegal copying and sharing of PC and computer game programming. The two examples further build-up copyright assurances for a variety of imaginative works [40].
- **Electronic government, public sector**: Central and neighborhood governments can utilize blockchain-based authoritative services for a variety of assignments: official record-keeping, online electronic democratic, schooling, tax collection, public protection, new and environmentally friendly power the executives, and related exchanges. For real explicitly, the services that will profit from such upgrades are sale and registration, records of reproduction, and repair. In all cases, blockchain innovation will assume an extraordinary part in cost decrease. With such administrative changes, society can get more attractive and more straightforward, leading to greater convenience and personal satisfaction for all residents [41].
- **Eco-friendly management system sector**: By utilizing artificial intelligence and large information advancements running on Internet of Things (IoT) blockchain stages, it is achievable to refine flow techniques for overseeing and improving water and air contamination records. One such advanced utilization of blockchain technology interlocks sensors for light, air temperature, moistness, and natural poisons like fine residue, ozone harming substances, soil pollutants, and wastewater [42].
- **Sharing economy sector**: As key participants in the sharing economy, Uber and Airbnb have both made critical changes in the transportation and convenience ventures, separately. In like vein, one influencer on the clinical business is Folding

Coin, a task wherein members trade the handling force of their computer to help cure cancer and different infections as a trade-off for cryptographic money coins [43]. From this model, it is normal that blockchain technology will catalyze the speed increase of substances in the sharing economy. For instance, each person who produces esteem added products (for example electric force, work of art) will turn into a business operator. Accordingly, it is normal that this will introduce a new time of worldwide dealing and distribution of items and products. As more stable exchange systems are established, the need for digital forms of money will likewise rise.

- **Other fields for blockchain technology utilization**: Different uses of blockchain technology incorporate the constant and permanent management of dependable official statements, the accelerated development of notices and travel items to draw in tourists, hyperspectral imaging [43] the enlargement of a reliable and utilizable rental industry, the minimization of debilitating and superfluous lawful questions or philosophical contentions, and widespread culture.

6 Conclusion

The above book chapter describes the basic to advance levels of blockchain technology and the future trends associated with the blockchain. According to the author's point of view, there is the following observation that the author found out while writing this book chapter. As the author found out several observations.

- Blockchain is the term that is used to describe distributed-ledger technology that uses smart contracts to share multi-party transactions with the member organizations of a business network. It helps to build up trust in these networks by providing cryptographic proof over a set of facts. The information that one participant sees is the same as what another participant sees. This idea is powerful and is enabling a new generation of transactional applications.
- It's also a simple idea, and it's good to remember when you get started with blockchain. You generally don't need to understand, *for example*, the tree data structure that underpins a blockchain, or how consensus works between nodes on the network. That information is best left to the people who are working to advance blockchain technology. Instead, consider using a blockchain to reliably share transactions between organizations. The network will handle how the information is distributed.
- Blockchain technology is a revolutionary innovation that can make many traditional systems safer, distributed, transparent, and collaborative while empowering users. Combining with all the aforementioned aspects, the immediate, low-cost trust assurance provided through blockchain technology can trigger customer innovation by enabling any provider and every producer to immediately, mutually start trading business.
- A typical manufacturing supply chain modal with Walmart's case study and give out a back-end architecture and build a blockchain upon it. It addresses blockchain

as a tool for solving their daily task-related work and helps companies understand the significance of smart contracts in blockchain to them and then concentrate on their concept proof in this study.

- Organizations involved in the blockchain system can better understand how their transactions are triggered and further explained throughout the supply chain. The feedback level can be used to improve marketing, production, and sales accounts, etc.
- Smart contracts can be better integrated into this system for the flexibility of transactions. And the application is a front-end tool for testing the usability and performance of the back-end architecture while enabling users to interact with directly.
- The application of blockchain technology across the manufacturing supply chain space is endless. The target users are looking forward more other functions according to the real business demand such as Audit trails, real-time negotiation, supply chain visibility, and traceability, collect data from IoT, IP management in product development based on the built back-end blockchain system and the computation mechanism inside the "black box".
- People from various areas are expecting the pace of disturbing innovation of blockchain to accelerate in near future. As the absolute predominance blockchain can bring to manufacturing.

Acknowledgements The book chapter entitled "The Future of blockchain technology, Recent advancement and challenges" has been written as a part of the Author[1] Master program and Author[1] Ph.D. Program. The author would like to thanks the dept of computer science at Shoolini University for providing the lab space for working and phrasing the concept of the book chapter. The author has dedicated this book chapter to the anonymous author who reviews this book chapter.

Author Contribution All the author has the equal participation as the Ajay and Vandana has prepared the menu script and all the assessment and the correction was done by the Dr. Varun.

Conflict of Interest There is no any conflict of interest in the book chapter. The author has taken care of proper citation of work with the proper acknowledgment and all other things related to the conflict.

References

1. Nakamoto, S.: Bitcoin whitepaper. https://bitcoin.org/bitcoin.pdf (Дата обращения: 17.07. 2019) (2008)
2. Cachin, C.: Architecture of the hyperledger blockchain fabric. In: Workshop on Distributed Cryptocurrencies and Consensus Ledgers, Chicago, IL (2016)
3. Zheng, Z., et al.: An overview of blockchain technology: architecture, consensus, and future trends. In: 2017 IEEE International Congress on Big Data (BigData Congress). IEEE (2017)
4. Hamida, E.B., et al.: Blockchain for enterprise: overview, opportunities and challenges. In: The Thirteenth International Conference on Wireless and Mobile Communications (ICWMC 2017) (2017)
5. Ateniese, G., et al.: Redactable blockchain–or–rewriting history in bitcoin and friends. In: 2017 IEEE European Symposium on Security and Privacy (EuroS&P). IEEE (2017)

6. Urquhart, A.: The inefficiency of Bitcoin. Econ. Lett. **148**, 80–82 (2016)
7. Nadarajah, S., Chu, J.: On the inefficiency of Bitcoin. Econ. Lett. **150**, 6–9 (2017)
8. Böhme, R., et al.: Bitcoin: economics, technology, and governance. J. Econ. Perspect. **29**(2), 213–238 (2015)
9. Muzammal, M., Qu, Q., Nasrulin, B.: Renovating blockchain with distributed databases: an open source system. Futur. Gener. Comput. Syst. **90**, 105–117 (2019)
10. Andreev, R., et al.: Review of blockchain technology: types of blockchain and their application. Intellekt. Sist. Proizv. **16**(1), 11–14 (2018)
11. Prajapati, S., et al.: Introduction to blockchain. Int. J. Sci. Res. Rev. **7**, 1800–1805 (2019)
12. Tang, H., Shi, Y., Dong, P.: Public blockchain evaluation using entropy and TOPSIS. Expert Syst. Appl. **117**, 204–210 (2019)
13. Dinh, T.T.A., et al.: Blockbench: a framework for analyzing private blockchains. In: Proceedings of the 2017 ACM International Conference on Management of Data (2017)
14. Li, Z., et al.: Consortium blockchain for secure energy trading in industrial internet of things. IEEE Trans. Industr. Inf. **14**(8), 3690–3700 (2017)
15. Gai, K., et al.: Privacy-preserving energy trading using consortium blockchain in smart grid. IEEE Trans. Industr. Inf. **15**(6), 3548–3558 (2019)
16. Sagirlar, G., et al.: Hybrid-IoT: hybrid blockchain architecture for internet of things-pow sub-blockchains. In: 2018 IEEE International Conference on Internet of Things (iThings) and IEEE Green Computing and Communications (GreenCom) and IEEE Cyber, Physical and Social Computing (CPSCom) and IEEE Smart Data (SmartData). IEEE (2018)
17. Li, Z., et al.: A hybrid blockchain ledger for supply chain visibility. In: 2018 17th International Symposium on Parallel and Distributed Computing (ISPDC). IEEE (2018)
18. Ali, O., et al.: A comparative study: blockchain technology utilization benefits, challenges and functionalities. IEEE Access **9**, 12730–12749 (2021)
19. Ølnes, S., Ubacht, J., Janssen, M.: Blockchain in government: benefits and implications of distributed ledger technology for information sharing. Elsevier (2017)
20. Sukhodolskiy, I., Zapechnikov, S.: A blockchain-based access control system for cloud storage. In: 2018 IEEE Conference of Russian Young Researchers in Electrical and Electronic Engineering (EIConRus). IEEE (2018)
21. Mukhopadhyay, U., et al.: A brief survey of cryptocurrency systems. In: 2016 14th Annual Conference on Privacy, Security and Trust (PST). IEEE (2016)
22. Christidis, K., Devetsikiotis, M.: Blockchains and smart contracts for the internet of things. IEEE Access **4**, 2292–2303 (2016)
23. Bartoletti, M., Pompianu, L.: An empirical analysis of smart contracts: platforms, applications, and design patterns. In: International Conference on Financial Cryptography and Data Security. Springer (2017)
24. Croman, K., et al.: On scaling decentralized blockchains. In: International Conference on Financial Cryptography and Data Security. Springer (2016)
25. Belotti, M., et al.: A vademecum on blockchain technologies: when, which, and how. IEEE Commun. Surv. Tutor. **21**(4), 3796–3838 (2019)
26. Golosova, J., Romanovs, A.: The advantages and disadvantages of the blockchain technology. In: 2018 IEEE 6th Workshop on Advances in Information, Electronic and Electrical Engineering (AIEEE). IEEE (2018)
27. Syed, T.A., et al.: A comparative analysis of blockchain architecture and its applications: problems and recommendations. IEEE Access **7**, 176838–176869 (2019)
28. Yuan, Y., Wang, F.-Y.: Towards blockchain-based intelligent transportation systems. In: 2016 IEEE 19th International Conference on Intelligent Transportation Systems (ITSC). IEEE (2016)
29. Yuan, Y., Wang, F.-Y.: Blockchain and cryptocurrencies: model, techniques, and applications. IEEE Trans. Syst. Man Cybern. Syst. **48**(9), 1421–1428 (2018)
30. Li, J., Wu, J., Chen, L.: Block-secure: Blockchain based scheme for secure P2P cloud storage. Inf. Sci. **465**, 219–231 (2018)
31. Wang, Y., Kogan, A.: Designing confidentiality-preserving Blockchain-based transaction processing systems. Int. J. Account. Inf. Syst. **30**, 1–18 (2018)

32. Hammi, M.T., et al.: Bubbles of trust: a decentralized blockchain-based authentication system for IoT. Comput. Secur. **78**, 126–142 (2018)
33. Zikratov, I., et al.: Ensuring data integrity using blockchain technology. In: 2017 20th Conference of Open Innovations Association (FRUCT). IEEE (2017)
34. Fang, W., et al.: Digital signature scheme for information non-repudiation in blockchain: a state of the art review. EURASIP J. Wirel. Commun. Netw. **2020**(1), 1–15 (2020)
35. Gupta, A., Gupta, S.: Blockchain technology: application in Indian banking sector. Delhi Bus. Rev. **19**(2), 75–84 (2018)
36. Li, M., et al.: Crowdbc: a blockchain-based decentralized framework for crowdsourcing. IEEE Trans. Parallel Distrib. Syst. **30**(6), 1251–1266 (2018)
37. Gatteschi, V., et al.: To blockchain or not to blockchain: that is the question. IT Professional **20**(2), 62–74 (2018)
38. Yoo, S.: Blockchain based financial case analysis and its implications. Asia Pacific J. Innov. Entrep. (2017)
39. Ahmed, S., Tarique, K.M., Arif, I.: Service quality, patient satisfaction and loyalty in the Bangladesh healthcare sector. Int. J. Health Care Q. Assur. (2017)
40. Holland, M., Nigischer, C., Stjepandic, J.: Copyright protection in additive manufacturing with blockchain approach. Transdiscip. Eng. Paradig. Shift **5**, 914–921 (2017)
41. Warkentin, M., Orgeron, C.: Using the security triad to assess blockchain technology in public sector applications. Int. J. Inf. Manag. **52**, 102090 (2020)
42. Sharma, A.: Smart agriculture services using deep learning, big data, and IoT (internet of things). In: Smart Agricultural Services Using Deep Learning, Big Data, and IoT, IGI Global, pp. 166–202 (2021)
43. Sharma, A., et al.: Application and analysis of hyperspectal imaging. In: 2019 5th International Conference on Signal Processing, Computing and Control (ISPCC). IEEE (2019)

Printed in the United States
by Baker & Taylor Publisher Services